American Cities & Technology

American Cities & Technology

wilderness to wired city

Gerrylynn K. Roberts and Philip Steadman

ROUTLEDGE

in association with

The Open University

Coventry University

A catalogue record for this book is available from the British Library
A catalog record for this book is available from the Library of Congress

ISBN 0 415 20083 0 hardback
ISBN 0 415 20084 9 paperback

at308bk3i1.1

This book is part of the Cities and Technology series listed on the back of the first page. The series has been
prepared for the Open University course AT308 *Cities and Technology: from Babylon to Singapore*. Details of
this and other Open University courses can be obtained from the Course Reservations and Sales Office,
PO Box 724, The Open University, Milton Keynes MK7 6ZS, United Kingdom; tel: + 44 (0)1908 653231.

Much useful course information can also be obtained from The Open University's website: http://www.open.ac.uk

Contents

Introduction

The Cities and Technology series

This textbook is part of a series about the technological dimension of one of the most fundamental changes in the history of human society – the transition from rural to urban ways of living. The series[1] is intended first and foremost as a contribution to the social history of technology; the urban setting serves above all as a repository of historical evidence with which to interpret the historical relations of technology and society. The main focus, though not an exclusive one, is on the social relations of technology as exhibited in the physical form and fabric of towns and cities.

The main aims of the series are twofold. The first is to investigate the extent to which major changes in the physical form and fabric of towns and cities have been stimulated by technological developments (and conversely how far urban development has been constrained by the existing state of technology). The second aim is to explore within the urban setting the social origins and contexts of technology. To this end the series draws upon a number of disciplines involved in urban historical studies – urban archaeology, urban history, urban historical geography and architectural history. In so doing, it seeks to correct an illusion created by some past historical writing – the illusion that all major changes in urban form and fabric might be sufficiently explained by technological innovations. In brief, the series shows not only how towns and cities have been shaped by applications of technology, but also how such applications have been influenced by, for example, politics, economics, culture and the natural environment.

The wide chronological and geographical compass of the series serves to bring out the general features of urban form which differentiate particular civilizations and economic orders. Attention to these differences shows how civilizations and societies are characterized both by their use of certain complexes of technologies, and also by the peculiar political, social and economic pathways through which the potentials of these technologies are channelled and shaped. Despite its wide sweep, the series does not sacrifice depth for breadth: case-studies of technologies in particular urban settings form the bulk of the material in the Readers associated with each textbook.

Definitions

The series calls upon a diverse set of interpretations, models and approaches from the social history of technology, from urban historical and geographical studies, and from archaeology and architectural history. In a wide-ranging series such as this, it seems appropriate to introduce theoretical issues when called for by a particular topic. There is, however, a place in this introduction for discussion of the series' two main variables.

In this series, 'technology' is interpreted broadly to cover all methods and means devised by humans in pursuit of their practical ends, thereby including relevant developments in science, mathematics, public health and medicine. But however broad this interpretation, there are still distinctions to be drawn

---

[1] The three textbooks and Readers in this Cities and Technology series – all published by Routledge, in association with The Open University – are listed on the back of the first page of this volume.

between, say, technology (the means of building) and the built environment (the product of building technologies); or between technology (the means of achieving human ends) and society (perhaps the most ambiguous term of all, but in one sense the summation of human ends in the form of a set of religious, moral and political values). It may turn out to be difficult to draw a sharp line between these concepts; but the same can be said of night and day, or indeed, the modern city and the countryside, and we cannot do without such concepts in our dealings with the world. Similarly, without the distinctions proposed between technology, society and the built environment, we cannot hope to think clearly about the questions this series raises. In practice, the series focuses on the implications for urban form and fabric of a well-defined range of innovations, above all those in: agriculture and food-processing; military technology; energy; materials; transport; communications; water supply, sanitation and other developments in public health and social medicine; production processes; and building construction, including representational and measurement techniques and engineering science.

The contribution of these technologies is analysed at differing angles of relevance to the history of urban form and fabric. Innovations in agricultural, military, industrial and transport technology are linked with broad developments in the history of urbanization, including the origins of urban settlements, the changing relationship between town and country, and the increasingly specialized nature of cities within systems of cities. At the core of the series are the technologies most intimately involved in the processes of city-building – construction techniques, intra-urban transport, energy systems, water supply and sanitation, and communications networks. The series also attends to developments in science, technology and public health stimulated by a variety of urban crises.

If 'technology' is a slippery term, so too is 'city', the other great historical variable of this series. It should be emphasized at the outset that no particular store is set by the distinction between cities and towns, a distinction often made in a culturally specific way – for example, the peculiarly British criterion that cities have cathedrals. In this series, cities and towns are seen as part of an urban continuum, and references to cities should usually be read as covering all settlements of an urban degree of complexity and specialization. This needs to be stressed, because in a series as wide-ranging as this, the largest cities of a given period and region tend to claim the most detailed attention. But this is already to presuppose a distinction between cities and villages, the type of permanent human settlement which developed alongside agriculture, beginning some 11,000 years ago. What differentiates towns or cities from villages, apart from the brute facts that the former are physically bigger and contain more people?

The very fact that 'cities' have been transformed throughout history is what makes them compelling objects of historical investigation; but it also makes any abstract definition elusive. In many earlier periods and places, the city, densely populated and built up within its defining defensive walls, was sharply demarcated from the surrounding countryside and its small agricultural settlements. In the modern era, the old fortification rings which defined the city, especially in Europe, became redundant, partly through developments in artillery. They sometimes metamorphosed directly into circular roads, becoming part of the transport system which facilitated the outward diffusion of the city into the countryside, to the point where some recent commentators have begun to question whether the term 'city' is becoming outmoded. Cities, clearly, are dynamic entities, subject to great changes, not only expanding but declining and contracting, and even at times being reduced to rubble or ashes. For all we know, they may be a phenomenon of certain successive configurations of human society only, and be destined to disappear, just as there were none in human history, it seems, until little more than 5,000 years ago. There is no neat definition of a city, as the following quotation indicates:

> the city may be defined initially as a community whose members live in close proximity under a single government and in a unified complex of buildings, often surrounded by a wall. Since, however, this definition would also cover many villages, military camps, religious communities and the like, the city may further be described as a community in which a considerable number of the population pursue their main activities within the city, in non-rural occupations. But other communities, such as a monastery or small factory surrounded by the dwellings of its workmen, might be similarly characterized. A third characterization may therefore be that the city is a community which extends at least its influence and preferably its control over an area wider than that simply necessary to maintain its self-sufficiency.
>
> (Hammond, 1972, p.8)

The problem with this process of definition and redefinition is that in order to cover marginal exceptions, the definition gets loaded with various historical attributes of certain cities that are not part of the *meaning* of words such as 'city' or 'urban'. The search for a perfect definition, intended not only to cover all types of city in history but also to expose those settlements with bogus urban credentials, is surely a futile one, and this series will eschew it. The main point that will be emphasized is that a significant or dominant proportion of the population is engaged in activities other than agriculture: government, religion, administration, law, education, finance, manufacture, commerce, entertainment and so on. As it turned out, those involved in the non-agricultural occupations made possible by improved agricultural productivity generally clustered in relatively densely populated and built-up settlements; the reasons for this are considered at appropriate points in the series.

But a criterion based on economic specialization seems too disembodied, too dissociated from the physical urban reality, to capture the full meaning of 'city'. It is important therefore to add that the activities which distinguish towns and cities from villages become associated with particular spaces and structures: forums, squares and parks, streets, bridges and railways, markets, factories and shopping malls, temples, town halls and theatres; and that these spaces and structures often become emblematic of a given town or city. Always remembering that the urban built environment is designed by human beings for human purposes, we suggest that it is, apart from geographical location itself, the most definitive aspect of a city's identity, however dramatically it may be transformed by war or disaster. Emphasis on the changing spatial form and physical fabric of cities, and the ways in which they emerge from human activities, is therefore no arbitrary selection from the multiple historical phenomena of urbanization.

Colin Chant
The Open University

American cities and technology

Unlike the other textbooks in this series, this volume focuses on a single country, the USA. Since the range of urban contexts of the applications of technologies is, in that respect, less varied, the authors have organized this volume by technologies rather than by urban region or city-biography. As a result, they have been able to explore in greater depth the diverse technological experiences of US cities. It will be clear that even if there have been fewer barriers within the USA to the interurban traffic of technological innovations, local policies and circumstances have often been crucial in determining their implementation and effects in a given city. After an introduction to the pattern of urbanization in the USA from the time of the first US census in 1790, five chapters deal with the development of specific technologies in relation to the form and fabric of a range of cities up to about the time of the Second World War. Developments in New York, Chicago and Los Angeles are featured, but examples are also drawn from many other cities. The final two chapters analyse relations between technologies and cities in a broader, more comparative manner, developing these themes up to very recent times and looking appropriately beyond the borders of the USA. Although the approach here is technology-led, the focus is not narrowly technological. As with the other textbooks in this series, it is concerned with the social history of technology. Nor is this volume a social history of US technology in general, for which readers are referred to recent works by Ruth Schwartz Cowan (1997) and Carroll Pursell (1995); the aim is to examine the social history of technology in relation to the form and fabric of many diverse US urban environments.

Conventions and acknowledgements

Chapters 2–6 have appended extracts from printed primary sources. These are readily identifiable by the different typeface used and the grey rectangle beside the page number. Apart from their intrinsic value as historical texts, they have been selected to introduce students of the relations of cities and technology to a variety of types of relevant sources. These extracts have been included without critical comment so as not to pre-empt their use and analysis in a teaching context. In some cases bibliographical references have been edited out of them without ellipses in the text; ellipses do appear where other portions of text have been omitted.

We have sought to avoid what are now widely accepted as sexist expressions in our own writing, but these remain without editorial comment where they occur in quotations from existing sources. As a general rule, we have used British rather than US terminology and measures have been given in imperial units followed by a rough metric conversion. Where non-metric measures appear in existing sources, metric equivalents have not been added, but the conversion chart below should prove helpful. In addition, we list below the current, official abbreviations for US states.

The authors of a textbook such as this, which provides part of the print backbone of a mixed-media Open University course, are indebted to an unquantifiable degree to innumerable colleagues in all areas of the institution – first of all to their Course Team colleagues who authored the other volumes in the series and commented on this one. Special thanks are also due to Denise Hall, the course manager, without whose unsurpassed ability in her role, and total commitment to the course, there would be no series of books; and also to her successor, Andrew Ferguson, who picked up work on this volume in its final hectic stages. Special thanks too, to Linda Camborne-Paynter, Erleen Pilkington and their colleagues in Course Management, without whose expertise in the conversion of authors' drafts to first-rate electronic text, the textbook deadlines could never have been met; to Jonathan Davies and Sarah Hofton of The Open University's Design Studio for their skill and imagination in the exterior and interior graphic design, and to their colleagues Ray Munns and Michael Andrew Whitehead for their painstaking and creative work on numerous maps and illustrations; to picture researchers Paul Smith and Anthony Coulson, for their unstinting efforts to track down transatlantic illustration archives; and last but by no means least, to the dedicated team of Open University editors, Hazel Coleman, John Pettit and Jane Wood, whose close and rigorous reading of the textbook chapters has saved readers from numerous opacities, solecisms and awkward expressions.

Gerrylynn Roberts would like especially to acknowledge the assistance of Professor Joel Tarr as external assessor for Chapters 1–6. His deep engagement with the subject, electronic good humour and ready sharing of references, reprints and sometimes even preprints, have made a great difference to the book. She would also like to thank Robert Bruegmann, Harold Platt, Paul Sprague and Carol Willis for helpful discussion of various points. Philip Steadman would like to thank Mike Batty, Scott Bottles, John Naughton and Alex Reid for useful suggestions; Glen Cass, Andreas Schafer and Melvin Webber for copies of illustrations and papers; Bill Mitchell for sight of the manuscript of his forthcoming book *E-topia*; and Stephen Potter for long loans from his personal library of transport literature. Furthermore, the Course Team has been fortunate indeed to have the benefit of the appropriately broad knowledge and experience of its external assessor, Anthony Sutcliffe. Responsibility for any remaining errors rests with the authors.

Gerrylynn K. Roberts and Philip Steadman
The Open University

References

COWAN, R.S. (1997) *A Social History of American Technology*, Oxford, Oxford University Press.

HAMMOND, M. (1972) *The City in the Ancient World*, Cambridge, Mass., Harvard University Press.

PURSELL, C. (1995) *The Machine in America: A social history of technology*, Baltimore, MD, The Johns Hopkins University Press.

Conversion table: imperial/metric units	
To convert[†]	**Multiply by**
inches to centimetres	2.54
feet to centimetres; to metres	30; 0.3
miles to kilometres	1.61
acres to hectares	0.4 (2.5 hectares to the acre)
square miles to square kilometres	2.59
gallons (US)[‡] to litres	3.79
barrels (US liquid[§]) to cubic metres	8.38
pounds (avoirdupois) to kilograms	0.45 (2.2 pounds to the kilogram)

[†] *In the text, we have used rather rough conversions, to give a feel for metric magnitudes, rather than precise mathematical equivalents.*

[‡] *The US gallon is four-fifths the volume of an imperial gallon.*

[§] *There are 31.5 US gallons to the barrel. This does not apply to oil; there are 42 US gallons to the barrel of oil.*

Abbreviations of US states

Abbreviation	State	Abbreviation	State
AK	Alaska	MT	Montana
AL	Alabama	NC	North Carolina
AR	Arkansas	ND	North Dakota
AZ	Arizona	NE	Nebraska
CA	California	NH	New Hampshire
CO	Colorado	NJ	New Jersey
CT	Connecticut	NM	New Mexico
DE	Delaware	NV	Nevada
FL	Florida	NY	New York
GA	Georgia	OH	Ohio
HI	Hawaii	OK	Oklahoma
ID	Idaho	OR	Oregon
IL	Illinois	PA	Pennsylvania
IN	Indiana	RI	Rhode Island
IO	Iowa	SC	South Carolina
KS	Kansas	SD	South Dakota
KY	Kentucky	TN	Tennessee
LO	Louisiana	TX	Texas
MA	Massachusetts	UT	Utah
MD	Maryland	VA	Virginia
ME	Maine	VT	Vermont
MI	Michigan	WA	Washington
MN	Minnesota	WI	Wisconsin
MO	Missouri	WV	West Virginia
MS	Mississippi	WY	Wyoming

Chapter 1: THE GROWTH OF CITIES

by Gerrylynn K. Roberts

The technology of urbanization in the United States has had much in common with that of Europe, and in many cases European precedents were adopted. However, the pattern of urbanization and the built environments of individual cities have been quite distinct from European models. That this is the case is perhaps hardly surprising, given the size[1] and geographical diversity of the USA, with its abundant resources. The timing of city-building in relation both to emerging nationhood and to developments in technology was also very different from that of Europe. For students of the relationship between cities and technology, this immediately raises questions about the role of technology in the development of US cities. How did the role of technology vary from place to place in relation to local cultural, economic, political, organizational and social circumstances, including the constraints of physical geography, and individual perceptions and preferences?

In comparison with other books in this series, this volume covers a relatively short time-span. Although this feature eliminates certain historiographic challenges, the sheer scale, pace and variety of developments in the USA present other problems of analysis. Reflecting the diversity of US cities, writings on the history of the technology of the built environment in the USA tend to be either of a very general nature, aiming to comprehend broad patterns of development, or of a very particular focus, exploring individual technologies in specific locations in precise periods. Finding an accessible middle way is not straightforward. Our aim is to explore the technologies involved in the building of US cities in relation to a variety of contexts in which they were developed and applied, thus approaching the subject matter in a broader manner than a case-study approach would allow. The histories of many cities will be touched upon, but none explored in detail. This short introductory chapter will outline certain general features of the history of urbanization in the United States, and open out issues involved in considering the role of technology in city development there.

1.1   The pattern of urbanization

In 1790 the first Census of Population was taken, shortly after George Washington had become the first president of the United States of America under its new Constitution.[2] This census showed that about 5 per cent of the new nation's inhabitants lived in areas defined as 'urban' – that is, having a population of more than 2,500. These colonial urban centres occupied a relatively

---

[1]   It is roughly 1,500 miles (2,400 kilometres) from north to south, and 3,000 miles (4,800 kilometres) from east to west.

[2]   The original thirteen colonies had declared themselves to be independent of Britain in 1776. With the enactment of the Articles of Confederation in 1781, during the War of Independence, they had formed a federated nation of thirteen independent republics (states). Subsequently, a federal government was established at the national level under the Constitution of 1788. The first Congress was convened in 1789, the year that Washington was elected president.

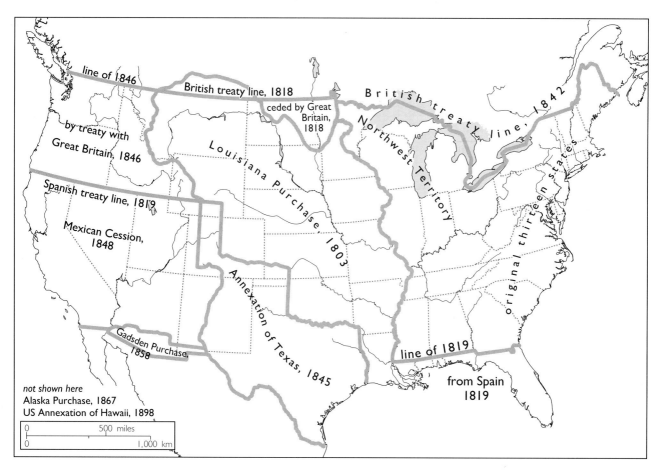

Figure 1.1 Territorial growth of the USA. The original territory of the United States in 1783 included the land that would become the original thirteen states, plus the area shown here as the 'Northwest Territory'. There was major expansion in 1803 with the Louisiana Purchase – territories bought from France by Thomas Jefferson. By mid-century, the United States had extended its territory to stretch from the Atlantic Ocean to the Pacific, and defined its northern and southern borders

narrow strip along the eastern seaboard. By the end of the nineteenth century, during which the area of the USA tripled (see Figure 1.1) and its population exploded from 5.3 million to 75.9 million (roughly quadrupling the density, even given the larger territory), the percentage of urban dwellers had increased to about 40 per cent. By then, the settled area, populated mainly by the movement of people from the east coast, extended well beyond the Mississippi River, in pockets as far as the Pacific Coast, as the 'centre of population' shifted westwards (see Figure 1.2).

The trend towards concentration into urban areas continued in the twentieth century, with 50 per cent – the figure generally taken as a benchmark for defining a nation as 'urban' – being reached by the time of the 1920 census (see Table 1.1, p.4). Fifty years later, the 1970 census revealed that – following a spurt of urbanization after the Second World War, particularly in the West – 73.5 per cent of the US's population (which now totalled more than 203 million) lived in cities (see Table 1.1). Despite the enduring image of 'wide open spaces' and rural, frontier spirit, the USA had become a very urban nation.

The enduring image is, in fact, understandable when we look at the geographical pattern of more recent urbanization in the western USA; see Figure 1.3. The 'wide open spaces' of the West have indeed persisted, despite the fact that the region, accounting for some 25 per cent of the nation's population in 1990 (but only 9 per cent in 1940), became the most heavily urbanized part of the country, with 86.4 per cent of its population living in

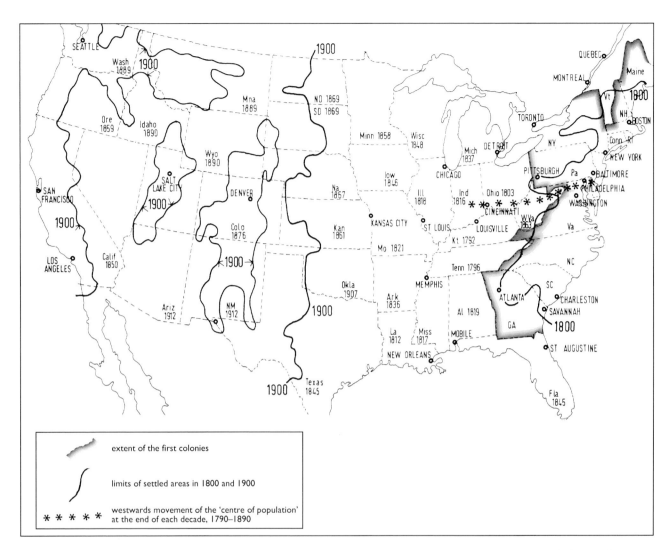

Figure 1.2 Population distribution up to 1900. 'Settled areas' in this case refers to contiguous settlement largely by people moving from the east, rather than occupation by Amerindians or colonial ventures of Spain and France. The asterisks indicate that the population as a whole, as well as the limits of settlement, was shifting west. The dates in smaller type indicate the years of statehood (reproduced from Morris, 1994, p.329; by permission of Addison Wesley Longman)

centres greater than 2,500 people, although these were very dispersed (Abbott, 1993, pp.xii, xix; United States Census Bureau, 1995c).

Furthermore, there are concentrations of population: certain states are far more urbanized than others. Table 1.2 (p.6) shows the westwards drift of the location of the nation's largest cities. In 1990, in the West Region for census returns, the states of Arizona, Colorado, Nevada, Hawaii and Utah were all more than 80 per cent urbanized, with California more than 90 per cent urban and Texas, which borders the West Region, also more than 80 per cent urbanized. Elsewhere in the country, the figures for Florida, Illinois, Maryland, Massachusetts, New Jersey, New York and Rhode Island were also over 80 per cent, with Connecticut at 79 per cent. While, as has been noted elsewhere in this series, the figure of 2,500 is rather a low threshold for defining an urban area, other criteria also indicate a rapidly urbanizing population.

Table 1.1 Urban population of the USA, 1790–1990 (adapted from: United States Department of Commerce, Bureau of the Census, 1975, A43–56, A57–72; United States Census Bureau, 1995a and 1995b)

Year	Total number of 'urban areas', defined as settlements with more than 2,500 inhabitants (> 2,500)	Urban areas with 2,500–50,000 inhabitants	Urban areas with 50,000–100,000 inhabitants	Urban areas with 100,000–250,000 inhabitants	Urban areas with 250,000–1,000,000 inhabitants	Urban areas with > 1,000,000 inhabitants	Population (in urban areas > 2,500) as % of total US population	Population (in urban areas > 100,000) as % of total US population
1790	24	24	–	–	–	–	5.1	–
1800	33	32	1	–	–	–	6.1	–
1810	46	44	2	–	–	–	7.3	–
1820	61	58	2	1	–	–	7.2	1.3
1830	90	86	3	1	–	–	8.8	1.6
1840	131	126	2	2	1	–	10.8	3.0
1850	236	226	4	5	1	–	15.3	5.1
1860	392	376	7	6	3	–	19.8	8.4
1870	663	638	11	7	7	–	25.7	10.7
1880	939	904	15	12	7	1	28.2	12.4
1890	1,348	1,290	30	17	8	3	35.1	15.4
1900	1,737	1,659	40	23	12	3	39.7	18.7
1910	2,262	2,153	59	31	16	3	45.7	22.1
1920	2,722	2,578	76	43	22	3	51.2	25.9
1930	3,165	2,974	98	56	32	5	56.2	29.6
1940	3,464	3,265	107	55	32	5	56.5	28.9
1950	4,284*	4,052	126	65	36	5	64.0†	29.4
1960	5,419‡	5,087	201	81	45	5	69.9‡	28.6
1970	6,435§	6,039	240	100	50	6	73.5§	27.8
1980	7,749			102	52	6	73.7	24.7
1990	8,510			131	56	8	75.2	24.9

* Using the 1950 definition of 'urban area'; using the 1940 definition, the figure is 4,023

† Percentages for 1950, 1960 and 1970 are based on figures that include some entities smaller than 2,500, but that are defined as belonging to urban areas. Excluding such places gives figures of 57.4, 64.0 and 65.7 per cent respectively

‡ Excludes Alaska and Hawaii

§ Includes Alaska and Hawaii

As can be seen from Table 1.1, and as we have already noted, by 1920 over 50 per cent of the population lived in areas defined as urban on the criterion of a population of 2,500. The data also show that about half of these urban dwellers (25.9 per cent of the total US population) lived in cities numbering more than 100,000 inhabitants. By 1880, New York City had already reached one million inhabitants. A decade later, the census revealed that Philadelphia and Chicago (whose population had quintupled in the previous thirty years) had also passed the million mark. Between 1880 and 1930, the percentage of people living in cities with a population greater than one million increased from 3.4 per cent to 13.3 per cent, which was to be an all-time peak (declining to 9.2 per cent by 1970 and 7.7 per cent by 1980).

This remarkable growth was fuelled in the nineteenth century by the great waves of European immigration, which were to give cities a particular character, and afterwards by rural migration. In 1930, roughly 30 per cent of the population still lived on farms, whereas in 1980 the figure was 3 per cent. Another distinctive feature of the urban population was its mobility, both between cities and their own suburbs, and between different cities. It is estimated that most US cities lost more than half of the individuals from their population every decade, but they continued to achieve rapid rates of growth none the less because the loss was compensated so rapidly by incomers. Nineteenth-century populations, then, were not very stable (Monkkonen, in The Open University, 1985, p.78).

One mechanism by which cities grew during the nineteenth century in both population and territory was annexation (of unincorporated land) or consolidation (of neighbouring municipalities). For example, in 1854 Philadelphia expanded from 2 to 130 square miles (5 to 337 square kilometres), increasing its population by a factor of four. Chicago added 133 square miles (344 square kilometres) in 1889. In 1898 Greater New York City was created, increasing from 44 to roughly 300 square miles (114 to 777 square kilometres), adding about two million people in the process. This consolidation took in Brooklyn, the fourth largest city in the country at the time. Such dramatic increases were motivated by a combination of civic pride and belief that a larger size would make urban services more efficient by allowing the better regulation of business and social affairs. Civic competition was also a motive. Chicago became the nation's second city with its 1889 expansion. Apparently, one of the motives for the expansion of New York City a decade later was concern that Chicago might overtake it (Jackson, 1985, pp.138–48). Table 1.2 shows that during the twentieth century New York remained by far the most populous city in the USA, with Chicago in second place until the 1990 census.

Urbanization in the USA was also characterized by a phenomenal amount of city-building on 'green-field' sites. It can be calculated from Table 1.1 that, between the first census (1790) and 1840, roughly two 'cities' were founded per year; over the next forty years, the average figure was twenty new cities per year; and over the fifty years from 1880 to 1930, forty-four per year (Conzen, 1981, p.321). From 1930, with the onset of the Great Depression, the number of new cities per year fell back to around thirty, but the twenty years from 1950 to 1970 saw an explosion, with more than one hundred new cities per year. Ninety-eight per cent of the cities with populations greater than 2,500 in 1900 had not even existed in 1800; of the US cities with populations greater than 100,000 in 1970, only 15 per cent had existed in 1800 (Monkkonen, 1988, pp.74–5). Clearly, starting new cities was easy in the United States. In terms of size, however, the main thrust of more recent urbanization has been the rise of medium to large cities, rather than huge new political entities.

Table 1.2 Rank-size of the ten largest US cities, 1790–1990 (adapted from: Goldfield and Brownell, 1979, pp.14, 16–17, 19; United States Census Bureau, 1995a)

	1790		1830		1870		1920		1970		1980		1990	
	Rank	Population: 1,000s	Rank	Population: 1,000s	Rank	Population: 1,000s	Rank	Population: 1,000s	Rank	Population: 1,000s	Rank	Population: 1,000s	Rank	Population: 1,000s
New York, NY	1	33.1	1	202.5	1	942.3	1*	5,620.0	1*	7,894.8	1*	7,071.6	1*	7,322.6
Philadelphia, PA	2	28.5	3	80.4	2	674.0	3	1,823.8	4	1,948.6	4	1,688.2	5	1,585.6
Boston, MA	3	18.3	4	61.3	7	250.5	7	748.0	16	641.0	21	563.0	20	574.3
Charleston, SC	4	16.3	6	30.2	†		†		†		‡		‡	
Baltimore, MD	5	13.5	2	80.6	6	267.4	8	733.8	7	905.7	10	786.7	13	736.0
Salem, MA	6	7.9	14	13.8	†		†		†		‡		‡	
Newport, RI	7	6.7	†		†		†		†		‡		‡	
Providence, RI	8	6.3	11	16.8	19	68.9	†		†		99	156.8	106	160.7
Gloucester, MA	9	5.3	†		†		†		†		‡		‡	
Newburyport, MA	10	4.8	†		†		†		†		‡		‡	
New Orleans, LA	–§		5	46.0	9	191.4	17	387.2	19	593.4	22	557.9	25	496.9
Cincinnati, OH			7	24.8	8	216.2	16	401.2	†		32	385.4	45	364.0
Albany, NY	16		8	24.2	†		†		†		161	101.7	191	101.1
Brooklyn, NY	12	4.4	9	20.5	3	420.0	*		*		*		*	
Washington, DC	–		10	18.8	13	109.2	14	437.6	9	756.5	17	638.4	19	606.9
St Louis, MO	–		†		4	310.9	6	772.9	18	622.2	27	452.8	34	396.7
Chicago, IL	–		†		5	299.0	2	2,701.7	2	3,366.9	2	3,005.1	3	2,783.7
San Francisco, CA	–		–		10	149.5	12	506.7	13	715.6	13	679.0	14	724.0
Detroit, MI	–		†		17	79.6	4	993.7	5	1,511.4	6	1,203.4	7	1,028.0
Cleveland, OH	–		†		16	92.8	5	796.8	10	750.9	18	573.8	24	505.6
Pittsburgh, PA	†		13	15.3	11	139.3	9	588.3	†		30	424.0	40	369.9
Los Angeles, CA	–		–		†		10	576.7	3	2,816.0	3	2,969.0	2	3,485.4
Houston, TX	–		–		†		†		6	1,232.8	5	1,595.1	4	1,630.6
Dallas, TX	–		–		†		†		8	844.4	7	904.6	8	1,006.9
San Diego, CA	–		–		†		†		14	696.7	8	875.5	6	1,111.1
Phoenix, AZ	–		–		–		†		20	582.0	9	789.7	9	983.4
San Antonio, TX	–		–		†		†		15	654.1	11	785.9	10	935.9

* New York included Brooklyn (from 1898)
‡ Population below 100,000

† Rank below 20
§ The dashes indicate that either the city did not exist, or it was not part of the USA at the time

1.2 Periodization of city development?

Using figures such as those in Table 1.1, historians do not always agree on which dates mark the beginning and end of significant periods in the history of US urbanization. Their arguments for differing 'periodizations' depend on the objectives of their investigation. For example, examining the spatial evolution of the US urban system of cities in terms of economic history, the historical geographer Michael P. Conzen points to the census figures from 1840, 1880 and 1930

> as useful dividing lines that separate periods of significantly different levels of urbanization, of new town founding, and of shifts in the proportional importance of the non-agricultural labour-force.
>
> (Conzen, 1981, p.321)

As the nation became urbanized and industrialized, the non-agricultural labour-force increased significantly as a percentage of the total labour-force (see Table 1.3), and by 1930, and the onset of the Great Depression, accounted for over three-quarters of the total (Conzen, 1981, pp.320–21). Conzen characterizes the first fifty years of nationhood as a period 'in which European colonialism was replaced by continental colonialism' (p.323). In other words, the great mercantile cities of the east coast shifted their focus from Europe to developing their own hinterlands and to establishing metropolitan manufacturing. During the period 1840–80, a continental network of cities with increasingly differentiated economic functions developed, while regional networks of cities also emerged, linked by increasingly complex railway connections. However, the major east-coast centres remained dominant (Conzen, 1981, pp.331–5). The period after 1880 saw the continued dominance of the east, but also extensive western urbanization, and 'the final emergence of a complex industrial belt covering nearly all the northern states' (pp.335–9; on p.338).

Eric Monkkonen, a social historian of urbanization, looking at the same figures, sees 1830 as the beginning of a full century of unhindered, continuous expansion of cities until the Great Depression. Concentrating on political and social developments during that century, he argues that 'cities began to work out their new mode of providing services, acting positively in local affairs, and doing so as competitive entrepreneurs'. When the Depression made certain activities of individualistic, local city governments unsustainable, 'a reluctant federal government stepped in ... creating new [federal] bureaucracies to accomplish the new services' (Monkkonen, 1988, p.6). Monkkonen stresses that this pattern of US city development was underpinned by virtually continuous population growth, which was closely linked not only to city prosperity but also to policy development and expansion, a situation well understood by contemporaries:

> Without the economic consequences of rising property values for generation after generation, local governments would not have been quite so able to combine conservative taxation with generous indebtedness policies and get the same benefits. These economic prospects loosened up otherwise cautious property owners, who knew that increases in land value compensated for the costs of city expansion.
>
> (Monkkonen, 1988, p.7)

Table 1.3 Percentage of non-agricultural labour-force (Conzen, 1981, pp.320–21)

Year	Non-agricultural labour-force: % of total labour-force
1840	36.9
1850	45.2
1880	48.7
1890	57.3
1930	78.4

By contrast, in *The Urban Wilderness: a history of the American city*, the urban historian Sam Bass Warner Jr adopts a periodization that prioritizes technological development in US urbanization, though his analysis is rather more fine-grained than the apparently 'determinist' quotations below would indicate:

> beginning in the late eighteenth and early nineteenth centuries, a rapid and accelerating succession of changes in transportation and technology disrupted the stability of the past and brought about several almost total reorganizations within the urban system ...
>
> To trace the patterns made by the impact of transportation and of technological innovation upon the system of American cities, it is convenient to divide our modern history into three periods, 1820–70, 1870–1920, 1920– ... In each we shall examine the state of technology and transportation, because it is from a particular technological climate and a particular configuration of transportation that the form of our cities and our business institutions inevitably takes shape.
>
> (Warner, 1995, pp.59–60)

This periodization is explicitly adapted from that put forward in classic works from the 1930s by a US journalist and commentator on urban planning, Lewis Mumford, who drew in turn on the ideas of the pioneering Scottish urban thinker Patrick Geddes. According to Mumford's scheme,[3] the first period was characterized by the use of hand-tools, water and gradually steam power – but, where there was industrialization, only slight mechanization. In terms of transport, Warner stresses the building of the canal network, and then the railway system that would link both coasts by 1869. The typical large cities of this period were regional economic centres.

The period 1870–1920 was characterized by the application of mature science and engineering to all aspects of manufacturing industry and, especially, the introduction of electricity. The railway network reached full maturity, while electric traction, the automobile and the lorry began to affect developments within cities. It was during this period, deploying the technologies of what is classically called the 'Second Industrial Revolution' when speaking of Europe, that the USA emerged as a major international industrial nation. According to Warner, the characteristic city type of this period – especially across the north of the country, stretching to the Midwest – was the metropolitan city with a specialized central business district and associated highly urbanized, large industrial regions.

The final period saw the rise of a consumer society and a transition from manufacturing to service industries, with the motorway, the aeroplane and long-distance pipelines being the new transport technologies. The characteristic urban unit was no longer the city but the 'megalopolis', a multicentred, multi-city urban region, most particularly in the Northeast, the Midwest and California (Warner, 1995, pp.60–63; see Figure 1.3).

That differing periodizations for US urbanization should be put forward – depending on whether the focus is economic history, political and social history, or technological history – underlines the complexity of the issues involved; several processes were occurring simultaneously. Our concern in analysing cities and technology in the USA will not be to prioritize any particular approach, but to look at the interplay between economics, politics, social circumstances and technological developments in shaping the built environment.

---

[3]   Mumford's terms for the three phases were 'eotechnic' (to describe the medieval city), 'paleotechnic' (to describe the industrial city) and 'biotechnic', sometimes 'neotechnic' (to describe the future city – or perhaps, now, 'post-industrial' city).

Figure 1.3 The USA at night, autumn 1985. This composite satellite photograph made dramatically clear the degree of urbanization in the late twentieth century. Note the contrast between the east – with linked urban concentrations on the north-eastern coast and along the rim of the Great Lakes, along with fairly widespread urbanization east of the Mississippi River – and the west, with very scattered concentrations of population (courtesy of Snow and Ice Data Center/Science Photo Library)

1.3 Characteristics of US cities

Although they inevitably drew on European precedents, US cities built in the nineteenth century were different in character from those of Europe (Conzen, 1980, p.120). The major difference – except for the case of New York City – was that US cities were low-density. Furthermore, their boundaries were less distinct; that is, the fluidity of development was such that there was seldom a sharp physical break between the rural and the urban. This was due partly to the absence of city walls – an important contrast with many European cities – and partly to the policy of annexation, which often took in large tracts of farmland within city boundaries to allow for future population expansion (Jackson, 1985, p.146).

Another distinctive feature was the widespread use of simple city designs on a grid plan; this made both the handling of land transactions and city expansion relatively straightforward. Importantly, the development of mass transportation occurred simultaneously with US urbanization, so it played a major role in shaping US cities and constituted a larger part of their built environments than in European cities. This led by the end of the century to a pattern of intensely concentrated central districts surrounded by a widely dispersed residential population, with industry moving from the centre to city fringes.

Underlying these specific characteristics was an important conjunction of conditions. First, a huge amount of good, adaptable land was readily available; it was both cheap and accessible. That availability – which could only be dreamed of in much of Europe with its formal land-based aristocracies – altered the status of land from having social meaning to having a more strictly economic function as a commodity for exchange between individuals. Second, those individuals were relatively unconstrained because of

> the prevailing principle that beyond establishing the most elementary frameworks for public order, the government was not to control or manage the processes of settlement and development but leave all to private initiative.
>
> (Meinig, 1993, p.255)

An ideology of individual social mobility underlay much of the city-building of the nineteenth century at a crucial formative period in the urban development of the USA.

Another major difference between the USA and Europe was that, unlike European countries, the United States had no single, 'primate' city serving as a focus of national life. This was a legacy of the colonial era, when a number of cities of more or less equal stature focused independently on Europe, rather than on each other. Furthermore, the federal structure of the nation allowed for considerable powers to be reserved to the individual states, which were competing entities and fostered their own, distinctive cities. Therefore, except arguably for New York City, which became the entry-point for the great waves of European immigration in the nineteenth century, there was no single 'magnet city', comparable to London or Paris, attracting population and enterprise.

1.4 *Industrialization and urbanization*

The role of cities in the development of the USA has been much debated by historians. In 1893, Frederick Jackson Turner, himself a son of rural Wisconsin, argued in his now-famous essay 'The significance of the frontier in American history' that until it ended in 1890, the process of development along an advancing frontier in the USA was unique, and definitive of both the American character and political outlook. That is, the West was the source of new ways of life. Driven by the availability of free land, pioneers moved relentlessly westwards, opening up new lands to the plough and developing a spirit of self-reliance and independence manifested in an individualistic form of democracy. Cities had a particular place in Turner's analysis; he argued that

> the United States lies like a huge page in the history of society. Line by line as we read this continental page from West to East we find the record of social evolution. It begins with the Indian and the hunter, it goes on to tell of the disintegration of savagery by the entrance of the trader, the pathfinder of civilization; we read the annals of the pastoral stage in ranch life; the exploitation of the soil by the raising of unrotated crops of corn and wheat in sparsely settled farming communities; the intensive culture of the denser farm settlement; and finally the manufacturing organization with city and factory system.
>
> (quoted in Meinig, 1993, p.258)

Thus Turner saw cities, linked to industrialization, as the final stage of an inexorable evolutionary process led by Europeans moving westwards. Quite apart from the fact that it overlooked the pluralism of US society (including its native peoples), and regional development, the overall model was unilinear, deterministic and projected as progressive. It has been superseded by models stressing wider cultural and contextual contingencies, along with the actions of specific peoples in continual, complex interactions between disparate societies, driven by competition and concern for prosperity (Meinig, 1993, p.259). Richard C. Wade pointed out that, far from coming at the end of the sequence of development of a moving frontier, cities were often in the van of that development; there was an urban west as well as a rural west:

> The towns were the spearheads of the frontier. Planted far in advance of the line of settlement, they held the West for the approaching population. Indeed, in 1763, when the British threw the Proclamation Line along the Appalachians to stop the flow of settlers, a French merchant company prepared to survey the streets of St Louis, a thousand miles through the wilderness. Whether as part of the activity of the French and Spanish from New Orleans or of the English and Americans operating from the Atlantic seaboard, the establishment of towns preceded the breaking of soil in the transmontane [trans-Appalachian] west.
>
> (Wade, 1996, p.1)

One implication of this is that the beginnings of US urbanization preceded industrialization. And the converse, that early (and even some later) industrialization was not a particularly urban phenomenon, is also true (Vance, 1977, pp.324–41). However, the very rapid urbanization from 1830 onwards was closely related to industrialization and the domestic demand created by so rapidly increasing a population:

> This urbanization was spurred not only by the extension of a national system of commercial towns along the major continental trade routes but also by immense growth in indigenous manufacturing that for market reasons and scale economies became increasingly concentrated in cities ... The concentration of industry was not only urban but also regional as a manufacturing belt emerged first between Baltimore and New England and later extended through the states bordering the Great Lakes.
>
> (Conzen, 1980, p.121)

Once urban industrialization was established, urbanization and industrialization became mutually reinforcing processes (Cowan, 1997, p.167).

1.5 Technology and the development of cities

During the period from the mid-nineteenth to the mid-twentieth century, what has been called the 'modern networked city' arose – that is, a city dependent on complex technological infrastructures (Tarr and Dupuy, 1988, preface). In *The Machine in America: a social history of technology*, Carroll Pursell describes the technological entity that was the modern US city by 1925:

> First, every city had an extensive (and expensive) engineering infrastructure of streets, sewers, water supply and electrical lines that formed the skeleton and nerves of the urban body. Second, no part of this infrastructure was more important than those transportation facilities that had been constructed over the years to serve first the commercial and then also the manufacturing activities of the city. Third, this manufacturing itself was only able to exist in an industrial form through the use of transport to bring in raw materials and take out finished goods, and through the application of steam and other forms of mechanical power. Fourth, many of the amenities that made city life exciting and rewarding – for example, newspapers, and parks and sports venues – were made possible on a large scale by modern invention and engineering. And fifth, in contrast to the older, inherited organic city, a newer planned city appeared, in the form of the company towns that often sprang up around new industries or in the form of suburbs, often laid out to take advantage of cheaper land and newly installed rapid transit lines. If the American city by 1920 was much more than a machine for making money, it was also unthinkable without that machine.
>
> (Pursell, 1995, pp.131–2)

The role of specific technologies in the development of US cities, especially from 1830 onwards, is both obvious and subtle. For example, the close relationship between transport and city development has already been stressed by several historians above. To take a specific instance: as in Europe, the new transport technology of railways greatly affected the built environment of cities – with railway stations, rights of way, goods sheds, level crossings, bridges – and the emerging industrial economy. To some extent, railways facilitated processes that were already under way, such as the post-colonial refocusing of eastern cities on their hinterlands. Railways were also influential in the siting of cities, but this is because they were already a known technology by the period of rapid new-city establishment in the USA (illustrated in Tables 1.1 and 1.2 above; see also Figure 2.9). City promoters could thus take advantage of the technology (Monkkonen, 1988, p.81). Although they were exceedingly influential in the development of the built environment, railways *per se* did not determine that environment. Individuals, influenced by many considerations and operating in a complex decision-making process, did that.

Within that framework, then, this volume will explore several of the technologies of urbanization in the USA. It will examine:

transportation: road, water, rail (both inter- and intra-urban), the motor car and the lorry;

building types: the balloon-frame house, the skyscraper, the suburban home;

other infrastructural and service technologies: lighting systems, communications technologies, sewage and water systems, waste and public health systems, fire-prevention technologies.

Although we shall treat them separately for convenience of organization, that these technologies did not develop independently will be a theme of this book. In common with the other volumes in the series, this book will not construe technology as a 'given', but examine it in the light of the complex of social, historical, economic, political and other contingencies that are the context for the development of a range of urban built environments in the United States.

References

ABBOTT, C. (1993) *The Metropolitan Frontier: cities in the modern American West*, Tucson, University of Arizona Press.

CONZEN, M.P. (1980) 'The morphology of nineteenth-century cities in the United States' in W. Borah, J. Hardoy and G.A. Stelter (eds) *Urbanization in the Americas: the background in comparative perspective*, Ottawa, National Museum of Man, pp.119–28.

CONZEN, M.P. (1981) 'The American urban system in the nineteenth century' in D.T. Herbert and R.J. Johnston (eds) *Geography and the Environment: progress in research and applications*, New York, John Wiley and Sons, vol.4, pp.295–347.

COWAN, R.S. (1997) *A Social History of American Technology*, Oxford, Oxford University Press.

GOLDFIELD, D.R. and BROWNELL, B.A. (eds) (1979) *Urban America: from downtown to no town*, Boston, Houghton Mifflin.

JACKSON, K.T. (1985) *Crabgrass Frontier: the suburbanization of the United States*, Oxford, Oxford University Press.

MEINIG, D.W. (1993) *The Shaping of America: a geographical perspective on 500 years of history*, vol.2, *Continental America, 1800–1867*, New Haven, Yale University Press.

MONKKONEN, E.H. (1988) *America Becomes Urban: the development of US cities and towns, 1780–1980*, Berkeley, University of California Press.

MORRIS, A.E.J. (1994, 3rd edn) *History of Urban Form before the Industrial Revolutions*, Harlow, Longman Scientific and Technical.

THE GROWTH OF CITIES

PURSELL, C. (1995) *The Machine in America: a social history of technology*, Baltimore, Johns Hopkins University Press.

TARR, J.E. and DUPUY, G. (eds) (1988) *Technology and the Rise of the Networked City in Europe and America*, Philadelphia, Temple University Press, pp.xiii–xvii.

THE OPEN UNIVERSITY (1985) A317 *Themes in British and American History: a comparative approach*, c.*1760–1970*, Focus Point 6, 'Cities and the social order, c.1850–1970', Milton Keynes, The Open University, pp.77–83.

UNITED STATES CENSUS BUREAU (1995a) '1980 and 1990 census counts for cities with 1990 population greater than 100,000', Table 1 <http:/www.census.gov/population/censusdata/c1008090.txt> (31 July 1998).

UNITED STATES CENSUS BUREAU (1995b) 'Population: 1790 to 1990; United States urban and rural', Table 4 <http://www.census.gov/population/censusdata/table-4.pdf> (3 December 1997).

UNITED STATES CENSUS BUREAU (1995c) 'Urban and rural population: 1900 to 1990', Table 1 <http://www.census.gov/population/censusdata/0090.txt> (31 July 1998).

UNITED STATES DEPARTMENT OF COMMERCE, BUREAU OF THE CENSUS (1975) *Historical Statistics of the United States from Colonial Times to 1970*, Washington DC, US Government Printing Office, Part 1.

VANCE, J.E. Jr (1977) *This Scene of Man: the role and structure of the city in the geography of Western civilization*, New York, Harper's College Press.

WADE, R.C. (1996) *The Urban Frontier: the rise of Western cities, 1790–1830*, Urbana, University of Illinois Press (first published 1959).

WARNER, S.B. Jr (1995) *The Urban Wilderness: a history of the American city*, Berkeley, University of California Press (first published 1972).

Chapter 2: TRANSPORT AND THE NINETEENTH-CENTURY CITY

by Gerrylynn K. Roberts

[I]n the explanation of that greatest of human features in geography – settlement –
transportation is a silent partner whose gift to humanity – geographical mobility –
is crucial. Settlements cannot exist unless people have some mobility. And shifts in
the availability of mobility provide, in all likelihood, the most powerful single
process at work in transforming and evolving the human half of geography …

We must be concerned with both the role transportation plays and the facilities
and technologies it employs at various times and under different circumstances.
But we always should view the transportation as a process brought into demand
by the needs of human occupation of the earth's surface – that is, by a human
geography whose greatest measure is to be found in settlement, particularly in its
most advanced form – in cities.

(Vance, 1986, pp.2–3)

Anyone who lives near a mechanized transport route may perhaps think that
'silent' is not the right adjective to describe the 'partnership' between transport
and cities in the United States. However, James Vance Jr's general point is an
important one: transport technologies profoundly affected the location, form
and built environment of cities. But it is important to recognize that, at the same
time, the development of transport systems was strongly influenced by the
requirements and thinking of city promoters, city-dwellers and people in
business. Politics, social and economic concerns (including land speculation)
and government action (or inaction) played a major role in that development
(Monkkonen, 1988, pp.163–4).

The pace of technological innovation in transport was rapid. During the
nineteenth century, transport passed through a broad trajectory – from animal
(including human) power and wind power to mechanical power, first steam-
based and later electrical. The nineteenth-century explosion of US city-building
occurred simultaneously with these transport innovations, some of which
originated outside the USA. Therefore entrepreneurs, enthusiastic city promoters
and governments were in a position to adopt the latest technologies while their
cities were being built, deploying them explicitly as means of development. This
point is meant to indicate that the opportunities were tremendous, not to imply
a lack of originality or ingenuity. Indeed, in some cases US adaptations of a basic
set of technological ideas originating from Europe were so fundamental as to
amount to a very different technological outcome.

This chapter will consider some of the built-environment issues arising from
the development of transport systems. Intercity systems such as roads, canals
and railways of course affected the environments and economies of the rural
areas through which they passed, and there are important historical questions to
raise about such systems as a whole. Our concern here, however, is principally
the cities that were linked by transport to those rural areas, and on whose urban
market-functions those areas relied. For cities, as entrepôts, accommodated both
the transport functions and the market functions, and this influenced their built
environments (Cronon, 1991). After examining some of the motives of urban
transport promoters for the development of *inter*city transport, this chapter will
look at the transition from animal power to mechanical power in *intra*city
transport over the nineteenth century.

2.1 The role of government

From the earliest years of the emergence of the United States as a new nation in the 1780s, transport was promoted – by entrepreneurs, and by the government at national, state and local levels – to foster settlement. At the same time, the focus of the former colonial mercantilist economy moved away from providing primary products for Britain towards a domestic economy developing the US hinterland and, later, towards emerging industrialization. Early motives were not only economic: many of the founding fathers, especially Thomas Jefferson, saw transport links as a crucial means of integrating the new nation, including the trans-Appalachian west. It was also seen to be important for defence, because the new country was surrounded on three sides by colonial territories of European powers – Britain in the north, France in the north and west, Spain in the south and west (see Figure 1.1 on p.2).

To say that various levels of government were involved is not to imply any overall transport strategy. Entrepreneurship and private enterprise were the principal agencies:

> turnpikes, bridges, canals, and, later, tramways and railroads were being built by private enterprise. Such projects had to be licensed by state authority, and their promoters lobbied relentlessly for subsidy and favor from governments at all levels, but such facilities as well as the actual systems of carriage – stagecoaches, wagon freighting, canal boats, steamboats, and railroad trains – were operated by private companies under minimal official regulation. Mail service and other essential government transport, such as shipments of ordnance and supplies to military posts, were normally undertaken by contract with private operators. This entire national infrastructure was shaped far more by private corporations in pursuit of profit than by government assessment of the basic needs of the nation, its regions and its local districts.
>
> (Meinig, 1993, p.252)

However, *ad hoc* government support was crucial to the success of many entrepreneurial transport schemes conceived for strictly local or private entrepreneurial reasons. This support was both financial, providing cash and/or favourable land grants, and practical – providing the expertise of army engineers (who constituted the principal technically trained body in the young country) and/or legislation.

The Land Ordinance of 1785

One way in which the fledgling federal government was particularly influential was through its method of dividing up, for distribution and sale, the unsettled public lands in the western areas of the thirteen states as far as the Mississippi River, a method eventually used for the rest of the country as new territories were acquired. The Land Ordinance of 1785 was the result of the efforts of the Continental Congress to balance the competing interests of speculative land companies, individual small farmers and its own desire to raise revenue:

> Prior to its sale, land was to be laid out in rectangular townships 6 miles square. Each township was to be divided into 36 square sections of 1 square mile or 640 acres [see Figure 2.1 overleaf]. Half of the land was to be sold by townships, the other half by sections. Sales were to be by auction in the eastern states, and a minimum price was established of $1 per acre plus survey costs of $1 per section or $36 per township. Terms of sale were cash at the time of purchase. The ordinance specified that the first seven ranges of townships should be laid out west of a line running directly north from the termination of the southern boundary line of Pennsylvania. From these townships one-seventh of the area was to be reserved by the Secretary of War to be exchanged for land certificates given in part payment for military service during the Revolution. Further, township sections 8, 11, 26 and 29 were to be reserved to the national government, and section 16 in every township was to be set aside for the maintenance of public schools.
>
> (Reps, 1965, p.216)

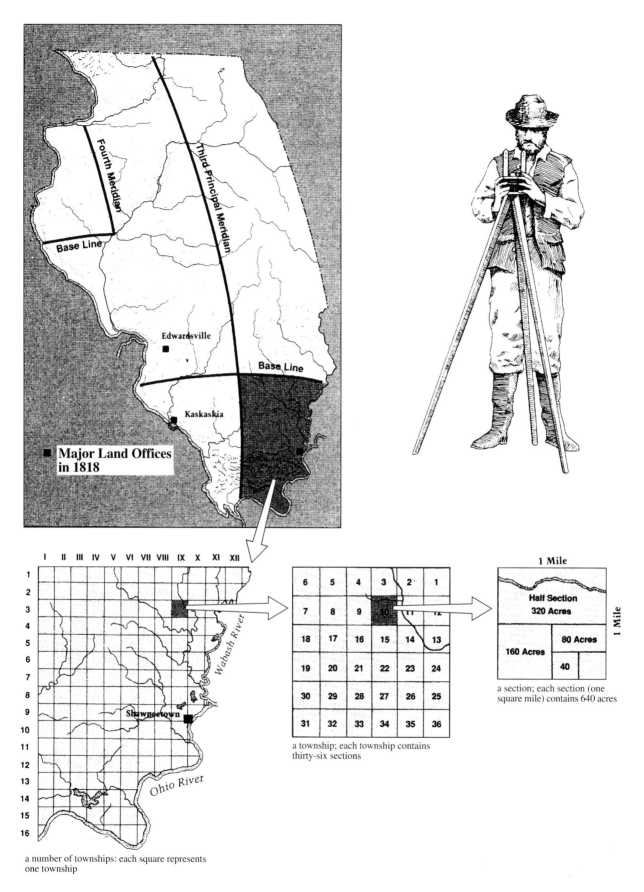

a number of townships: each square represents one township

a township; each township contains thirty-six sections

a section; each section (one square mile) contains 640 acres

Figure 2.1 The federal rectangular survey system, as described in the quotation from Reps on the previous page (adapted with permission from Buisseret, 1990; copyright © University of Chicago, all rights reserved)

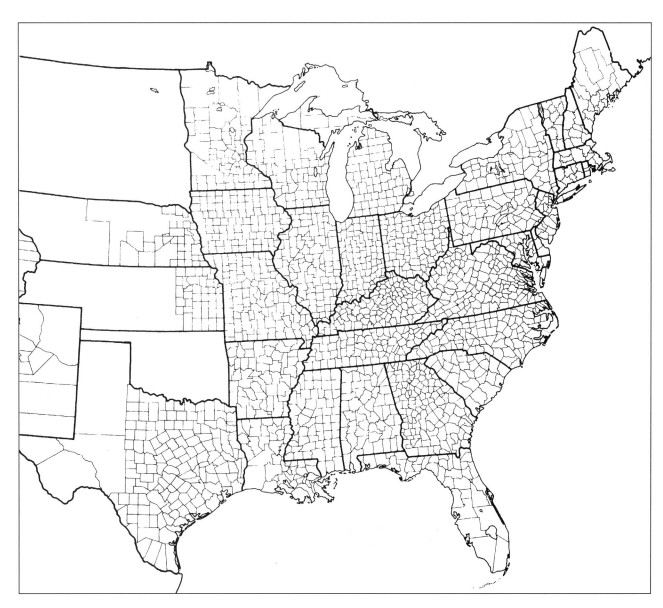

Figure 2.2 Boundaries of US counties, 1 June 1860. The influence of the federal rectangular survey system (Figure 2.1) is increasingly apparent in the more westerly areas, which were settled later. (In the USA, the hierarchy of government tiers is federal, state, county, local) (courtesy of the Library of Congress; Geography and Map Division, negative no.3261)

This method for laying out the land stimulated property speculation; it affected settlement patterns in rural sections, and also influenced the planning of towns and cities. It gave rise to the grid structure so characteristic of both town and country in the Midwest and West (see Figure 2.2).

> The effect on city planning was to reinforce the natural inclination for the gridiron street system, the easiest of all to lay out when speed or the desire for land speculation guide the hand of the surveyor. Section lines became rural roads. Where they intersected, small hamlets grew up slowly or were laid out with an eye to quick development stimulated by clever promotion and exaggerated claims of their advantageous locations. The original right-angle crossing served as base lines for new streets parallel and perpendicular to the section line roads. Later federal legislation establishing half-section townsites of 320 acres perpetuated the rectangular street system for the overwhelming majority of American cities.
>
> (Reps, 1965, p.217)

The Gallatin Plan of 1808

There was no official federal transport strategy. This was not because of lack of government attention, but a result of the way in which the implementation of the new US constitution was negotiated between the states and the federal government in the early years of the nineteenth century. Early in 1807, the secretary of the treasury, Albert Gallatin, was directed

> to prepare and report to the [US] Senate, at their next session, a plan for the application of such means as are within the power of Congress, to the purposes of opening roads, and making canals; together with a statement of the undertakings, of that nature, which as objects of public improvement, may require and deserve the aid of government; and also a statement of works of the nature mentioned, which have been commenced, the progress which has been made in them, and the means and prospect of their being completed.

(Gallatin, 1968)

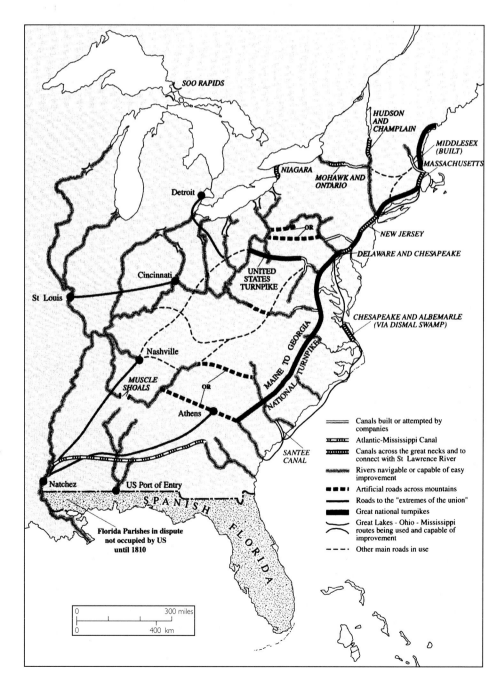

Figure 2.3 The Gallatin Plan, 1808. Gallatin proposed a number of major roads, including a north–south turnpike to complement the already planned east–west turnpike; his plan included a number of lesser roads and canals to link river navigations on opposite sides of the Appalachians, and a north–south coastal navigation system, using canals to bypass hazardous or lengthy routes (adapted from Meinig, 1993 by permission of Professor D.W. Meinig)

Extract 2.1 at the end of this chapter (p.43) is taken from Gallatin's 1808 *Report of the Secretary of the Treasury on the Subject of Public Roads and Canals.* Gallatin's report was in fact a national transport strategy, and he argued strongly that it should be funded extensively by the federal government. Figure 2.3 gives the main features of his plan.

Gallatin's analysis of the nation's transport needs was astute. He saw that a major north–south highway linking the coastal states – and various canal improvements to aid coastal shipping by eliminating hazardous passages (for example, around Cape Cod) – would help knit together the former colonies. Even more far-reaching, both literally and figuratively, were his proposals for east–west road and canal links. Their purpose was to bind to the nation the new states that were west of the Appalachian Mountains. These states – such as Ohio, Kentucky and Tennessee – were oriented naturally westwards towards the Mississippi rather than to the east, because of their river connections. Their geographical position raised fears that these states would not only bypass the East commercially, but might also ultimately leave the orbit of the eastern-based federal government. Many of the routes Gallatin proposed were eventually built (though not in a very systematic way) during the nineteenth century by entrepreneurs, often aided by – or even stimulated by – state or city governments expecting to benefit. They were not built as part of a federal plan, nor were they federally funded, for it was decided that matters of transport were not 'within the power of Congress' and must be left to individual states. The only major project to receive federal funding was the famous east–west National Road, known as the Cumberland Road, which had already been authorized in 1802. It is shown in Figure 2.3 as the United States Turnpike.

2.2   Intercity transport

'Misplaced' entrepôts

The aim of this chapter is not to explore the development of road, water (canal and steamboat) and railway systems in the US, which can be explored elsewhere (Cowan, 1997, Chapter 5; Vance, 1986). Instead this chapter aims to sketch in the principles underlying those systems as they were influenced by cities. As the Senate brief to Gallatin in 1807 recognized, a number of important transport projects had already been undertaken to serve specific local economic interests during the young country's existence. The transport system of the colonial mercantile economy had been water-based. Indeed the site along the Atlantic coast of the principal city of each of the thirteen colonies, which became the original thirteen states of the new nation, had been chosen for economic reasons on the basis of transport considerations – because of a good natural harbour, for optimal trading with England. Communications between colonies were also primarily maritime, but were not a major consideration before the revolutionary period.

When, after independence, transatlantic trade became less important than developing an internal market, not all cities were equally well placed. The Appalachian Mountain chain formed a natural western barrier for the full length of the settled area; those coastal ports that were at the mouths of rivers flowing down from the mountains were best able to establish trade with their own hinterlands. However, not all coastal ports were so well situated. Portland (ME), for example, had a fine natural harbour, but no river link to its hinterland. Other early coastal explorers had not proceeded sufficiently far inland to discover the extent, sometimes quite small, of rivers giving out to the sea (Boston; Baltimore).

Figure 2.4 Philadelphia: eighteenth-century waterfront (reproduced from Morris, 1994, p.340 by permission of Addison Wesley Longman Ltd)

In still other cases (Hartford, CT, for example), the upper reaches of some rivers were in fact in different states from their tidewater ports. Such ports were denied the opportunity for full commercial exploitation of their natural hinterlands. From the perspective of the developing economy of the new nation, Vance labels such sites 'misplaced cities', 'misplaced entrepôts' or 'badly located' mercantile centres in the sense that civic and political choices did not conform optimally to geography (1986, p.108; 1995, pp.21–6).

Coastal transport for both passengers and freight continued to grow, and it remained important until the First World War (Barsness, 1973/4, p.168). Structures relating to port activities (wharves, warehouses, merchant halls), with adjacent business districts, were characteristic of port cities; see Figure 2.4.

Canals

With the shifting of focus towards the domestic economy, transport development began to be deployed differently: in an emerging continental economy, the aim – as has already been stated – was to integrate the increasingly settled centres beyond the barrier of the Appalachians with the eastern coastal ports. Initially, eastern entrepreneurs concentrated on improving the natural system of river communications between the coastal cities and their hinterlands, notably by building canals to get around unnavigable stretches of river. Thus, while canal promoters in England built them to connect existing centres of economic activity, and helped to spur industrial development, many early canals in the US were built to allow a coastal port city to get access to its hinterland, thus promoting its development as a mercantile centre and helping it to compete with other cities.

From the first decade of the nineteenth century, steam power was applied to river and lake transport, making it possible to travel upriver against prevailing currents and to navigate the Great Lakes without depending on the wind. Robert Fulton's side-wheeled paddle-steamer, *Clermont,* regularly plied the Hudson from New York City to Albany from 1807. By the 1820s, there were regular steamship services on all the tidal rivers of the east coast and in the Chesapeake Bay. Steamboat services began on the Ohio and Mississippi Rivers in 1811; by 1855, more than 700 boats plied them regularly. Steamboats cut transport time and costs: their introduction on the major rivers stimulated inland travel, especially upriver and especially in opening up the west as river towns exploited the new possibilities.[1]

---

[1]   Meinig (1993, p.323) suggests that, despite the very great influence of the steamboat in the west, and the rise of New Orleans as an entrepôt with a population as large as Boston's by 1840, improved trans-Appalachian routes combining road and canal were still favoured by many in the west because they provided links to the established eastern entrepôts and what were seen as their favourable commercial terms.

In a bid to make New York City the principal east-coast trading centre, the state of New York sponsored the 364-mile (586-kilometre) Erie Canal, in which it was also a major shareholder. The Erie Canal was perhaps the grandest canal project of the early nineteenth century, linking the middle west, via the western New York State port of Buffalo on Lake Erie, with Albany on the Hudson River and hence with the eastern international port of New York City (see Figure 2.5). The Erie Canal was built during 1819–25, through the geographically most obvious crossing of the Appalachians along the Mohawk Valley (see Figures 2.6 and 2.7 overleaf). The Erie was the most successful of the trans-Appalachian canals, spawning an elaborate system of feeder canals from its profits; none of the ventures in other states that attempted to follow its example were able to match its success, nor did they enjoy such favourable geography.

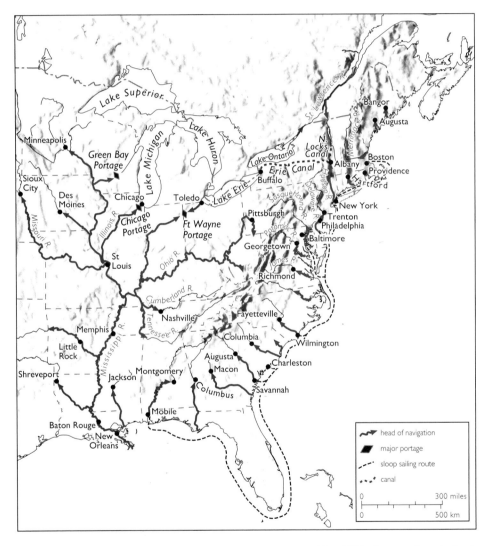

Figure 2.5 Potamic[2] phase of North American transport, 1607–c.1830; the river-based inland transport system formed a vast, but by no means integrated, network (based on material from Vance, 1995, p.71 by permission of The Johns Hopkins University Press)

---

[2] 'Potamic' means 'pertaining to rivers'.

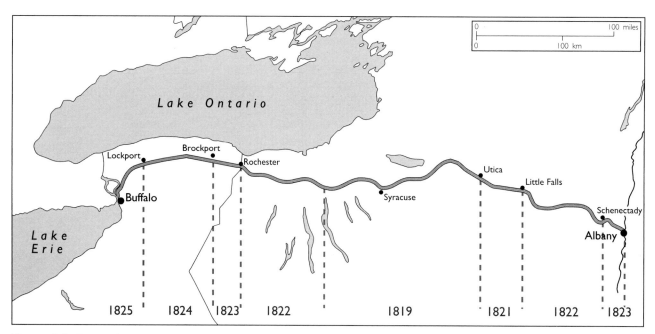

Figure 2.6 Dates of construction of the Erie Canal. The canal was built in stages, with revenues from earlier stages being used to finance subsequent construction; the string of towns that has grown up along the canal can be seen in Figure 1.3 (p.9) (adapted from Vance, 1986 by permission of Addison Wesley Longman Ltd)

Figure 2.7 Constructing the Erie Canal at Lockport. In terms of the amount of earth moved and the sheer scale of the construction, the canal set records for the period before the Civil War. Note the horse-powered crane (reproduced from Colden, C., 1825, *Memoir*, New York; Rare Books Division, The New York Public Library, Astor, Lenox and Tilden Foundations)

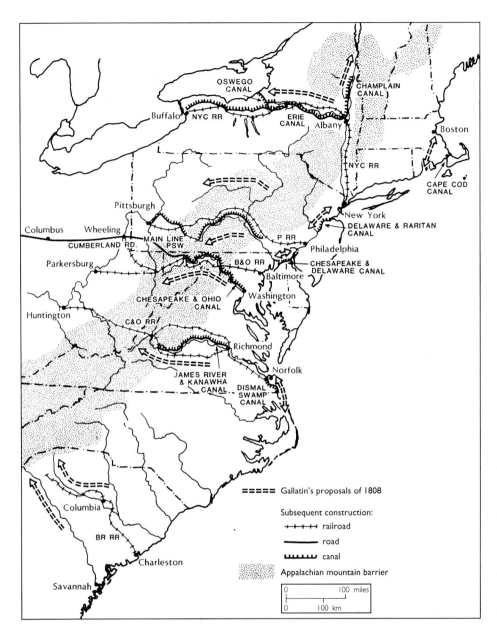

Figure 2.8 Routes proposed by Gallatin in 1808 (see Figure 2.3) that were eventually constructed. Much of what Gallatin proposed was actually achieved, not by means of roads or waterways but by using the new technology of the railway. In some cases, rail eventually superseded earlier canal lines – for example, the Erie Canal (reproduced from Vance, 1986 by permission of Addison Wesley Longman Ltd)

Railways

It was not practicable for all cities to be linked to their hinterlands by canals, and from the 1820s many entrepreneurs and city authorities eagerly grasped the opportunities provided by the new technology of railways (see Figure 2.8). Baltimore, for example, was particularly badly placed because 'its' river, the Susquehanna, ran most of its course in the neighbouring state of Pennsylvania to the north, where the city of Philadelphia was a fierce competitor for control of the trade (see Figure 2.5). Initially, helped by investment from the state of Maryland, Baltimore's city leaders promoted a canal northwards to the Susquehanna River. However, observing the beneficial effects of the westward link of the Erie Canal on New York City, the merchants of Baltimore decided in 1827 to open up their own route westwards by using a technological system recently introduced in Europe, the railway. It took twenty-four years to build the Baltimore and Ohio Railroad 200 miles (320 kilometres) across the mountains and reach the important western transport artery of the Ohio River

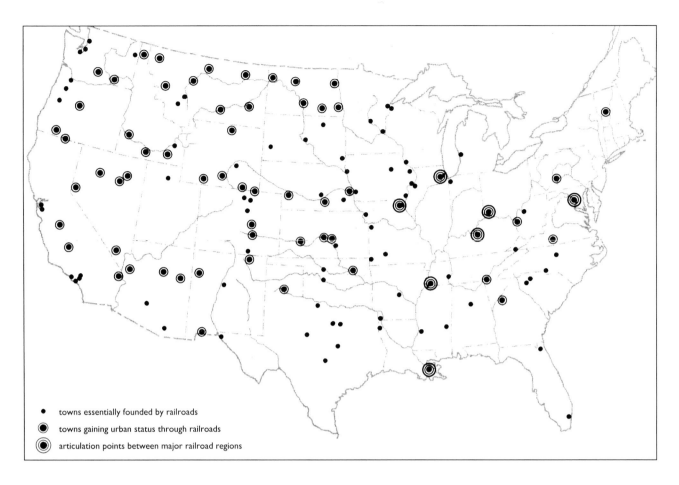

• towns essentially founded by railroads

◉ towns gaining urban status through railroads

◎ articulation points between major railroad regions

Figure 2.9 The railway-based
urban system (reproduced from
Vance, 1995, p.239 by
permission of The Johns
Hopkins University Press)

at Wheeling, now West Virginia. It is famous in railway history for having set
the standards that would define US railway practice by solving the practical
problems of crossing the Appalachians. The technology developed was, as a
result, quite distinct from European railway technology. Because of the rugged
geography, the US railway system had to tolerate steeper gradients, rougher
track-beds and tighter curves than British railways. This resulted in the
development of the swivelling front bogie-wheels for engines, which
accommodated the curves and dealt with the bumpy tracks. More powerful
engines than European ones coped with the gradients (Vance, 1995). As well
as establishing east–west connections for coastal entrepôts, such lines
stimulated the development of new towns in a corridor along their routes (see
Figure 2.9), as canals had also done. Being bypassed by the new line could be
disastrous for existing settlements. This phenomenon would subsequently be
exploited in settling the trans-Mississippi west.

By 1840, mileage of railways in the US exceeded that of canals, and was
greater than that of any other country. By the end of the 1850s, there were
more than 30,000 miles (48,000 kilometres) of track. However, this scarcely
amounted to a network: the competitive and idiosyncratic way in which the
system had been developed meant that a long-distance journey involved
several breaks, because of changes of company and gauge, and lack of urban
through-lines. However, by 1856 the Atlantic coast was connected with the
Mississippi River. Most of the early nineteenth-century development occurred
east of the Mississippi, but from the 1850s lines were built further west, joining
the Great Plains to Chicago, bringing agricultural products directly into the city
for processing and sale to the East and Europe – so establishing Chicago as a
new entrepôt, which would become dominant in the Midwest (Cronon, 1991).
The first through-route to the Pacific was completed in 1869 (see Figure 2.10).

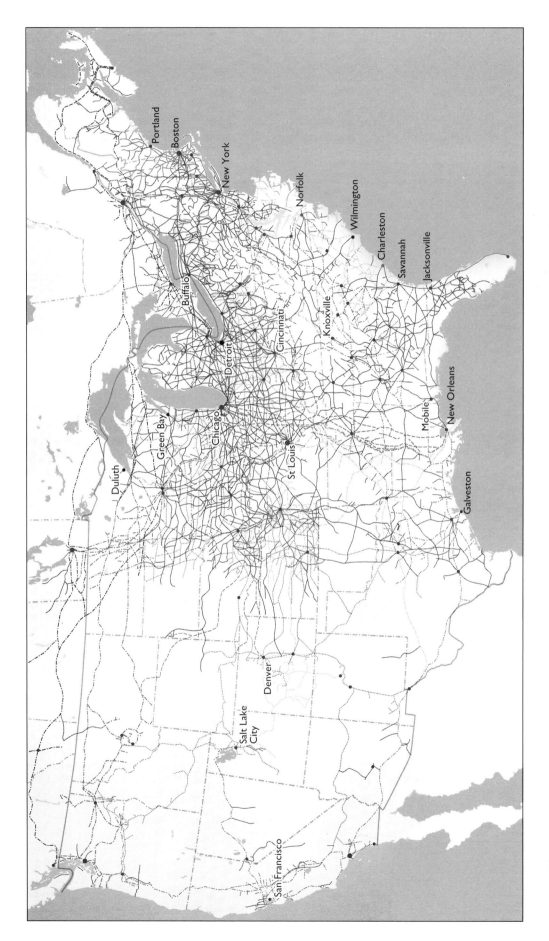

Figure 2.10 The North American railway network, 1899. By the end of the century there were some 200,000 miles (320,000 kilometres) of railway. It would be another twenty years before the network was complete (reproduced from Vance, 1995 by permission of The Johns Hopkins University Press)

Figure 2.11 Grand Central Terminal, New York City; constructed 1869–71. This was the first station in the US to cover the tracks with a vast, vaulted roof or 'balloon' shed along European lines (photograph: copyright © Collection of The New-York Historical Society)

Facilities of the built environment spawned by canals were the same as those in other types of port – wharves and warehouses, with business districts adjacent, and bridges to carry street traffic over them:

> Low bridge, everybody down!
> Low bridge, for we're comin' to a town!

as the well-known US folk-song, 'The Erie Canal', describes it (Boni, 1947, p.37). However, the railways brought new building types into cities, especially stations and train sheds, which provided architectural and engineering challenges (see Figure 2.11).

Cities also had to cope with the implications of fixed rails and rail traffic. Early railways ran into cities such as Baltimore and New York directly along their streets. For fear of accidents from collisions and exploding boilers, some cities introduced local by-laws prohibiting steam operation in central areas; consequently, horse teams were laid on to provide the power from the edge of the city to the terminus (see Figure 2.12).

Railways were very demanding of city space (see Figure 2.13). Already, by the 1860s, Chicago had become so important a rail junction that its central business area was virtually ringed by rails. No other city in the world had so many major stations at this time, an early indication of the concentrating effect of railway transport. This presented a serious problem because all lines, including those carrying goods, ran into the city

Figure 2.12 The New York and New Haven Railroad, after New York's by-law came into effect in 1832 (*The Illustrated London News*, 10 April 1852, p.285)

centre at street-level, creating great congestion and making cross-town travel both time-consuming and hazardous for goods, passengers and other street-users:

> Large freight trains posed especially awkward problems, blocking streets for such extended periods that the city passed ordinances and posted policemen to limit how long trains could remain in one place. To reduce the risk of collisions and injury to pedestrians ... by the late 1860s ... the companies had to station flagmen at busy intersections ... They had to slow trains to six miles per hour, so it took the better part of an hour just to reach city limits. And on very busy intersections, companies started building overpasses so that trains and street traffic could cross each other on different grades with no risk of interference.
>
> (Cronon, 1991, p.373)

All such measures involved costs to the railways, costs that rural customers resented. In the early 1880s, at least three rail 'beltways' were constructed around Chicago by different consortia of railway companies, with a view to easing the goods-traffic congestion by providing an alternative route for trains with destinations beyond the city. However, because of lack of co-operation among the competing railway companies, the beltways were under-utilized until well into the twentieth century. A different way of handling the transfer of goods around the city – involving a massive marshalling yard – was proposed in 1889 (see Figure 2.14). The entire system ground to a halt in the winters of 1909 and 1910, after which a uniform tariff for changing lines was finally agreed (Pinkepank, 1966).

Figure 2.13 View to the north-east across the Illinois Central yards on the downtown lakefront, Chicago, 1917. Dominating the Chicago lakefront adjacent to its central business district until the mid-1920s, this was one of more than a hundred goods yards that were part of the city's rail network (photograph: Illinois Central Railroad)

Figure 2.14 Clearing Yard, Chicago, 1923. In 1889 one railway entrepreneur proposed the construction of a massive marshalling yard to sort wagons for all the railways of the city so that they could be targeted to specific destinations, thereby eliminating much unnecessary traffic. Thirteen railway companies backed the project, and Clearing Yard was opened in 1902. Built on undeveloped land outside the city, Clearing Yard was 5 miles (8 kilometres) long. It was the biggest such yard in the world, and technologically 'state of the art'. Until the 1911 combined tariff agreement, it was used only briefly (photograph: copyright © Diggins Aerial Photo Company; reproduced from Showalter, 1923, p.382, courtesy of Cambridge University Library)

Yet in the early 1890s, 600 people per year (almost two per day on average) were killed, and even more were injured, at railway crossings in Chicago. The cynical comments of one visiting New York journalist, Julian Ralph, are often quoted. In *Our Great West* of 1893, he wrote:

> [railway companies believe] that they invented and developed Chicago, and that her people are ungrateful to protest against a little thing like a slaughter which would depopulate the average village in a year.
>
> (quoted in Cronon, 1991, p.374)

2.3   *Intracity transport*

Walking

During the early nineteenth century, as in earlier centuries, the principal means of intracity transport for most people was their own two feet, just as in Europe. This had implications for the practical size of cities and their spatial arrangement. The 'walking city', in Sam Bass Warner's phrase (1962), tended to be compact and defined by the distance an individual would be prepared to walk in a day: this gave a diameter of roughly three miles (see Figure 2.15). Keeping a horse and carriage, or even just a horse, was an option only for the very wealthy, as was 'commuting' from one of the small villages through which the new railway lines passed; such commuter railways were particularly numerous around Boston, Philadelphia and New York City (Jackson, 1985, pp.35–9).

Instead, individuals of all classes tended to live relatively close to their places of work. This would remain the case for the working classes in particular, despite the rapid development throughout the century of various intracity transport systems that would result in changes in the size and form of cities. In general, as late as 1914,

> the typical American averaged about 1,640 miles of total travel per year, and nearly 1,300 (that is, about 3.5 miles/day) of this was accounted for by walking. In other words, he travelled only about 340 miles per year with the aid of horse, cycle, or mechanical means.
>
> (Aldcroft, 1976, p.217)

It was in the twentieth century, with the internal-combustion engine, that the typical US resident stopped walking.

Transport and suburbanization

We tend to think of congestion as a recent phenomenon. However, cities have often been perceived as congested, as was the case in US cities throughout the nineteenth century, even at the time of the walking city. Rapid population increase from the second quarter of the century exacerbated this perception:

> Associated with this population rise was a nascent suburban movement; many wealthy families gave up residential locations close to the noisy and crowded marketplaces, opting instead for houses in smaller peripheral towns. These suburbanites maintained their connection with the larger population center by water ferry and steam railroad, or they assumed the expense of providing their own carriages ... Thus the residential movement away from the city center and into suburban areas predates the development of mass transit. Mass transportation innovation accentuated this thrust; it was not the seminal cause.
>
> (Holt, 1972, p.324)

Figure 2.15   The walking city: assuming that all points were equally accessible by foot, its characteristic shape was circular; see also Figure 2.30 (adapted from Monkkonen, 1988, p.178 by permission of University of California Press)

c .3 miles

The relationship between transport technology and suburbanization has aroused considerable scholarly debate, a debate focused in part on the issue of technological determinism. Some contemporary observers saw suburbanization as a solution to the problem of congestion. At the end of this chapter the views of a contemporary pioneering student of urbanization, Adna Ferrin Weber, are reproduced as Extract 2.2, which is taken from his chapter 'Tendencies and remedies'. Most scholars agree that US suburbanization had several 'causes', including aesthetic and emotional preferences for a rural ideal, differential land prices and entrepreneurial activity by developers, the location of natural resources and employment, anxieties about the ethnic and later the racial mix of cities, health concerns, a desire for privacy and an escape from city taxes (Jackson, 1985; Tarr, 1973).

Clearly, public-transport systems and suburbanization developed in parallel in the nineteenth century, though it is difficult to envisage suburbanization taking place in the way that it did without those transport systems. This is not to assert that transport caused suburbanization, but that the technologies of public transport were developed and deployed by city entrepreneurs and politicians to achieve ends that resulted in suburbanization. Though not a 'seminal cause', mass transport (operation along a particular route, to a schedule, for a single fare) was a key enabling factor for suburbanization, and profoundly influenced both the form of cities and their built environments.

Street paving

Paving was fundamental to transport developments. Seventeenth-century New York City and Boston did have some cobblestone paving, using river-worn stones set in sand or gravel; some southern cities were paved with crushed sea shells. According to McShane (1979),[3] as wagon traffic became more common towards the end of the eighteenth century, many cities began major paving programmes to accommodate it. Property developers began to provide paved streets to make their properties more accessible and therefore more valuable; cities soon required this of them as a civic duty. As paving programmes increased in the nineteenth century, the capital cost of paving established streets was met, not from the funds of impoverished municipal governments but by assessing property owners according to the length of their street frontages. This could lead to rather random quality, because those with abutting property ('abutters') often asserted their views by agreeing to assessments only sufficient for inadequate, cheap solutions. This was partly to keep their streets free of the heavy traffic that required more substantial, and therefore more expensive, surfaces. From the 1850s, paving was increasingly introduced; squared granite blocks, at first acquired from European ships (for which they had been provided as ballast), and then locally quarried with improved methods, became the preferred heavy-duty paving. Block paving, with its grooves, was thought to give horses a better grip than some of the smoother surfaces tried, such as concrete and iron (see Figure 2.16 overleaf). Where less robust surfaces were required, graded dirt, gravel or macadam was used. However, by 1880 roughly half of all city streets remained unpaved (McShane, 1979, p.280), and that was well into the era of mass-transport systems.

[3]   An abridged version of this article is reprinted in Roberts (1999), the Reader associated with this volume.

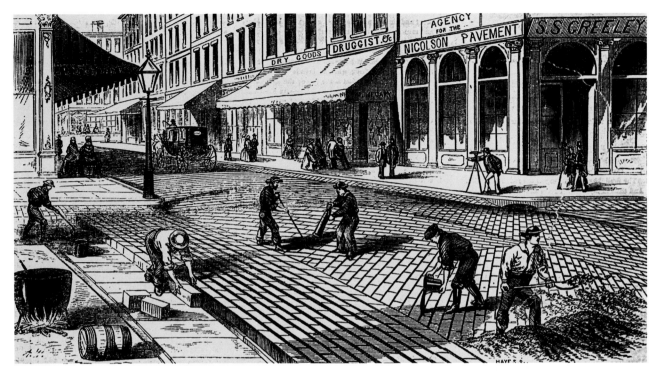

Figure 2.16 Laying Nicholson (wood-block) paving in Chicago, 1859; creosote-soaked pine blocks were laid like bricks over two layers of tarred planking on a bed of sand, then infilled with pitch and gravel. This was said to give a quiet surface, ideal for horses and pedestrians. By the time of the infamous Chicago fire of 1871, the city had 50 miles (80 kilometres) of wood-paved streets, and this is thought to have contributed to the fire's rapid spread (reproduced from *Ballow's Pictorial Drawing Room Companion*, 15 October 1858, Boston; courtesy of Chicago Historical Society, ICHi 01955)

Ferries

Ferries were the earliest form of commuter transport in the USA. Ferry services to Manhattan began in 1814, crossing the East River from Brooklyn; in 1821 a service began across the Hudson from New Jersey. Long Island land speculators were quick to buy up farm land and subdivide it into building plots for sale to people commuting by steam-ferry to lower Manhattan. Enterprising ferry owners and subdividers could argue that Brooklyn was just as close to the city's business district as residential areas further up Manhattan Island. Already, by 1870, before the completion of the Brooklyn Bridge in 1883, almost one million passengers were carried each week (Jackson, 1985, p.28).

Omnibuses

As in Europe, the first land-based form of intracity public transport was the omnibus, a merging of the idea of the stage-coach and the hackney cab, the former established to serve eighteenth-century turnpike roads, and the latter a more individualized conveyance for hire from the early nineteenth century. Initiated in France, the omnibus was soon introduced to New York City (in 1829), and quickly to other eastern cities. By 1853, New York had 683 licensed omnibuses belonging to twenty-two firms and carrying some 120,000 passengers per day. They were horse-drawn rectangular boxes, accommodating up to twenty passengers on two lengthways benches. Requiring little capital, they were often one-person operations working competitively in lucrative areas, with no constraint on their numbers or requirement to organize into a system that would serve all areas of the city. However, they did travel on fixed routes, though not on fixed schedules, stopping wherever they were flagged down.

Omnibuses were apparently the 'boneshakers' of their day – uncomfortable vehicles that jolted over cobbled streets or rutted dirt roads – but they could travel at 5–6 miles an hour (see Figure 2.17). However, congestion often made them much slower, no faster than walking, so their importance at the time has to be queried. In northern cities, snow would cause the seasonal suspension of service, though some lines substituted sledges. Fares put them out of reach of the poor, but omnibuses did make it possible for the better-off who could not afford personal conveyances to live further from city centres. Nationally, omnibus numbers did not peak until after the end of the Civil War in 1865, as the idea spread to new and growing cities; but their role in promoting change seems to have been more to establish the idea of commuting, or the 'riding habit', than to help shape particular cities. They also established the principle of fixed routes along busy streets between identified points, one end of the route usually being in a bustling city centre (Holt, 1972, p.327).

Figure 2.17 Chicago omnibus, for twenty passengers. Omnibus services began in Chicago in 1850, taking over vehicles that had formerly been used to transport guests between hotels and boat or train stations. Railways generated increased intra-urban traffic of this kind because of the need to transport passengers and goods to and from stations (photograph: courtesy of Chicago Historical Society, ICHi 09304)

Horse trams

Street railways, or horse trams (in effect omnibuses drawn along fixed rails by horses), appeared in the USA almost simultaneously with the omnibus because of the prohibition noted above on steam-engines entering Baltimore and New York. The horse-drawn extensions into those cities generated profitable local traffic from the 1830s. However, they were unpopular because their rails projected above street-level, making travel difficult and dangerous for other street-users. The development by an expatriate Frenchman in New York, in 1852, of a grooved rail that lay flush with the street, made the horse tram a very practical conveyance, and omnibuses were soon relegated to secondary and feeder lines. From that time, entrepreneurs began to bid to city governments for the rights to run strictly local street railways. Boston had a line in 1856; Philadelphia, Cincinnati, Pittsburgh and Chicago all had services by 1859 (see Figures 2.18 and 2.19 overleaf). But street railways did not spread widely to other cities until after the Civil War. By 1886 there were 525 street railways in the USA, using 100,000 horses (Rowsome, 1956, p.17).

Figure 2.18 Horse trams on State Street, Chicago, c.1870. State Street was one of the principal commercial streets of pre-fire Chicago. Numerous horse trams can be seen in this photograph; note that the street was only partially paved (photograph: Carbutt, Chicago Historical Society, ICHi 04739)

Figure 2.19 Interior of a Boston horse tram; date unknown. This was quite a plush tramcar, well decorated and with oil lamps, but straw on the floor was still the principal means of keeping passengers warm, as it had been in omnibuses. It also made the car easier to keep clean, as it could collect street dirt from boots and easily be swept out. This line apparently also sold advertising space (reproduced from Rowsome, 1956, p.18)

Rails made it possible for a single horse to pull double the load of an omnibus, up to forty passengers, and to travel at a steady 6–8 miles per hour, that is 30 per cent faster than an omnibus. People willing to commute for half an hour could now live three miles from work, rather than two, and therefore the potential residential area of a city increased (McShane, 1994, p.15). Horse-tram fares were beyond the reach of the poor, and so wealthier areas tended to be more favoured with lines, as tram investors sought a return on their substantial capital outlays (Holt, 1972, p.329). Indeed, lines for horse trams were often built by property entrepreneurs intending to make their property investment accessible and hence saleable (see Figure 2.20).

The laying of rails involved far more capital than running an omnibus line, so horse tramlines were typically developed by investment consortia rather than individuals. Rights of way for installing track had to be sought from city authorities, and this required good political connections. Generally city franchises set fares as well. Typically, fares of the more efficient horse trams were set at a much lower rate than had been the case for omnibuses, 50–80 per cent less (McShane, 1974, p.4). Most cities came to require the *quid pro quo* that horse-tram companies pave at least the area between tracks, if not the whole street, and that the width of the paved area be sufficient to accommodate wagons in order for all vehicles to move smoothly.

In addition to rails and tramcars, the major investment was in horses and their care. According to one estimate, the 'care and feeding of horses made up 40 to 50 per cent of operating costs and about 40 per cent of total investment' (J.H. White Jr; quoted in McShane and Tarr, 1997, p.120). Horses generally worked between four and six hours per day, depending on the terrain, weather and traffic conditions, and were stabled for the remainder. They covered 12–15 miles (19–24 kilometres) per shift. The number of horses kept by a line depended on, among other things, the stresses of the route; for example, horses were worked in teams to cope with hilly routes. One line in Troy (NY) had a ratio of about nine horses per car, whereas the ratio in Boston was of the order of 5:1. The working life of a horse was four or five years, after which it would be sold off for lighter work at 75 per cent of its original cost (Rowsome, 1956, p.25).

Figure 2.20 First horse tram to reach Humboldt Park, a new subdivision of Chicago; c.1886. This is an estate agent's promotion car; a poster advertising lots and cottages 'on new street car line' is difficult to see. The building in the background is a restaurant, put up to cater for those visiting the otherwise as yet undeveloped site (photograph: courtesy of Chicago Historical Society, ICHi 05499)

Figure 2.21 Manhandling a tramcar during the Great Epizootic of 1872 (reproduced from Rowsome, 1956, p.29)

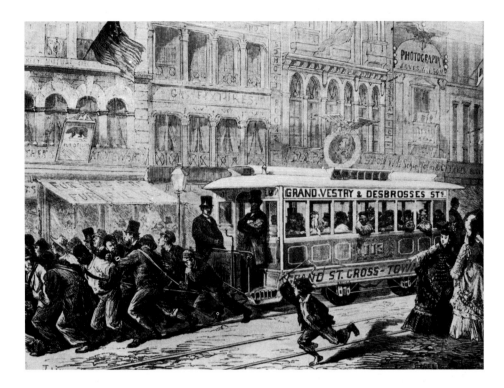

Because of the investment in stables, which had to be near the route, most lines remained discrete and fairly short rather than combining with other companies to create complex route systems. Centralized stabling was not practicable, and would not result in economies of scale. However, in cities such as New York – where, by 1860, horse trams carried 100,000 passengers per day on 142 miles (229 kilometres) of track – horse trams were a crucial link in the transport system, connecting with railways, ferries and omnibuses (Jackson, 1985, p.41). Horses were vulnerable to extremes of temperature, and also to disease. The Great Epizootic of 1872, probably an equine flu, rapidly spread throughout eastern Canada and the USA. Over two thousand horses died in Philadelphia alone in just three weeks. Eighteen thousand horses were out of action in New York City, and there was serious concern that its economy would collapse; unemployed men were rounded up into gangs to pull tramcars (see Figure 2.21). Boston suffered a devastating fire because the disease laid low the motive power for its fire engines (McShane and Tarr, 1997, p.105). And undeniably, the manure produced (15–35 pounds, or 7–16 kilogrammes, per day per horse), both from the street and from stables, presented a major problem for disposal as well as a public health nuisance (see Chapter 5).

The horse-tram companies developed policies that would characterize US public transport well into the twentieth century. They were more concerned with the number of riders than with the more subtle set of relationships between distance, cost and fares. A cheap flat-fare policy meant that it was in fact a bargain to travel farther to the wealthier suburbs, so poorer passengers with shorter journeys effectively subsidized the new developments and longer journeys of their wealthier contemporaries. Hence the horse tram favoured suburbanization and the expansion of cities' built-up areas. However, the effect of the low flat fare on suburbanization should not be exaggerated. Cheap land was a stronger motive, as was the fact that suburban living offered the better-off physical separation from the poor, from immigrants and from industry. The tram 'acted as a separating agent that allowed middle- and upper-class citizens to move to their own fashionable neighbourhoods' (Holt, 1972, p.328).

Cable-cars

The Great Epizootic emphasized the vulnerability of a horse-based transport system. The maximum load that horses could pull limited them to drawing a single tramcar, and there was a maximum number of cars that a line could accommodate. Mechanical substitutes were actively sought. For a long time some steam railways had had suburban stops on their intercity routes, but steam locomotion – even with smaller engines – was generally rejected for intracity travel for fear of explosions, because of objections to smoke and soot, and because it was not very efficient for constant stopping and starting. The effective minimum interval for stations was rather large for intra-urban transport purposes, about one mile (1.6 kilometres). In some cities, so-called steam dummy engines were used, partly to disguise the boiler and partly to help contain any possible explosion (see Figure 2.22).

Figure 2.22 Steam dummy in suburban Chicago, c.1860. So called because the boiler was masked by side-panels, this steam dummy operated at the suburban end of a horse tramline for over two decades. Steam dummies were accepted for regular use only in a very few other places. According to Rowsome (1956, p.36), one inventor reckoned that people were simply prejudiced in favour of horses (photograph: Lilian M. Campbell Memorial Collection; courtesy of West Side Historical Society, Legler Branch Library)

Andrew Halladie, a California wire manufacturer, devised a system for using a stationary steam-engine located in a power house to drive continuously a cable loop to which tramcar operators could hook their vehicles with a special 'grip' mechanism (see Figure 2.23 overleaf). He introduced this in San Francisco in 1873, where it was ideally suited to that city's grid pattern of streets climbing steep hills; it still operates as a tourist attraction. The cable-car was technologically excellent for such circumstances because, barring incidents, the cable travelled at a constant speed, untroubled by the terrain. Moreover it was effective wherever cities laid out with straight streets had sufficiently concentrated centres of business and population to guarantee a return on the heavy financial investment in the necessary infrastructure (see Figures 2.24 and 2.25). In the period 1882–90 the cable-car was taken up by most major cities other than Boston and Detroit. In 1890 cable-cars carried some 375 million passengers; cable-car mileage was roughly 5 per cent of horse-tram mileage, but cable-cars carried 20 per cent of the total number of passengers carried by horse trams (McShane, 1974, p.8). It was in Chicago that the cable-car was adopted most enthusiastically: at its greatest extent, in 1893, the Chicago system had 34 cables, 710 grip cars (plus trailers) and 82 miles (132 kilometres) of cable track run by three companies. It was the cable-car loop that gave Chicago's central business district its famous name, The Loop.

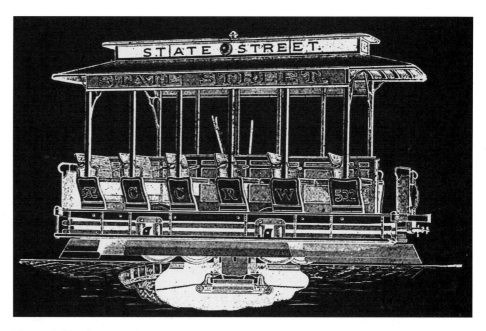

Figure 2.23 Cutaway of a grip car, showing one type of mechanism for attaching the car to a cable. Through a slit in the road, the mechanism extended below the car to grip the moving cable, which ran below ground through a specially designed and installed iron yoke (reproduced from *The Chicago Surface Lines Album*, 1882, Chicago, p.7; courtesy of Chicago Historical Society, ICHi 05493)

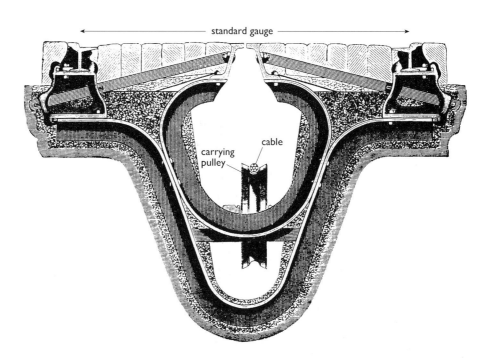

Figure 2.24 Cross-section of the Chicago City Railway roadbed. The distance between the rails was 4 feet, 8.5 inches (1.435 metres), the standard railway gauge; the iron yokes were firmly set in concrete (adapted with permission from *Cable Railways of Chicago*, p.12; courtesy of Chicago Historical Society, ICHi 29273)

In Chicago, just as the horse tram had relegated the omnibus to the role of feeder line, so in turn the cable-car relegated the horse tram to a subsidiary role and indeed took over many routes in the centre. The fact that it could add trailers, as passenger demand required, gave the system flexibility (see Figure 2.26). Furthermore, with speeds of 12 miles (19 kilometres) per hour, the cable-car doubled the distance that people could commute in half an hour.

Although the cable-car was very successful and, especially on main routes in larger cities, displaced the horse, or at any rate the horse tram for public

Figure 2.25 The Caldwell patent concrete mixer as used to lay the North Chicago Cable Road, c.1888; engraving in *The Cable Line Illustrated*. Special construction equipment was devised for setting the yokes in concrete (courtesy of Chicago Historical Society, ICHi 29272)

Figure 2.26 Cable-cars on State Street, Chicago. After the fire of 1871, State Street was the principal shopping district. In addition to the Boston Store on the left, Marshall Field's department store can be seen on the right-hand side (it has a clock on the corner, above the awning). Field led a consortium of businessmen who promoted the cable-car for Chicago. It was important to them that women residents of the new suburbs be able to get into the city centre easily to shop. An estimated 100,000 passengers per day passed his shop in the 1890s (photograph: Barnes-Crosby, Chicago Historical Society, ICHi 19268)

transport, the cable-car itself was inherently inefficient. A study of San Francisco in 1888 revealed that 57 per cent of the power generated was used to move the cable, 39 per cent to move the cars and only 4 per cent to move the passengers (Hilton, 1954, p.40). Furthermore, cables were liable to break or tangle, which would bring all the cars on that cable to a halt. In cities with severe weather, such as Chicago, snow and ice in the cable tubes were a special hazard. Elaborate steam-thawing devices were developed, but were very expensive to operate. Once the electric tram was perfected, cable-cars were quickly phased out. Some cable traction persisted in Chicago until 1906, but this was only because of the city's reluctance to allow overhead electric wires on the streets of The Loop.

The electric tram

The electric tram was a great technological success story. In the early 1880s, central electricity generation for lighting systems became commercially viable for cities. Various inventors in Europe and the USA, including Thomas Edison, had worked on aspects of electrical propulsion since the 1840s and with a degree of success (Rowsome, 1956). The availability of centrally generated electricity spurred further work on a motor for electric traction, and this, together with various other developments, made an electric-tram system technically possible (Nye, 1992, pp.86–9). It was Frank Sprague, an engineer trained at the US Naval Academy, who – after spending some time in London at the 1882 Crystal Palace Electrical Exhibition, and then working with Edison – put the elements together and developed them into a system; this was first introduced in Richmond (VA) in 1888. The principle was simple: a four-wheeled cage, or troller (hence 'trolley car'), attached to a pole, was pulled along an overhead electric wire by electric motors mounted in the trolley cars; the overhead wire gave current to the motors, and the return was through the rails.

The system in Richmond itself did not prosper, but Sprague used it to solve numerous teething problems – such as making the various components of his system more robust and arranging a technical *modus operandi* with telephone companies whose lines suffered from interference when electric trams were near by (Rowsome, 1956, p.88). The experiment was widely reported and Sprague's reputation was assured. Famously, Henry M. Whitney, who developed the wealthy Boston suburb of Brookline, rejected a proposed cable-car system for his new project in favour of electric traction, after seeing Sprague's system at work. Until well into the twentieth century, electric traction was the single biggest user of electricity in cities (Platt, 1991). In turn, it stimulated the development of large-scale central electricity-generating plants in many cities, and hence electrification more generally (Nye, 1992, p.92).

The electric tram was far more reliable than the cable-car, and much cheaper to install and run. It was adopted with astonishing speed: within a year of Sprague's Richmond experiments, 200 electric-tram systems were being installed in cities throughout the USA, either by Sprague himself or by others using his patents. By 1890 electric-tram track was already more than twice as extensive as cable-car track. Three years later, 60 per cent of the country's 12,000 miles (19,300 kilometres) of track was electrified. Electric transport continued to expand rapidly over the next decade with 30,000 miles (48,300 kilometres) of track in 1903, 98 per cent of it electrified (Jackson, 1985, p.111). In comparison with Europeans, US residents quickly took to the new means of transport. Passenger numbers in the United States, already double those of Europe in 1890, continued to increase, roughly in proportion to the increase in length of track, into the early years of the twentieth century (Nye, 1992, p.94; Weber, Extract 2.2). In Chicago, all new tramlines after 1892 were electrified, and by 1893 there were more than 200 miles (322 kilometres) of track, with some 200 million fares per year. It was not just large cities that acquired electric trams:

> The trolley represented progress and technological achievement; no community that thought well of its future could afford to be thought backward and unpromising. The electric street car was a source of pride; the very symbol of a city.
>
> (Jackson, 1985, p.111)

Not only did the electric tram enable a longer journey to work, leisure use of it also altered the lives of those who did not move to the suburbs. In the very early years of electric traction, weekend and holiday use far exceeded that of workdays. (This had also been the case with Chicago's cable-cars.) Tram companies contributed to this by building leisure attractions at the ends of their lines. The amusement park, often aglitter with new electric lighting and featuring electric-powered attractions, was a phenomenon of the late nineteenth century, capitalizing on the new importance of leisure to urban families. Though some contemporaries worried about the mixing of classes, sexes and strangers in such close proximity, Nye points to the new experience of the ride itself:

> The ride itself had a kinetic impact, not only in the movement of the car but in the varying experience of weather. Compared to walking or a buggy ride the trolley was dry if it rained, warm in cold weather, and cooled by a refreshing breeze when it was hot. In effect, it improved the climate without eliminating contact with it. The rider also became slightly detached from the passing scene, as the city became an unwinding panorama. In the same years that Americans saw their first kinetoscopes and short films, they began to experience the urban landscape as a passing spectacle framed by a moving window. Electric streetcars were speeding sightseeing platforms, and could be used to tour the city; they could cover a vast area in a few hours.
>
> (Nye, 1992, p.106)

Elevated railways and underground systems: separation of traffic levels

As streets in major cities became congested, the possibility of improving vehicle-flow through densely developed areas, by creating separate levels with clear rights of way for different types of traffic, was explored. It was on the geographically restricted island of Manhattan that such solutions were first tried in the USA. A short, and very expensive, underground line was built there in 1864, but further proposals were blocked politically, apparently owing to the influence of horse-tram owners (Vance, 1986, p.372). Among the recommendations in 1867 of a New York State legislative committee, which investigated the city's congestion problem, was a trial length of elevated line to be carried 14 feet (4.2 metres) above street-level on vertical columns placed a minimum of 20 feet apart. The cable-powered experimental line started at the southern tip of the island and ran northwards for half a mile. The experiment was so successful that immediate plans were made to extend the line to the northern end of Manhattan; four lines were built by 1880. Construction costs per mile were half those of the earlier underground. From 1871, steam power was substituted for cable traction; the system was not electrified until 1902. The 'elevated' was noisy and dirty, and it robbed light from streets and abutting buildings, but it was well used. For example, in 1879 the Third Avenue Line carried more than 45 million passengers (Vance, 1986, p.373). By 1890, New York had the most comprehensive mass-transport system in the world, each New Yorker taking 300 trips per year, compared with Londoners' 74 (Hood, 1993, p.56).

Businesspeople in Boston and Chicago also built elevated lines towards the end of the nineteenth century, and one was started in Philadelphia in 1903 – all cities with densely built-up urban cores. The Chicago 'El' (*ele*vated) was

Figure 2.27 The Chicago El
in The Loop, 1890s
(photograph: Barnes-Crosby,
Chicago Historical Society, ICHi
19001)

Figure 2.28 Bird's-eye view of the Chicago Business District, c.1898; lithograph published by Poole Brothers. The marked rectangle is the route of the Chicago El in The Loop (courtesy of Chicago Historical Society, ICHi 13554)

opened in 1892 as part of the city's preparations for the World's Columbian Exposition of 1893 (see Figures 2.27 and 2.28). It had been calculated that the existing cable-cars and horse trams would not be adequate to carry the anticipated crowds. The Chicago El was built by Charles Yerkes, entrepreneur extraordinary and master of shady deals and political manœuvres (Platt, 1991), who unified the public-transport system by using all his considerable powers of manipulation. He began with the cable-car system in the 1880s and was very quick to substitute electric power. He then developed the elevated Loop, initially powered by steam and then by electricity, to replace the cable-car loop in the central business district. The Loop line was integrated with Chicago's suburban light railway system.

Once electrified, the Chicago El companies were quick to adopt Frank Sprague's invention of multiple-unit control, which made it possible to control several electric motors from a single location. Thus each electric-powered car on an El train had an individual motor, but the whole train was run by a single driver. Cars could therefore easily be added or removed as required. It was this that made possible the efficient functioning of later underground and elevated lines. Sprague's technical case for the replacement of steam power by electric traction for elevated systems is argued in Extract 2.3, 'Application of electricity to propulsion of elevated railroads', at the end of this chapter.

The other way of removing traffic from street-level was to run it underground, but this was far costlier. Boston, whose underground system opened in 1898, was the only US city to have one in the nineteenth century; but that was effectively an underground tramcar service. An underground for New York was discussed almost continuously from the 1860s onwards, but the first line was not opened until 1904. A major role was played by a municipal body, the Rapid Transit Railroad Commissioners (RTC), which had been set up to authorize routes for elevated lines (Hood, 1993, pp.50, 66). The RTC financed the building of the New York underground system and then leased it to a private operator to run (Hood, 1993, p.71). The contractor had travelled to London to study English methods of construction and operation; after observing the electrified City and South London underground line, he opted for electric traction because its operating costs were lower than those of London's steam-powered lines. Manhattan's geology posed particular problems for deep tunnels, so instead of deep tubes, the cut-and-cover method was used to construct a shallow railway for more than half of the system. The underground framework was made of steel columns embedded in concrete. Both tunnels through hard rock and open viaducts had to be used in places to keep the tracks level (Hood, 1993, pp.84–7). An important operating innovation which made the system very efficient was to have four tracks, so that there could be a local service and an express service in each direction.

2.4 Mass transport and the late nineteenth-century city

As McShane put it:

> transit innovation was the product of a conscious search by a group of engineers drawing on a well-established body of knowledge to meet a social need. They had active support from those with the greatest financial stake in the fulfilment of that need.
>
> (McShane, 1974, p.17)

The need was to cope with the rapid expansion of city populations (see Figures 2.29 and 2.30 overleaf). The electric tram could travel at up to 20 miles (32 kilometres) per hour, and thus it increased again the settlement area from which people could commute to work in half an hour. (See Weber's explanation in Extract 2.2, p.48.)

Figure 2.29 Electric tram operating in an as yet sparsely settled suburb of Chicago, 1890. Electric tramlines were built far out into the countryside in advance of settlement. As with horse trams, the middle classes benefited, but skilled workers did also by this time (Nye, 1992, pp.93, 96–7). Elevated lines and underground systems were also used to open up new areas to settlement (Hood, 1993; photograph: courtesy of Oak Park Public Library, Oak Park Illinois)

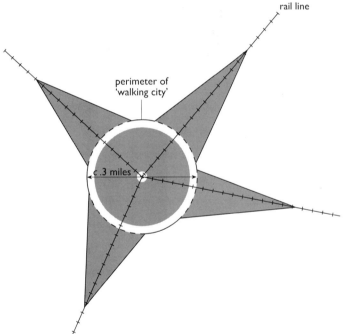

Figure 2.30 The rail-based city. Rail-based transport resulted in a characteristic 'star' shape for cities, as people settled further out along the radial lines but within walking distance of the tracks, keeping travel times constant (adapted from Monkkonen, 1988, p.178 by permission of University of California Press)

In the case of Chicago, as in many other US cities, efforts to spread out the residential population acted to concentrate the density of business activity ever more in the centre (Weber, Extract 2.2; see Figure 2.31). Radial transport lines enabled more people to live further out from the centre, but they also enabled more people to come daily into the centre where the increasing number of workers was accommodated in the new skyscraper office buildings (see Chapter 4). It was into this transport context that the key transport innovation of the early-twentieth century, the motor car, would be introduced.

And of course, public transport only dealt with passengers. At the end of the nineteenth century, goods were still transported across the city by horse and wagon. It was a simple matter for Chicago teamsters (wagon drivers) to bring the city completely to a halt in 1902, by unhitching their horses in concert. Over the last third of the nineteenth century, the number of teamsters in the ten largest US cities grew by more than 300 per cent, while the population, though expanding dramatically, grew by 100 per cent. Thus traffic was increasing steadily, as was the horse population – despite the displacement of horses from passenger transport. On average, the largest cities had one horse for every twenty-three people (McShane, 1994, pp.43–4). It was almost four times as costly per mile to move goods within cities in the 1890s as to convey them long distances. In the case of goods, too, the transformation would come with the deployment of another technology, that of the internal-combustion engine.

Figure 2.31 A junction in
The Loop, Chicago, c.1905
(photograph: Chicago Historical
Society, ICHi 04191)

Extracts

2.1   Gallatin, A. (1968) *Report of the Secretary of the Treasury on the Subject of Public Roads and Canals,* **Library of Early American Business and Industry, XIX, New York, Augustus M. Kelley, pp.5–8, 66–75 (first published 1808)**

The general utility of artificial roads and canals, is at this time so universally admitted, as hardly to require any additional proofs … The general gain is not confined to the difference between the expenses of the transportation of those articles which had been formerly conveyed by [a natural] route, but many which were brought to market by other channels, will then find a new and more advantageous direction; and those which on account of their distance or weight could not be transported in any manner whatever, will acquire a value, and become a clear addition to the national wealth. Those and many other advantages have become so obvious, that in countries possessed of a large capital, where property is sufficiently secure to induce individuals to lay out that capital on permanent undertakings, and where a compact population creates an extensive commercial intercourse, within short distances, those improvements may often, in ordinary cases, be left to individual exertion, without any direct aid from government.

There are however some circumstances, which, whilst they render the facility of communications throughout the United States an object of primary importance, naturally check the application of private capital and enterprize, to improvements on a large scale.

The price of labor is not considered as a formidable obstacle, because whatever it may be, it equally affects the expense of transportation, which is saved by the improvement, and that of effecting the improvement itself. The want of practical knowledge is no longer felt: and the occasional influence of mistaken local interests, in sometimes thwarting or giving an improper direction to public improvements, arises from the nature of man, and is common to all countries. The great demand for capital in the United States, and the extent of territory compared with the population, are, it is believed, the true causes which prevent new undertakings, and render those already accomplished, less profitable than had been expected ...

The present population of the United States, compared with the extent of territory over which it is spread, does not, except in the vicinity of the seaports, admit that extensive commercial intercourse within short distances, which, in England and some other countries, forms the principal support of artificial roads and canals. With a few exceptions, canals particularly, cannot in America be undertaken with a view solely to the intercourse between the two extremes of, and along the intermediate ground which they occupy. It is necessary, in order to be productive, that the canal should open a communication with a natural extensive navigation which will flow through that new channel. It follows that whenever that navigation requires to be improved, or when it might at some distance be connected by another canal to another navigation, the first canal will remain comparatively unproductive, until the other improvements are effected, until the other canal is also completed ...

The general government can alone remove these obstacles ...

The early and efficient aid of the *federal* government is recommended by still more important considerations. The inconveniences, complaints, and perhaps dangers, which may result from a vast extent of territory, can no otherwise be radically removed, or prevented, than by opening speedy and easy communications through all its parts. Good roads and canals, will shorten distances, facilitate commercial and personal intercourse, and unite by a still more intimate community of interests, the most remote quarters of the United States. No other single operation, within the power of government, can more effectually tend to strengthen and perpetuate that union, which secures external independence, domestic peace, and internal liberty ...

Recapitulation and resources

The improvements which have been respectfully suggested as most important, in order to facilitate the communication between the great geographical divisions of the United States, will now be recapitulated [see Figure 2.3]; and their expense compared with the resources applicable to that object.

I   From north to south, in a direction parallel to the sea coast:

	Dollars
1 Canals opening an inland navigation for sea vessels from Massachusetts to North Carolina, being more than two-thirds of the Atlantic sea coast of the United States, and across all the principal capes, cape Fear excepted,	3,000,000
2 A great turnpike road from Maine to Georgia, along the whole extent of the Atlantic sea coast,	4,800,000
	7,800,000

II   From east to west, forming communications across the mountains between the Atlantic and western rivers:

	Dollars
1 Improvement of the navigation of four great Atlantic rivers, including canals parallel to them,	1,500,000
2 Four first-rate turnpike roads from those rivers across the mountains, to the four corresponding western rivers,	2,800,000
3 Canal around the falls of the Ohio,	300,000
4 Improvement of roads to Detroit, St Louis and New Orleans,	200,000
	4,800,000

III   In a northern and north-westardly direction, forming inland navigations between the Atlantic sea coast, and the great lakes and the St Laurence:

brought forward	*12,600,000*	*Dollars*
1 Inland navigation between the North river and lake Champlain,	800,000	
2 Great inland navigation opened the whole way by canals, from the north river to lake Ontario,	2,200,000	
3 Canal around the falls and rapids of Niagara, opening a sloop navigation from lake Ontario to the upper lakes, as far as the extremities of lake Michigan,	1,000,000	
		4,000,000
Making together,		$16,600,000

… Amongst the resources of the union, there is one which from its nature seems more particularly applicable to internal improvements. Exclusively of Louisiana, the general government possesses, in trust for the people of the United States, about one hundred millions of acres fit for cultivation, north of the river Ohio, and near fifty millions south of the state of Tennessee. For the disposition of those lands a plan has been adopted, calculated to enable every industrious citizen to become a freeholder, to secure indisputable titles to the purchasers, to obtain a national revenue, and above all to suppress monopoly. Its success has surpassed that of every former attempt, and exceeded the expectations of its authors … It is believed that nothing could be more gratifying to the purchasers, and to the inhabitants of the western states generally, or better calculated to remove popular objections, and to defeat insidious efforts, than the application of the proceeds of the sales to improvements conferring general advantages on the nation, and an immediate benefit on the purchasers and inhabitants themselves. It may be added, that the United States, considered merely as owners of the soil, are also deeply interested in the opening of those communications, which must necessarily enhance the value of their property. Thus the opening of an inland navigation from tide water to the great lakes, would immediately give to the great body of lands bordering on those lakes, as great value as if they were situated at the distance of one hundred miles by land from the sea coast. And if the proceeds of the first ten millions of acres which may be sold, were applied to such improvements, the United States would be amply repaid in the sale of the other ninety millions …

In the selection of the objects submitted in obedience to the order of the Senate, as claiming in the first instance the aid of the general government, general principles have been adhered to, as best calculated to suppress every biass [sic] of partiality to particular objects. Yet some such biass, of which no individual is perfectly free, may without being felt, have operated on this report. The national legislature alone, embracing every local interest, and superior to every local consideration, is competent to the selection of such national objects …

All which is respectfully submitted.
ALBERT GALLATIN
Secretary of the Treasury
Treasury Department
4th April 1808

2.2   Weber, A.F. (1965) *The Growth of Cities in the Nineteenth Century: a study in statistics*, Ithaca, Cornell University Press, pp.457–75 (first published 1899) [4]

… Is there, then, 'no remedy until the accumulated miseries of overgrown cities drive the people back to the country?' One remedy is to admit the harmful tendencies of city life, to fight city degeneration on its own ground, and free city life from as many ills as possible. This work is now proceeding on a vast scale, and in a vast number of ways. Private philanthropy and public supervision go hand in hand. Not only complete drainage, paving, water-supply, inspection of food, etc., are required from the municipality, but also small parks, playgrounds, public baths and laundries, and a variety of other institutions. A vast deal has been accomplished in this line, and the work is only begun. Much may be expected from the progress of invention and discovery and the growth of capital. Prof. Marshall indicates how certain improvements (some of which have already been made in American cities) would

> enable a large part of the population to live in towns and yet be free from many of the present evils of town life. The first step is to make under all the streets large tunnels, in which many pipes and wires can be laid side by side, and repaired when they get out of order, without any interruption of the general traffic and without great expense. Motive power, and possibly even heat, might then be generated at great distances from the towns (in some cases in coal mines) and laid on whenever wanted. Soft water and spring water, and perhaps even sea water and ozonized air, might be laid on in separate pipes to nearly every house; while steam-pipes might be used for giving warmth in winter, and compressed air for lowering the heat of summer; or the heat might be supplied by gas of great heating power laid on in special pipes, while light was derived from gas especially suited for the purpose, or from electricity; and every house might be in electric communication with the rest of the town. All unwholesome vapors, including those given off by any domestic fires which were still used, might be carried away by strong draughts through long conduits, to be purified by passing through large furnaces and thence away through huge chimneys into the higher air. [†]

But while much is to be expected in this direction in the near future, the most encouraging feature of the whole situation is the tendency, heretofore alluded to in the present essay, toward the development of suburban towns. The significance of this tendency is that it denotes, not a cessation in the movement toward concentration, but a diminution in the *intensity* of concentration. Such a new distribution of population combines at once the open air and spaciousness of the country with the sanitary improvements, comforts and associated life of the city …

American cities are not so compactly built as European cities. [Across the Atlantic], where it is still the practice to live in rooms connected with one's store or workshop, the density of population is remarkable in comparison with American cities, thus:

	No. of cities considered	Population per acre
United States	28	15:2
England	22	38.3
Germany	15	25.9

Almost as many English urbanites dwell on 230,000 acres as Americans on 638,000 acres. The German percentage is somewhat more favorable, until we restrict the comparison to the building area. Then the population per acre in fifteen American cities is 22 as compared with 157.6 in thirteen German cities. [‡]

Paris is one of the most densely populated cities in the world. Upon an area slightly smaller than Kansas City's (20,774 acres), Paris concentrates about two and one-half million persons as contrasted with the latter's 133,000. The following comparative table of individual cities will further illustrate the point here insisted upon: [§]

[4]   For ease of reference, some data have been omitted from some of the tables in this extract.

[†]   *Principles of Economics*, 3rd edn, p.305n.

[‡]   *11th Cen., Social Statistics of Cities*, p.14.

[§]   Ibid., p.13.

	Population per acre			Population per acre
London	56.4	Liverpool	99.4
present New York	13.0	Manchester	39.5
Paris	126.9	Hamburg	30.7
Berlin	100.8	St Louis	11.5
Chicago	10.7	Boston	18.5
Philadelphia	12.6	Baltimore	23.0
Brooklyn	44.6	Birmingham	51.1

These figures are to be used cautiously, as they depend somewhat upon the amount of suburban territory recently annexed: but on the whole, they demonstrate that population is spread out over a larger territory in American than in European cities. It has sometimes been urged that this is largely a result of the development of the electric street railway in America, but the causal connection is not apparent. The first street railway using electric propulsion was opened in 1886, and the number of miles in operation at the time of the latest census was not only small in the aggregate, but was restricted for the most part to smaller cities than those at present under consideration. It should rather be said that the American *penchant* for dwelling in cottage homes instead of business blocks after the fashion of Europe is the cause, and the trolley car the effect. Philadelphia was the 'city of homes' long before *rapid* transit. Philadelphia in 1880 led all other American cities in length of horse-car lines, but the horse-car is too slow to carry the majority of workingmen to and from their work each day. Hence the comparative figures of mileage and number of rides per inhabitant of American and European cities are indications of low or high density of population, which may be regarded as the cause of street railway buildings.†

A better index of suburban travel is the number of commuters carried by the steam railways and their percentage in the total number of passengers. Such statistics are furnished by the Eleventh Census (*Social Statistics of Cities*, pp.49–50) and it appears that, on the whole, suburban traffic increases in the same ratio with the magnitude of cities:

Cities	No. of cities	Commuters	Ratio of commuters to all passengers
10,000–25,000	75	4,764,884	28.6
25,000–50,000	29	6,667,220	29.3
50,000–100,000	23	3,956,938	31.1
100,000+	24	79,945,182	52.1
Total	151	95,334,224	46.4

† The 11th Cen. Rep. on Transportation by Land (p.685) contains comparative figures of the average number of rides annually per inhabitant:

New York	297	Buffalo	65
Kansas City	286	London	74
San Francisco	270	Liverpool	51
Boston, Lynn and Cambridge	225	Glasgow	61
Brooklyn	183	Berlin	87
Chicago	164	Hamburg	78
Philadelphia	158	Vienna	43
St Louis	150	Budapest	37

Berlin, with the best street railway system in Europe, would rank twenty-second among the twenty-eight American cities of 100,000+.

In the matter of street railway travel, Chicago and Philadelphia ranked far below New York. Evidently this is connected with the fact that they have a larger traffic on the regular railway lines. But the palm for suburban travel belongs to Boston, which had almost as many commuters as New York and Chicago put together:

	Commuters	Ratio to all passengers
Boston	24,587,000	62.9
Chicago	16,903,000	85.9
Philadelphia	10,714,000	70.7
New York	8,643,000	26.9
Cincinnati	3,697,000	86.9
Pittsburgh	2,698,000	48.8
San Francisco	2,367,000	37.3
St Louis	2,164,000	75.8

It is clear that we are now in sight of a solution of the problem of concentration of population. The trolley car and the bicycle may serve the purpose in middle-sized cities or even in the less populous cities of the first class. But when the city attains a population of a quarter of a million, more rapid transit than the electric surface railway can furnish is imperatively demanded. A surface railway cannot well run cars at a speed of more than nine miles an hour, and the legal limit in New York State is ten miles an hour. But, as few workingmen can afford to spend more than half an hour in going to their work, they would then be compelled to dwell within three or four miles of the factory and could not have homes in the open country or suburbs, which are at least seven miles beyond the center of the large city. Even the elevated system would not serve the purpose with its regular trains, which cannot be run at a speed in excess of 12 miles an hour. On the other hand, the regular railway lines with fast suburban trains are too few in number to serve a large territory. The sole remedy is the multiplication of steam railroads or the building of elevated and underground four-track systems, thus providing for express trains with a speed of at least 25 miles an hour. Then the workingman could establish his dwelling in the suburbs anywhere within a radius of ten miles of the center of the city. Moreover, by virtue of the geometrical proposition that the areas of two circles are to each other as the squares of their radii, you will quadruple the area for residences every time you double the distance travelled. If in the first case (3 miles radius), the land within the circle affords room for 20,000 dwellings, when you double the speed (ordinary elevated system) you will have an area large enough for 80,000 dwellings. Double it again (the underground railway supposition) and you will have ground for 320,000 houses …

New York still lags behind, but Chicago and more especially Boston (where the legislature has aided the public) are developing a first-rate system of suburban communication. It cannot be doubted that the extremely satisfactory housing of Bostonians as compared with European urbanites is due not less to the fostering of suburban travel by the steam railways than to the development of the trolley system …

Attention has often been called to another encouraging tendency favoring suburban growth, namely, the transference of manufacturing industries to the suburbs. The local advantages of a suburban town have been pointed out; they include not only a great saving in rent and insurance, but economy in the handling and storing of goods. All carting is avoided by having a switch run directly into the factory; saving to machinery is effected by placing it all on solid foundations on the first floor; and plenty of space is at hand for the storing of fuel and materials, so that these may be bought when the market offers the most favorable terms. Finally, the suburban employer is likely to secure a high grade of employees. On the one hand, he is not antagonized by the trade unions, who can treat with him as effectively as if he were in the city itself; on the other hand, his large workshops, and the prospect of a cottage and garden, and open-air life, attract operatives of the best class. Statistical data regarding the location of factories in suburbs are not available, but the strong tendency in that direction is familiar to all Americans …

Suburban growth as a result of this tendency cannot be forced; it must wait upon economic forces. But the growth of purely residential suburbs can be influenced a great deal by public policy. In the past it was chiefly the middle classes who could afford to dwell in the suburbs. But if society wishes to minimize the evils of concentration of population, it must abandon the hope of accomplishing great things by such palliatives as model tenements (which, if located in the city, often serve merely to prevent factories from moving to the suburbs), building laws, inspection of buildings, and the various other ameliorations already discussed. Four goals are of fundamental importance: (1) a shorter working day, which will permit the workingman to live at a distance from the factory; (2) associations for promoting the ownership of suburban homes by workingmen; (3) cheap transit; (4) rapid transit. The importance of the two latter policies has been urged in so eloquent words by Dr Cooley that they deserve quotation:

> Humanity demands that men should have sunlight, fresh air, the sight of grass and trees. It demands these things for the man himself, and it demands them still more urgently for his wife and children. No child has a fair chance in the world who is condemned to grow up in the dirt and confinement, the dreariness, ugliness and vice of the poorer quarters of a great city. It is impossible to think with patience of any future condition of things in which such a childhood shall fall to the lot of any large part of the human race. Whatever struggles manhood must endure, childhood should have room and opportunity for healthy moral and physical growth. Fair play and the welfare of the human race alike demand it. There is, then, a permanent conflict between the needs of industry and the needs of humanity. Industry says men must aggregate. Humanity says they must not, or if they must, let it be only during working hours and let the necessity not extend to their wives and children. *It is the office of the city railways to reconcile these conflicting requirements.*

The extent to which this function may be fulfilled is indicated by the progress already made in Boston, Sydney, etc. … The electric trolley car is helping in the transformation, and its influence will undoubtedly be apparent in the Twelfth Census.

The 'rise of the suburbs' it is, which furnishes the solid basis of a hope that the evils of city life, so far as they result from overcrowding, may be in large part removed. If concentration of population seems destined to continue, it will be modified concentration which offers the advantages of both city and country life …

2.3 Sprague, F.J. (1886) 'Application of electricity to propulsion of elevated railroads', *Electrical World*, vol.7, pp.27–44

When, only a few years since, against all the hue and cry of its opponents, but in answer to the growing demand for rapid transit, the elevated railroads in New York were projected and finally constructed, I doubt if their most ardent supporters looked forward to the fact that in 1885 the roads could not accommodate the number of passengers who wish to be carried. But so great has been the growth of the city, so rapidly has the business section pushed to the northward and forced large numbers of the people, especially the laboring classes, to seek homes at the upper limits of Manhattan Island, where rents, already increasing because of increased facility of access, are still lower than could be readily commanded for business purposes in the lower part of the city, that the roads at night during the commission hours are now taxed beyond their present resources, and the demand is urgent that their carrying capacity be increased, or that new routes be opened for the accommodation of the public. But new roads require time for construction … Most urgent, then, is the necessity for additional means of transportation, and, for the present at least, by the elevated railroads …

In considering how the capacity of the roads may be increased, it is necessary to note the character of the roads and their method of operation, and for that purpose we will take the Third avenue line as an example. This extends from the Harlem River to South Ferry, and has a length of about eight and a half miles. With the main features of the construction of the road-bed you are all familiar. It consists of a series of built-up wrought iron columns of various heights, the bases resting in solid stone foundations and supporting a superstructure of lattice girders, making up a very strongly braced web system, on which are laid the ties and rails of a complete double-track road, with all the necessary arrangements of switches and sidings …

In the seventeen miles of single track there are altogether fifty-two stations. A train weighing, as often happens, over ninety tons, must make the trip from South Ferry to Harlem River in about forty-two minutes, including stopping at twenty-six stations and discharging and taking on passengers. This is at the rate of only about ten traffic miles per hour. Surface railroads stopping at any street corner make six miles per hour. Hence it is evident that the elevated railroads cannot afford to make any slower time than at present. If they could run faster, their traffic would be very materially increased …

With cars of a seating capacity of forty-eight and room to accommodate forty more standing, it is evident that the capacity of each car of the train cannot well be increased, and the capacity of the trains can only be increased by increasing the number of cars. There are already sixty-three trains in operation at one time on this road during commission hours; they run at very close intervals and they must be under the most perfect control. Every car added increases the momentum of the train, rendering it more difficult to quickly and efficiently check the speed, and increasing the liability of and the danger from any derangement of the brake apparatus, especially where dependent upon one source. If the length of trains be increased, admitting that they can be handled without difficulty, the engines will be forced to a duty much beyond that for which they were constructed …

Furthermore, when the tracks are in a slippery condition, the weight of the engines does not now give sufficient rail adhesion to promptly get away from the stations. An additional car will render this delay more frequent and tend to reduce the traffic miles per hour run during commission hours. To get increased traction in the present system, it is absolutely necessary to increase the weight of the locomotives, but such increase of weight is useless unless it be effective on the drivers; but if effective on the drivers, then we have an increased dead weight applied to the superstructure of the road at one point …

How, then, is the capacity of this road to be increased? It certainly ought not to be done by increasing the traction of the locomotives. This is out of question, for leaving out of consideration the impracticability of a change of the locomotives, it would be impolitic and unwise to increase their weight … Any increased weight to increase the traction of the locomotive will increase the strain at the weakest point of the structure …

Two ways, however, remain by means of which the capacity of the roads can be increased: one, by increasing the number of cars in a train, provided it can be done without increasing the weight of the tractive engine, or if it can be distributed and the train remain under perfect control: another, by increasing the mean running speed … To get a higher speed of the trains it is probably necessary to increase the power, and at the same time to increase the traction. This increased traction however, must be distributed: it cannot be put at any one point.

It is not practicable to materially increase the number of trains during commission hours, because they run nearly as closely as can be now to make their running time.

We see, then, that steam locomotion does not present many opportunities to better the condition of the road. Hence we are obliged to turn to some other method of locomotion, and that which promises the most satisfactory solution is an electrical system.

I have for a long time been elaborating such a one, and am now convinced that this is the future method of propulsion for the trains of the elevated roads, and it is a near future.

In the substitution of one system by the other on these roads, what is the object sought? Ultimately, of course, it is to get a greater return for a given investment …

By a system of electrical propulsion the power can be distributed underneath the cars – every car, or two cars if need be, being a unit – and at the same time arrangements can be made for propelling five or six cars under simultaneous control.

By distributing the power under the car, the whole weight of the car and passengers can be made effective for traction, such traction weight being six times as great as is afforded by the present locomotives. This will enable the cars to be started more promptly, brought to speed more quickly, and stopped in shorter intervals, increasing the mean rate of speed, and thereby the capacity of the road.

Weight is the necessary practical adjunct for traction. The elevated roads present a peculiar problem. To attempt to solve that problem by replacing the present locomotives by electric locomotives of lighter weight, or even of the same weight, is to shut our eyes to plain mechanical and engineering truths, and does not advance by one single step such solution.

The making of cars individual units of locomotion will enable the intervals between trains to be made one-third of the present schedule for a large part of the time. This would greatly increase the number of passengers during the day and night who would make use of the roads, and this too, without materially increasing the running expenses.

Another important advantage will be the great reduction in the vibration and wear and tear of the superstructure by distributing the weight so much more evenly … Slipping or skidding, such as is now common both in starting and stopping, would be unknown.

The objects thus far indicated as attained are important, and some of them would be very instrumental in the accomplishment of that most immediate and pressing object, the increasing of the carrying capacity of the road without reducing the time schedule. This must be done, even if the operating expenses are increased. I am glad to say that it can be done with a decrease of these. It is a problem requiring some thought.

References

ALDCROFT, D.H. (1976) 'A new chapter in transport history: the twentieth-century revolution', *Journal of Transport History*, n.s., vol.3, pp.217–39.

BARSNESS, R.W. (1973/4) 'Maritime activity and port development in the United States since 1900: a survey', *Journal of Transport History*, n.s., vol.2, pp.167–84.

BONI, M.B. (ed.) (1947, 18th edn) *The Fireside Book of Folk Songs*, New York, Simon and Schuster.

BUISSERET, D. (1990) *Historic Illinois from the Air*, Chicago, University of Chicago Press.

COWAN, R.S. (1997) *A Social History of American Technology*, Oxford, Oxford University Press.

CRONON, W. (1991) *Nature's Metropolis: Chicago and the Great West*, New York, W.W. Norton.

GALLATIN, A. (1968) *Report of the Secretary of the Treasury on the Subject of Public Roads and Canals*, Library of Early American Business and Industry, XIX, New York, Augustus M. Kelley (first published 1808).

HILTON, G.W. (1954) *Cable Railways of Chicago*, Chicago, Electric Railway Historical Society (Bulletin 10).

HOLT, G.E. (1972) 'The changing perception of urban pathology: an essay in the development of mass transit in the United States' in K.T. Jackson and S.K. Schultz (eds) *Cities in American History*, New York, Alfred A. Knopf, pp.324–43.

HOOD, C. (1993) *722 Miles: the building of the subways and how they transformed New York*, Baltimore, Johns Hopkins University Press.

JACKSON, K.T. (1985) *Crabgrass Frontier: the suburbanization of the United States*, Oxford, Oxford University Press.

McSHANE, C. (1974) *Technology and Reform: street railways and the growth of Milwaukee, 1887–1900*, Madison, State Historical Society of Wisconsin.

McSHANE, C. (1979) 'Transforming the use of urban space: a look at the revolution in street pavements, 1880–1924', *Journal of Urban History*, vol.5, pp.279–307.

McSHANE, C. (1994) *Down the Asphalt Path: the automobile and the American city*, New York, Columbia University Press.

McSHANE, C. and TARR, J.A. (1997, 2nd edn) 'The centrality of the horse in the nineteenth-century American city' in R.A. Mohl (ed.) *The Making of Urban America*, Wilmington, S.R. Books, pp.105–30.

MEINIG, D.W. (1993) *The Shaping of America: a geographical perspective on 500 years of history*, vol.2, *Continental America, 1800–1867*, New Haven, Yale University Press.

MILLER, D.L. (1996) *City of the Century: the epic of Chicago and the making of America,* New York, Simon and Schuster.

MONKKONEN, E.H. (1988) *America Becomes Urban: the development of US cities and towns 1780–1980*, Berkeley, University of California Press.

MORRIS, A.E.J. (1994, 3rd edn) *History of Urban Form before the Industrial Revolution*, Harlow, Longman Scientific and Technical (first published 1972).

NYE, D.E. (1992) *Electrifying America: social meanings of a new technology, 1880–1940*, Cambridge, Mass., MIT Press.

PINKEPANK, J.A. (1966) 'How the belt came to be', *Trains*, vol.26, pp.42–9.

PLATT, H.L. (1991) *The Electric City: energy and the growth of the Chicago area, 1880–1930*, Chicago, University of Chicago Press.

REPS, J.W. (1965) *The Making of Urban America: a history of city planning in the United States*, Princeton, Princeton University Press.

ROBERTS, G.K. (ed.) (1999) *The American Cities and Technology Reader: wilderness to wired city*, London, Routledge, in association with The Open University.

ROWSOME, F. Jr (1956) *Trolley Car Treasury: a century of American streetcars, horsecars, cable-cars, interurbans and trolleys*, New York, Bonanza Books.

SHOWALTER, W.J. (1923) 'America's amazing railway traffic', *National Geographic*, vol.43, pp.353–404.

SPRAGUE, F.J. (1886) 'Application of electricity to propulsion of elevated railroads', *Electrical World*, vol.7, pp.27–44.

TARR, J.A. (1973, 2nd edn) 'From city to suburb: the "moral" influence of transportation technology' in A.B. Callow Jr (ed.) *American Urban History: an interpretative Reader with commentaries*, London, Oxford University Press, pp.202–12.

VANCE, J.E. Jr (1986) *Capturing the Horizon: the historical geography of transportation since the transportation revolution of the sixteenth century*, New York, Harper and Row.

VANCE, J.E. Jr (1995) *The North American Railroad: its origin, evolution, and geography*, Baltimore, Johns Hopkins University Press.

WARNER, S.B. (1962) *Streetcar Suburbs: the process of growth in Boston, 1870–1900*, Cambridge, Mass., Harvard University Press and MIT Press.

WEBER, A.F. (1965) *The Growth of Cities in the Nineteenth Century: a study in statistics*, Ithaca, Cornell University Press (first published 1899).

Chapter 3: TRANSPORT IN THE TWENTIETH-CENTURY CITY – AUTOMOBILITY

by Gerrylynn K. Roberts

3.1 Introduction

The new technology of the motor car became central to the development of cities during the twentieth century in the USA, where it was adopted far more rapidly and extensively than anywhere else in the world. At the start of the twenty-first century, when the motor car is such a dominant part of our lives, be they rural or urban, it might be difficult to view this as other than inevitable. In technological terms, the motor car differed importantly from its mechanically propelled transport predecessors. In the first place, the motor car was independent in that it did not have the constraint of following routes fixed by rails, and in that it carried its own fuel supply. In addition, although the basic technology was eventually applied to public transport in the form of the motorized bus, the motor car was projected and immediately understood as a device for personal, rather than public transport. In social terms, it gave individuals the possibility of independence not only from prescribed routes but also from rigid timetables. That is, it made possible the personal convenience of temporal as well as spatial flexibility, while assisting with physical burdens. And it was private. Based on these differences, politicians, planners, manufacturers, businessmen and individuals made decisions which affected profoundly the physical fabric of cities and the daily lives of all city-dwellers and users.

As was the case with the technologies of public transport in the nineteenth century, although the technology of the motor car made possible important changes in cities, the motor car itself was not an autonomous agent of change:

> Because so much of our initial knowledge of a city derives from a transportation perspective, we often err in our analyses of the relationship between the two. Because automobiles, for instance, structure how we visualize mid-twentieth-century US cities, we begin to think that the automobile has actually structured them. Consider, for instance, how urban observers have viewed the relationship between the automobile and the city. In the early twentieth century they saw the automobile as the solution to urban problems from slum housing to traffic congestion; by the 1960s, it had become the cause of urban problems. Both positions are wrong, but their wrongness suggests the power which we willingly attribute to transportation technology.
>
> (Monkkonen, 1988, p.159)

This chapter will explore some of the political, social, economic, technological and other circumstances which help to explain why and in what ways the motor car came to have so strong an influence on the built environment of US cities during the twentieth century.

3.2   The emergence of the motor car[1]

There were numerous attempts in Europe and the USA to develop self-propelled vehicles during the nineteenth century. Steam, electricity and the internal-combustion engine were all investigated as power sources for them (see Figure 3.1). However, steam-engines were large, prone to explosion, required long warm-up periods and access to top-ups of bulky fuel and water. The use of early viable, steam-powered, self-propelled vehicles was discouraged by regulatory constraints resulting from fears for public safety (McShane, 1994, pp.97–8). Later in the century, after the development of electric storage-batteries in the 1880s, experiments were also made with electric propulsion. But the power-to-weight ratio of electric batteries was low and they had only a short range before needing expensive recharging. Refined versions of both steam and electric vehicles were in successful commercial production at the end of the nineteenth century and enjoyed considerable vogue in the US (McShane, 1994; Villalon and Laux, 1979). However, by that time, motor cars with petrol-powered internal-combustion engines were also in production and, with none of the drawbacks of steam- or electric-powered vehicles, soon prevailed.

The internal-combustion engine was developed in Europe and first applied to self-propelled vehicles there. The first US-produced cars did not appear until the 1890s; they drew on European models. Methods of publicity, such as staging races, were also derivative. The first, high-profile, US automobile race, sponsored by the *Chicago Times Herald*, took place on Thanksgiving Day, 1895, in freezing temperatures shortly after a blizzard had deposited a foot (30 cm) of snow, bringing telegraphic communications, railways, cable and tramcars to a standstill (*Scientific American*, 1895, p.357). After the race, no less a figure than the pioneering developer of electric lighting, Thomas Edison, predicted that all vehicles in large cities would soon be powered by motors.

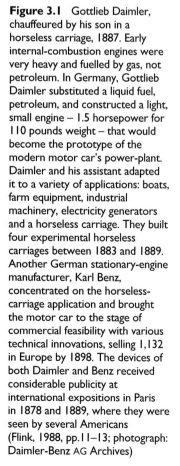

Figure 3.1 Gottlieb Daimler, chauffeured by his son in a horseless carriage, 1887. Early internal-combustion engines were very heavy and fuelled by gas, not petroleum. In Germany, Gottlieb Daimler substituted a liquid fuel, petroleum, and constructed a light, small engine – 1.5 horsepower for 110 pounds weight – that would become the prototype of the modern motor car's power-plant. Daimler and his assistant adapted it to a variety of applications: boats, farm equipment, industrial machinery, electricity generators and a horseless carriage. They built four experimental horseless carriages between 1883 and 1889. Another German stationary-engine manufacturer, Karl Benz, concentrated on the horseless-carriage application and brought the motor car to the stage of commercial feasibility with various technical innovations, selling 1,132 in Europe by 1898. The devices of both Daimler and Benz received considerable publicity at international expositions in Paris in 1878 and 1889, where they were seen by several Americans (Flink, 1988, pp.11–13; photograph: Daimler-Benz AG Archives)

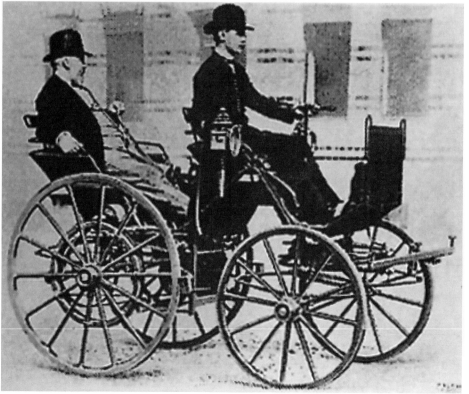

[1]   This section draws heavily on Cowan (1997, Chapter 10) and Flink (1988, Chapters 1–4).

Table 3.1   US Motor vehicle registrations, 1900–1975* (United States Department of Commerce, Bureau of the Census, 1975, A73, G152–G155)

Year	Total: 1,000s	Motor cars: 1,000s	Trucks: 1,000s	Buses: 1,000s	Persons per car	Number of cars to trucks
1900	8.0	8.0	–	–	9,499†	–
1905	78.8	77.4	1.4	–		55
1910‡	468.5	458.3	10.1	–	201	45
1915	2,490.9	2,332.4	158.5	–		15
1920	9,239.1	8,131.5	1,107.6	–	13	7
1925	20,068.5	17,481.0	2,569.7	17.8		7
1930	26,749.8	23,034.7	3,674.5	40.5	5	6
1935	26,546.1	22,567.8	3,919.3	58.9		6
1940	32,453.2	27,465.8	4,886.2	101.1	5	6
1945	31,035.4	25,796.9	5,076.3	162.1		5
1950	49,161.6	40,399.0	8,598.9	223.6	4	5
1955	62,688.7	52,144.7	10,288.8	255.2		5
1960§	73,868.6	61,682.3	11,914.2	272.1	3	5
1965	90,357.6	75,257.5	14,785.7	314.2		5
1970	108,407.3	89,279.8	18,748.4	379.0	2	5
1975		106,731.0	25,755.7			4††

*   *1975 figures from Jackson, 1985, pp.162–3*
†   *It is estimated that there were some 18,000 people per car in 1898 (Showalter, 1923, p.337)*
‡   *State registration was not required until 1913, after which the figures are rather firmer*
§   *Alaska and Hawaii are included from here on*
††   *Jackson (1994, p.167) suggests that the 1990 figure was about 2*

He was himself developing an electric delivery-van. In the same year, the *Horseless Carriage*, the first US motoring-trade journal, appeared and recorded some 300 companies and individuals working on motor cars. They were not lone tinkerers but tended to be involved in other businesses such as bicycle or carriage manufacture, or engineering of some variety. For the most part, US car-makers were assemblers of parts gathered from a variety of sources, rather than manufacturers, so start-up costs were low. By 1899, thirty companies had entered into commercial production, making some 2,500 vehicles (Cowan, 1997, p.227). The early years of the new century would see rapid expansion of the use of automobiles, especially during the 1910s and 1920s. Except for dips in the early years of the Depression and subsequently in the early years of the Second World War, motor-car use increased steadily and rapidly thereafter, especially from 1950 when post-war prosperity brought its purchase price within the range of US workers. So important did the motor car become in the USA that over the period 1947 to 1983, the demand for it rose at three times the rate of increase in income (Bowden and Offer, 1994, p.743).

By the eve of the First World War, US production had far outstripped that of Europe and accounted for 80 per cent of motor cars world-wide. One reason for this was a shift in the USA from producing individually crafted cars for the luxury market to mass-producing motor cars at affordable prices for a much wider market. Though perhaps the most famous and persistent, Henry Ford was not the first entrepreneur in this market. Ransom Olds, whose Oldsmobile marque, like Ford, still exists, pioneered the idea of a mass-produced down-market motor car in 1900 and had already sold some 6,500 by the time the car's popularity was demonstrated by its being featured in a hit song in 1905 (see Figure 3.2 overleaf).

Figure 3.2 Songsheet cover page for an early automobile hit from Tin Pan Alley, 1905:

'Come away with me, Lucille
In my merry Oldsmobile,
Over the road of life we'll fly,
Autobubbling you and I,
To the church we'll swiftly steal,
And our wedding bells will peal,
You can go as far as you like with me,
In our Merry Oldsmobile'

(lyrics by Gus Edwards, quoted in Jackson, 1985, p.159). From very early in its history, the motor car was considered to be 'fast' in more senses than one (courtesy of The Lester S. Levy Collection of Sheet Music, Special Collections and Archives, Milton S. Eisenhower Library, The Johns Hopkins University)

Figure 3.3 The moving-chassis assembly line for Ford's Model T, Highland Park, Michigan, c.1913. The moving assembly line was first used for making magnetos, motors and transmissions in mid-1913. The output from these sub-assembly lines threatened to flood the final line, so a moving-chassis assembly line was introduced that autumn, cutting overall assembly-time by three-quarters (photograph: Henry Ford Museum and Greenfield Village)

In 1908, Ford, having previously produced more luxurious cars, introduced his basic, box-like Model T, a model that was available only in black and would persist until 1927. Innovative production methods allowed it to be manufactured in unprecedented quantities and eventually at an unprecedentedly low price. On the eve of the First World War in Europe, Ford's firm introduced the moving assembly line for motor-car manufacture (see Figure 3.3).

The price of the Model T plummeted: it cost $850 in 1908, $360 in 1916 and $290 in 1927. In 1916 alone, Ford produced 577,000 Model Ts, reaching a peak of two million in 1923, a production figure that would not be topped until the 1970s, by the Volkswagen Beetle (see Figure 3.4). By 1927, the final year of Model T production, more than 15 million had been sold. One in every two cars in existence anywhere was a Model T (Jackson, 1985, pp.161–2).

Figure 3.4 Thomas Edison with the Model T given to him by Henry Ford in 1916, subsequently updated, photographed at Fort Myers. Early in his career, Ford worked for the Detroit Edison Company through which he met Edison in 1896. At the time, Edison was experimenting with battery cars and Ford with a petrol-driven car. Apparently, Edison advised Ford to stick with it, for the petrol car had the advantage of carrying its own fuel. Edison was reportedly enamoured of his Model T, and Ford – his neighbour in Florida during the winter months – made sure that it was well maintained and updated (photograph: courtesy of the Edison and Ford Winter Estates, Fort Myers, Florida)

Figure 3.5 1923 Dodge Custom coupé, the first car with an all-steel, closed body. The closed car was introduced in the late 1910s. The all-steel body was stronger and allowed for automatic welding, thus increasing durability and lowering labour costs. The continuous strip mill for rolling sheet steel and the continuous-process technique for manufacturing plate glass both helped to make the closed car affordable and profited from the demand of the motor-car industry (photograph: Chrysler Historical Collection)

But by 1927, Ford's market share was falling because of the success of the strategy of his competitor, General Motors. Instead of offering a universal car for a mass market, General Motors, supported by heavy lifestyle advertising, aimed to sell more expensive cars with annual model changes to a more mature motor-car market that was ready to 'trade up'. Also, in the 1920s, the closed car became the preferred purchase, marking a change among urban users from leisure to more regular motor-car use in all weathers. By 1927, more than 80 per cent of motor cars manufactured in the US were closed; that is, enclosed by a roof (Flink, 1988, p.213; see Figure 3.5).

James Flink points to a number of reasons for the rapid adoption of the motor car in the early years of the twentieth century: utilitarian advantages of economy and efficiency over the horse; public-health benefits of horseless cities; the breaking down of rural isolation; and the enhancing of US values of individualism and mobility (Flink, 1988, p.28). Furthermore, people in the USA had higher per-capita incomes and experienced a more equitable distribution of income than was the case in Europe. In a vast domestic market, more people were able to afford motor cars in the USA than in Europe, especially with the hire-purchase schemes introduced first by General Motors and then by Ford. '[A]utomobility permitted a restructuring of American society, via technical efficiency along lines dictated by traditional cultural values' (Flink, quoted in Pursell, 1995, p.239). Extract 3.1, 'The automobile industry', written in 1923 by William Joseph Showalter, a staff reporter on *The National Geographic Magazine*, is an enthusiastic contemporary view of the spread of the motor car in the USA (see pp.82–4 below).

3.3   *Good roads*

The successful adoption of the automobile was dependent on the existence of roads on which it could be driven. Up until about 1910, the motor car in the United States was largely an urban vehicle, for that is where there was paving (Monkkonen, 1988, pp.175, 293; McShane, 1994, p.106). By 1880, roughly 50 per cent of urban streets were paved; the figure would reach only 58 per cent by 1909. By that date, however, 30 per cent of paved streets had a superior asphalt surface suitable for pneumatic tyres, as compared with 2.5 per cent in 1880 (McShane, 1979, table 1).[2] The Good Roads movement for the improvement of rural roads was the result of a joint campaign by bicyclists and reforming farmers. The bicyclists' organization, the League of American Wheelmen, was formed in 1880 with strong backing from bicycle manufacturers. In 1892, it founded the *Good Roads Magazine* and spearheaded the lobbying of Congress for better roads, resulting in the setting up in 1893 of a national government body with responsibility for them. This was a major departure from the usual principle that roads were a local responsibility. In general, however, farmers were reluctant to support the funding of roads both because of resistance to higher taxation and because of concern that they would be funding outsiders' transport. Not until they were made 'auto-mobile' by Henry Ford and the Model T did farmers back the movement. In fact, Ford had recognized the problem of inadequate roads, and among the reasons for the great success of his Model T was the car's suitability for the unsatisfactory rural roads of the period, making it attractive to farmers. Ironically, the improvement in roads generally was among the reasons for the decline in the Model T's popularity: people no longer had to put up with the car's basic design and rough ride.

As early as 1888, New Jersey was the first state to legislate for the funding of roads through state and county taxation. By 1913, twenty-six more states had followed New Jersey's lead. From having almost no paved rural roads in 1910, the nation moved rapidly to a position of having 250,000 miles (400,000 kilometres) of surfaced roads by 1914 and a system of interconnected concrete highways by 1930 (Flink, 1988, pp.169–71). From 1921, 50 per cent federal funding became available for state road-building projects. Petrol tax was introduced by the state of Oregon in 1919, and by 1929 every state imposed one as a source of revenue for road-building. The section on 'Highways and Highway Utilization' reproduced as Extract 3.2 at the end of this chapter (p.86) discusses the development of a national road system in the United States (see Figure 3.6).

---

[2]   An abridged version of this article appears in the companion Reader to this volume (Roberts, 1999).

Figure 3.6 US Route 1 in New Jersey. Finance from the Federal Highway Act helped to fund good four-lane roads such as US Route 1, which would eventually stretch from northernmost Maine to the southern tip of the Florida Keys, linking up the major eastern cities (photograph: *American City Magazine*)

The bicycle was also important for the history of the motor car in that bicycle manufacturers provided technical expertise for some early developments. Furthermore, as the first form of widely affordable individual transport, the bicycle created an awareness of the attractions of personal transport which made it easier for the public to understand the motor car in such terms, much as the horse car had prepared the ground for later, mechanized, public-transport systems. In addition to promoting road surfacing and literally smoothing the way for the later motor car, the bicycle lobby also won some path-breaking judicial decisions that would metaphorically smooth the way. During the 1870s and 1880s, several US cities banned the bicycle, but the last of the bans was lifted in 1885. There was an urban cycling boom in the 1890s, by which time cycling was defined as a common right. '[Bicycles] are not an obstruction to, or an unreasonable use of the streets of a city, but rather a new and improved method of using the same, and germane to their principal object as a passageway' (George B. Clementson, *The Road Rights and Liabilities of American Wheelmen*, 1895 quoting an 1892 judgement in Kansas, in McShane, 1994, p.117).

The proper use of urban streets was a matter of controversy. Clay McShane (1979) argues that the late nineteenth century was marked by a major change in perception from urban streets being seen as recreational and social spaces to their acceptance as arteries for fast-moving transport. Often this difference in perception was one between residents and municipal engineers. As noted in Chapter 2, surfacing of urban streets was a local government responsibility, funded, until well into the twentieth century, by taxing abutting property owners whose particular interests might differ from those of the city authorities. However, municipal engineers and city authorities had many reasons for promoting smooth surfaces beyond responding to cyclists' pressures. As sanitation became a major issue in the closing decades of the century, smooth surfaces were favoured because they were easier to sweep clean of horse droppings (see Chapter 5). Smooth streets could also be treated to keep dust down, a concern for influential businessmen in central Chicago, for example (Barrett, 1983, p.51). Thus for a mixture of reasons, at least some urban roads were ready for the motor car (Monkkonen, 1988, p.176).

3.4 Urban car ownership

If the motor car was mainly an urban phenomenon up until about 1910, by 1920 the Model T had made it ubiquitous. Rising rural incomes up to the First World War and a comparative lack of rail transport in rural areas made the robust Model T attractive to farmers (see Figure 3.7), and rural car ownership was very high by the mid-1920s (Flink, 1988, p.132).

Figure 3.7 Cartoon postcard showing multiple uses for the Model T on the farm. Transport was not the motor car's only rural use. Farmers recognized it as an all-purpose power-provider for running farm (and in some cases domestic) machinery; eventually the industry responded to this type of user initiative by selling adapter kits (photograph: Henry Ford Museum and Greenfield Village, 'Just in Time Images', Graphics Collection)

Although car ownership was ubiquitous by 1927, it was not universal. In that year, 56 per cent of US families owned cars. Given the high rural car ownership of the 1920s, that means that a lower percentage of urban families owned them. By 1950, the figure had increased to only 59 per cent of households owning cars, although there was one car registered for every four people in the country, indicating that many families had more than one car. (This applied to 18 per cent of families already in 1927.) In general, the extent of urban car ownership and use varied inversely with city size, and there was class variation as well. In the Pittsburgh area, for example, 1934 data show that the wealthiest areas had high levels of car ownership, as high as 96 per cent, while the figure for working-class mill areas was 38 per cent. Usage also varied depending on the availability of transport alternatives and economic level. Pittsburgh's middle classes had both higher rates of car usage for travel to work and higher rates of usage of public-transport alternatives. Only 12 per cent of the workers in Pittsburgh's mill areas travelled to work by motor car; 72 per cent of them walked (Tarr, 1978, pp.36–7). Joseph Interrante suggests that 20 to 30 per cent of daily traffic into large cities was car-borne in 1930; the figure for medium-sized cities was 50 to 60 per cent, while cities of about 10,000 in population were totally dependent on the motor car (Interrante, 1983, p.94). Certainly, the middle-class citizens of Middletown (Muncie), Indiana, where Robert and Helen Lynd conducted their famous social surveys in 1924–5 and 1933, were heavily car-dependent and remained so throughout the Depression, prioritizing the car over other possessions when money was tight.[3]

In 1937, the auto was covered by the Lynds in a chapter on 'Spending leisure', underlining the fact that much use of the motor car by urban owners was for leisure purposes. And much leisure involved travel out to the

---

[3]   See Extract 3.3 at the end of this chapter, which is taken from the second of Robert and Helen Lynd's survey publications, *Middletown in Transition* (1937).

Figure 3.8 A roadside picnic in New York State in the 1920s (reproduced from Showalter, 1923, p.392)

countryside. Henry Ford, for example, saw the motor car as a means of escape from the city: 'We shall solve the city problem by leaving the city' (quoted in Flink, 1988, p.139).

> I will build a motor car for the great multitude.
> It will be large enough for the family, but small enough for the individual to run and care for. It will be constructed of the best materials, by the best men to be hired, after the simplest designs that modern engineering can devise. But it will be so low in price that no man making a good salary will be unable to own one – and enjoy with his family the blessings of hours of pleasure in God's great open spaces.
>
> (Ford in 1909, quoted in Jackson, 1985, p.160)

And those who could afford to do so certainly did take the road to the wide open spaces, travelling both locally and much further afield (see Figure 3.8).

Although leisure use predominated, already in the 1920s a noticeable minority of people began to commute to work by car, even in large cities. McShane suggests that the rapid spread of the closed car is evidence of commuting use (McShane, 1994, p.191). By 1928, the car-commuting figure for industrial workers in Baltimore (which had excellent public transport) was 8 per cent; in Milwaukee it was 12 per cent, and in Washington, more than 20 per cent (Barrett, 1983, p.162).

3.5 Accommodating the motor car in the city

Even though the largest cities were the least car-dependent, the rapid increase in the number of motor cars had considerable consequences for them. For example, comparing the data in Table 3.1 (p.55) and Table 3.2 overleaf indicates that, around 1910, when motor car ownership was principally an urban phenomenon, Chicago's number of inhabitants per motor car was comparable to the national figure. Subsequently, the Chicago figure was notably higher than the national figure. In 1920 there were almost twice as many inhabitants per car in Chicago than nationally, and about one and a half times as many in 1930. However, by that later date, Chicago's population was more than three million, so, even at eight persons per car, there was a huge number of cars to accommodate.

Table 3.2   Chicago motor cars (Condit, 1973, table 7)

Year	Registered cars	Persons/car
1910	9,963	219
1915	34,441	71*
1920	106,500	25*
1925	289,948	11
1930	406,916	8

*   *Barrett (1983, p.130) gives 61 for 1915 and 30 for 1920*

Gradually, both regulatory and physical measures for handling such quantities of vehicles were devised. City governments, planners, engineers, entrepreneurs and residents deliberately took measures to accommodate the motor car both in terms of facilitative traffic control and new types of facilities and roads. In this way, the private motor car came to be seen as a public responsibility in cities and developed into a form of urban mass transport with considerable public support (Barrett, 1983). Before the First World War, the principal stimulus to regulation was the rising accident rate (McShane, 1994, Chapter 9). Especially after the war, however, the motor car became a major contributor to urban congestion, and relieving congestion became the main focus of urban efforts to deal with it.

Traffic control[4]

Regulation was the initial strategy for dealing with traffic and it tended to be the responsibility of the police. The first dedicated, traffic-police squad was designated in Philadelphia in 1904. New York City quickly followed suit and would have 6,000 officers by the 1920s. Baltimore, Chicago, Cleveland, Detroit and St Louis had traffic police from about 1911. And police forces themselves acquired mechanical personal transport. Boston had a steam-powered car in 1902 and New York City police obtained three bicycles in 1903. Chicago's traffic police were mounted on horseback as horse-drawn goods vehicles accounted for a substantial amount of the traffic congestion in the central business district. In that same year, even though motor-car traffic was only about 5 per cent of the total, the New York police issued the first ever comprehensive set of rules for the road in the USA. They were devised by an independently wealthy self-styled amateur traffic engineer, William Phelps Eno. He helped frame the pioneering Parisian traffic regulations and would later be a consultant on London's traffic. Among other regulations, slower traffic had to stay to the right (driving on the right had been the law in New York from 1804) and all overtaking was to be done on the left; wagons could no longer back up to the kerb (see Chapter 2, Figure 2.16), but had to off-load sideways to the kerb; all vehicles had to park next to the kerb and face the direction of traffic on their side of the street; parking was not allowed within ten feet (three metres) of intersections (and later, fire hydrants). Motor-car drivers were to give hand signals before stopping or turning. Furthermore, some attempt was made to define the relationship between the motor car and other traffic. Pedestrians were expected to cross only at corners, not mid-block. Trams were given priority in their tracks: motor cars were supposed to give way to trams when turning across their paths, and to leave the tracks free for trams wherever possible (see Figures 3.9 and 3.10).

Such regulations met considerable opposition from drivers. In New York and Chicago, as well as in other cities which attempted to regulate traffic before the First World War, motorists, both as individuals and via their quickly growing

[4]   This section draws heavily on McShane, 1994 and 1997.

Figure 3.9 The car versus the tram in fiction. That the Keystone Cops would stage such a stunt on film indicates that there was an issue of competition for road space. The first Keystone Cops film appeared in 1912, and comic car chases were frequently featured (reproduced from Rowsome, 1956, p.10)

Figure 3.10 The car versus the tram in reality, New York, 1925 (reproduced from Rowsome, 1956, p.10)

clubs, resisted what they interpreted as an infringement of their personal freedom. In this period, with motoring still an elite activity and accounting for a minority of traffic, motorists tended to favour those rules which enhanced their own mobility, rather than those which focused on public safety or the convenience of other road users (McShane, 1994, p.189). However, some of the early rules came to be observed as custom and practice for pragmatic reasons. Roundabout-type junctions came into use in New York and Washington; one-way systems were put in place in the narrow streets of Philadelphia and Boston. Various other types of regulations were tried: prohibiting turns across traffic and restricting the amount of time a car could park on the street in central business districts. These were largely ignored; the second quickly generated public protests when instituted in Chicago and Los Angeles in 1911.

Physical devices also began to be developed to cope with the motor car in cities. Already by 1905, the New York police painted lines on the road to guide pedestrians across the street, and they were required to use them. Painted safety-islands for pedestrians were placed at mid-road tram-stops, to be replaced eventually by more effective raised paved areas. Stop signs, where traffic on side streets crossed main arteries, were first introduced in Detroit in 1911. The painting of lines to mark road centres and traffic lanes was also first done near Detroit in that year. From 1912, when motorized traffic exceeded horse-drawn traffic there for the first time, New York City resorted to stationing a traffic policeman on every street corner in the central business district to keep traffic on the move. Enforcement was problematic as it was necessary to arrest traffic-regulation violators. That is, traffic offences were treated as crimes and the burden of proof was on the officer making the complaint. The creation of traffic courts in Chicago and New York in 1916 was ineffective as there was little legislation to act on and a reluctance to treat traffic offenders as criminals.

Towards the end of the First World War, many city governments began to appoint traffic engineers, whose methods would be different from the regulatory ones of the police. They began to use technical approaches to the problem of traffic control and there were early attempts at modelling traffic flows. This approach was favoured by the motoring clubs which viewed engineering solutions as facilitating motoring in contrast to what they saw as the restrictive approach of the police. Traffic signals were an early innovation by the engineers. Manually controlled semaphore signals had been used by traffic police at intersections from 1908 in Toledo and Detroit. At particularly busy junctions, elevated police boxes ('crows' nests') were installed to give police a better view, and manually operated electric signals came into use in 1912 in Salt Lake City and in 1914 in Cleveland (see Figures 3.11 and 3.12). Once an improved lens was developed which afforded better day-time visibility of coloured lights, electric signalling became the norm and allowed for much better synchronization.

Figure 3.11 A Detroit intersection with crow's nest topped by traffic lights, plus a traffic policeman at street level, c.1920. Note the central white lines on the road and the parked cars on either side of the street. This particular street seems to have had no horse-drawn traffic when the photograph was taken. Crows' nests were preferred by police as they left them in control, but they were ineffective in maintaining traffic flow through a series of junctions. The first crow's nest was introduced in Detroit in 1917, with three-colour traffic signals from 1918. From 1920, the colours were shown in four directions (photograph: Library of Congress)

Figure 3.12 Manually operated electric traffic signal, Chicago, 1922. This was a demonstration; 316 Chicago junctions had signal lights by 1926. Automatic timing to create phased traffic waves was introduced in that year after considerable discussion by traffic engineers and tram engineers. The latter group made an especially strong case for such controls in order to facilitate tram traffic. The system reportedly improved tram and motor-car speeds by 25 to 50 per cent (Barrett, 1983, p.158; photograph: Chicago Historical Society ICHi 05118)

Timed automatic traffic lights were first introduced in Houston in 1922 by a firm that manufactured railway signals, though there had been experiments in Salt Lake City a few years earlier. By 1929, New York City had 3,000 automatic traffic lights and had been able to reduce its number of traffic police from 6,000 to 500 (McShane, 1997, p.73). Vehicle-activated signals were first introduced in Baltimore in 1928; the currently used inductance loop became widely used from 1934.

By the mid-1920s, motor cars had become the predominant element in urban traffic jams (a term coined by the popular magazine, the *Saturday Evening Post*, in 1910). Twice-daily rush hours became a feature of big cities by 1914. In 1915, daily traffic of more than 25,000 vehicles and 142,000 pedestrians was recorded at the corner of Fifth Avenue and 34th Street in New York City. By 1923, 42,000 vehicles per day would be recorded just a bit further uptown on Fifth Avenue above 42nd Street – as many as 4,500 in a single hour (see Figure 3.13). In Manhattan in 1926, travel speed fell below 3 miles (about 5 kilometres) per hour; it was quicker to walk.

What all measures to deal with motor traffic and congestion had in common was a desire on the part of city governments to accommodate the private motor car. This had the effect of turning it into a means of mass transport. Ironically, considerable public money was thus put into providing an infrastructure for private transport. This was certainly the case in Chicago, where traffic congestion in the highly concentrated central business district, the Loop (see Chapter 2, Figure 2.28), was seen as a potentially disastrous hindrance to business. Parking on the street reduced considerably the amount of space available for the flow of traffic, so a parking ban was seriously canvassed in the 1920s and finally implemented in 1928 after surveys conducted in 1926 established that, contrary to businessmen's fears, customers did not object to paying a small fee for off-street parking (Barrett, 1983, p.159). To compensate, public land was devoted to privately run car parks. By 1926, Chicago had the most public land devoted to motor car storage in the nation (see Figure 3.14 overleaf). By 1931, there was off-street parking for more than 23,000 vehicles (Barrett, 1983, pp.134–5, 161). For many years, Chicago's was the only blanket parking ban in a US central business district. By decreasing congestion, the ban was thought to have encouraged more customers to enter the Loop.

Figure 3.13 Traffic on Fifth Avenue, north of 42nd Street, New York City, 1923 (photograph: Ewing Galloway)

Figure 3.14 Chicago Park District Car Park, 1923. This huge car park, laid out in 1921, was located between Lake Michigan and Michigan Avenue, which became a major thoroughfare from the Loop to wealthy northern residential areas during the 1920s. The car park was a short walk from the Loop and located on reclaimed ground along the lakefront. In the 1930s, many cities tore down as much as a third of their central business districts in order to provide for car parking and 'retrofit for the automobile' (McShane, 1997, photograph 8). Subsequently, buildings for parking cars became features of the urban built environment (Showalter, 1923, p.413) (photograph: Chicago Historical Society, ICHi 19437)

Street improvements[5]

During the 1920s, the approach in cities began to change from absorbing the motor car by means of regulations and traffic control to designing for it. Architects began to take account of the motor car when designing new buildings (see Chapter 4) and several cities designed new roads specifically to accommodate it. Specially designed street improvements were complementary to regulations and traffic-control measures for maintaining the flow of traffic. They could be used to maintain or alter existing traffic-patterns, by channelling traffic or creating distinct rights of way for each type of traffic. Already in the nineteenth century, Chicago, for example, had prohibited goods traffic from its parks and the system of boulevards that linked them. From 1863, when crossing the boulevards, side-street traffic had to defer to boulevard traffic. Trams were also forbidden on those streets. Not surprisingly, Chicago's boulevards proved especially attractive to leisure motorists; they became precursors of later limited-access roads that would segregate motor-car traffic. Park pedestrians would come to complain that the level of traffic was undermining their important urban amenity (see Figure 3.15).

Figure 3.15 Pedestrians trying to cross the road, Lincoln Park, Chicago, 1921 (photograph: Chicago Historical Society, ICHi 03451)

[5] This section draws heavily on Barrett (1983).

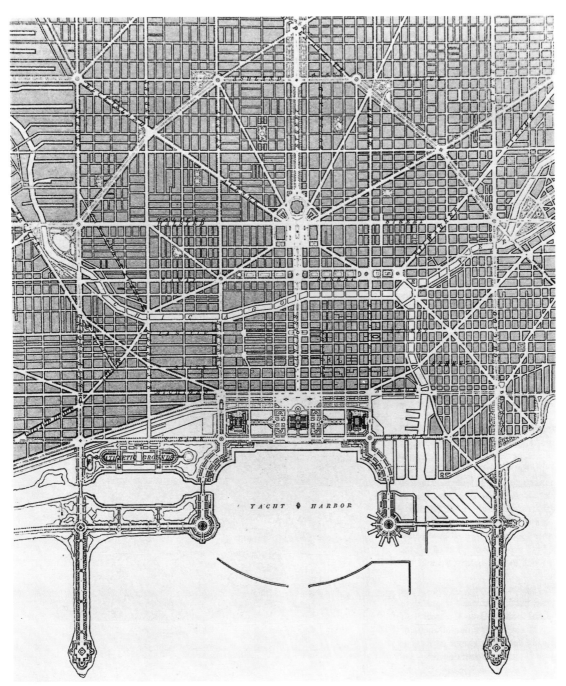

Figure 3.16 Plan of the complete system of street circulation (1909, from the *Plan of Chicago*, Chicago Historical Society, ICHi 03545, Commercial Club of Chicago)

Street improvements were featured in the influential Burnham Plan of 1909 (see Chapter 6, Extract 6.2). Quite apart from its much-studied aim to beautify the city, the Burnham Plan was in essence a transport plan, tackling the congestion problem. One recommendation was that the central business district should spread beyond the Loop, thus deconcentrating some of its activities and eliminating the need for so much traffic in the centre. Although the plan said little about mass transport and did not embrace the motor car, one of the means of dispersal that it proposed was to widen certain thoroughfares and to create new diagonal boulevards through Chicago's rectangular grid system. Thus, even though carried through in only a limited way, the features of the plan that were eventually implemented further facilitated the increasing numbers of motor cars (see Figure 3.16). Little was done before the First World War.

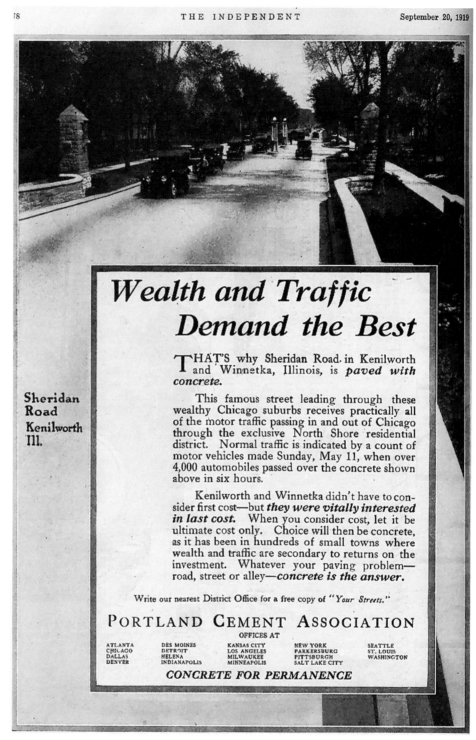

Figure 3.17 Concrete roads in the North Chicago suburbs, 1919. Wealthier suburbs could afford higher-quality surfaces to link them to the centre. Notice how the count of 'normal' traffic was made on a Sunday (reproduced from *The Independent*, 20 September, 1919, p.378)

A city government's ability to act depended on the authority it received from voters. In Chicago, where, from the beginning of the twentieth century, the city was expected to take responsibility for traffic problems, business interests were the key. Because business people favoured measures that would ease congestion in the Loop and keep business traffic flowing smoothly, the city authorities were able to take various actions. For example, to solve the problem of abutters' unwillingness to fund general street improvements, Chicago had been able to impose in 1908 a vehicle tax that at least funded repairs. This measure was of course not directed specifically at motor car users at this date, but it marked an important change of principle that would affect the way they were supported later: the measure shifted the financing of repairs from property owners to road users. However, implementing the Burnham Plan required funds far greater than could be generated by the usual methods. The city sought state permission to widen the base of property owners it could assess for street improvements beyond abutters, on the grounds that the improvements would benefit a much wider area. It also sought permission to purchase compulsorily more land than it needed physically for improvements, in order to raise money by selling the excess (along streets it improved) at a profit.

Some measures could be implemented quite inexpensively. A series of through-streets was created in Chicago in the 1920s, simply by defining them as such and by installing stop signs on streets that crossed them. The double-decked Wacker Drive, one of the earliest features of the Burnham Plan to be implemented, was an altogether more expensive undertaking (see Figure 3.18). Intended to be a through-route funnelling traffic from the north of the city, over a new, double-decked Michigan Avenue bridge, around the edge of the Loop, Wacker Drive was built on the site of a long-established produce market that was very costly to buy out and relocate. (Moving the market had the additional benefit of removing an anarchic form of goods traffic from the Loop; see Chapter 4, Figure 4.38.) The purpose of building two decks was to segregate passenger and goods vehicles, with the latter being on the lower deck. In some places, access was given to riverside parking and to loading-docks (see Figure 3.19 overleaf). Wacker Drive and the new tram-free through-routes allowed motor-car traffic to travel at the rate of 20 to 25 miles (32 to 40 kilometres) per hour. Further improvements were hindered by a property tax payers' revolt in 1927. By the time the issue was resolved, the Depression had begun and the city moved to seek outside funds. Rather than channelling traffic around the Loop, as Wacker Drive was intended to do, developments in the 1930s concentrated on establishing a new sort of

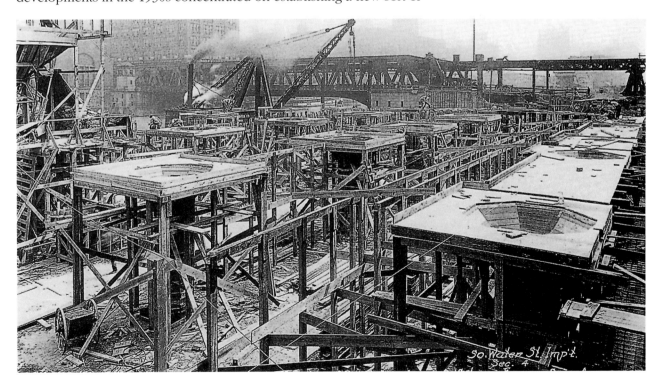

Figure 3.18 Wacker Drive under construction, 18 July 1925. At a width of 135 feet (40.5 m) on the lower level and 100 feet (30 m) on the upper level, Wacker Drive was constructed of reinforced concrete. Its flat-slab concrete framing was of a scale otherwise known only in railway viaducts. Flat-slab framing was independently developed in Europe and the US around the turn of the twentieth century. With this technique, the floor slab rests directly on the columns. At this time, it was distinguished by its flaring column capitals, which serve to spread out the cross-sectional area and thus reduce the shearing stress where the slab meets the column (Condit, 1982, pp.243–4; photograph: Kaufmann and Fabry)

Figure 3.19 Wacker Drive, 1929 (Chicago Historical Society, ICHi 04860, Kaufmann and Fabry)

roadway, the limited-access expressway right to the heart of the city, thus concentrating even more business in the Loop (see Figure 3.20).

It was New York City that pioneered this new form of road with the limited-access dual-carriageway Bronx River Parkway. Building such a road had been discussed from 1907 onwards, but it was finally opened only in 1923. Drawing on the ideas that Frederick Law Olmsted had implemented in Central Park in the heart of the city in the nineteenth century (see Chapter 6), the Bronx River Parkway was proposed as a leisure road to take city-dwellers through countryside that was to be retained as amenity parkland, to recreational reservoirs to the north. At the same time, it was meant to prevent further development along the Bronx River and thus control any increase in river pollution. Suburban property interests recognized that it would have the effect of making towns in Westchester County commutable to the city by road. Their donations of land proved crucial to the project's success. Their assumption was borne out; driving time to New York City was halved and the new road tripled the value of property at its northern end (McShane, 1994, p.222; see Figure 3.21).

Figure 3.20 Aerial view of the Outer Drive, built 1929–1937, looking north-eastwards (photograph 1956). This road paralleled the lakefront for the entire length of the city and included a new bridge over the Chicago River at its mouth. Eventually a link was provided to Wacker Drive along the south bank of the river. Its intention was to provide a way of bypassing the Loop on the eastern side of the city. While it did become the principal distributor highway for through-traffic, it also brought ever more motor cars ever more rapidly to the city. The number of motor cars in the city increased by 27 per cent between 1929 and 1939. By the latter year, Chicago accommodated a third of Illinois' motor vehicles (Sennott, 1993, p.65; photograph: Chicago Historical Society ICHi 16212, Copelin Commercial Photographers)

Figure 3.21 The Bronx River Parkway, 1923. According to Siegfried Giedion (1967, pp.824–32), the parkway concept humanized and beautified the highway by integrating it closely with the terrain. He argued that the principles of separation of pedestrians and traffic and limited access for abutters should also guide solutions to urban traffic problems (photograph: reproduced courtesy of the Westchester (NY) County Archives)

Robert Moses, a civil servant who was New York City's enterprising and single-minded Commissioner for Parks, obtained authority to build a series of highways around New York City in the 1920s and 1930s. The authority came from New York State legislation of 1924, which he helped to draft, giving him power to appropriate land (however wealthy and politically powerful the landowner) for road-building purposes (Hall, 1988, p.277). Financed in part by tolls, thus passing the cost on to users, and in part by the revenues generated from the increased land-values along the routes, these highways adopted the parkway principles of limited access and grade separation, but decreasingly included park-like landscaping. More than 300 miles (480 kilometres) of such highways were constructed in the Greater New York area before the Second World War (see Figure 3.22).

Some, like the Bronx River Parkway, were intended initially as amenities to link urban dwellers to rural or seaside recreational areas. However, the later highways became straightforward traffic-movers, now planned for suburban commuters. The low bridges of Moses' parkways excluded lorries and buses; they were for middle-class car owners (Hall, 1988, p.277). The parkways did not solve the city's congestion problems, but they did provide one route around them, making in effect a partial ring-road for Manhattan Island (Giedion, 1967, p.831; see Figure 3.23). This made it possible to accommodate even more vehicles.

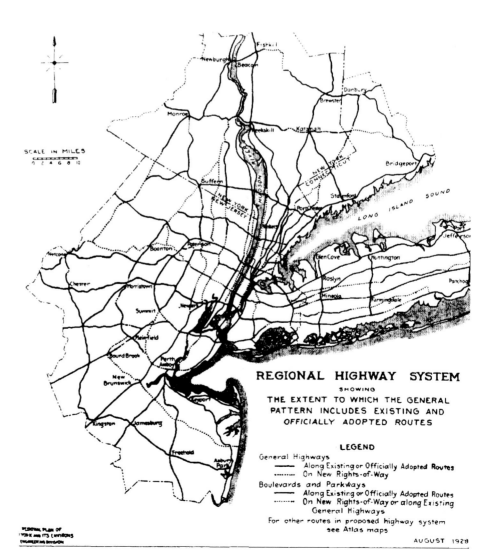

Figure 3.22 The proposed highway system, New York and environs, 1928 (reproduced from the *Regional Plan of New York and its Environs: The Graphic Regional Plan*)

Figure 3.23 The Henry Hudson Parkway, built 1934–7. The Henry Hudson led down from the northern suburbs as an extension of the Westchester County Park system right along the Hudson River to the tip of Manhattan. There it met the East River Drive which took drivers around to the access point to the Long Island system. Peter Hall calls this the 'first true urban motorway' (1988, p.277). Note the cloverleaf junction in the middle ground. The first one was built in Woodbridge, New Jersey in 1929 (reproduced from Giedion, 1967, p.830, figure 508; photograph: New York City Park Commission)

3.6 The motor car and suburbanization

As was made clear in Chapter 2, suburbanization was under way long before the motor car became a form of mass transport. However, during the 1920s – a decade of dramatic increase in motor car ownership – suburban populations grew twice as rapidly as those of central cities, on average 39 per cent in the suburbs and 19 per cent in the central cities (United States National Resources Planning Board, 1937, p.35), with as much as 1,000 per cent growth for some suburbs. The Depression and the Second World War halted such growth (indeed, as they slowed the rate of increase in car ownership). However, after the war, the trend to suburbanization increased dramatically, as did the trend to car ownership. By 1950, suburban populations were growing at ten times the rate of those in central cities. The 1960 census showed a 45 per cent increase, but an 11 per cent increase in central cities. Some long-established central cities, such as Boston and St Louis, even lost population during the 1950s, which was the beginning of a trend for such places (Hall, 1988, p.294). Meanwhile, the over-concentrated populations of New York City and Chicago had been steadily becoming less dense throughout the century because of the growth of suburbs (Jackson, 1985, p.185). According to Peter Hall:

There were four main foundations for the suburban boom. They were new roads, to open up land outside the reach of the old trolley and commuter rail routes; zoning of land uses, to produce uniform residential tracts with stable property values; government-guaranteed mortgages, to make possible long-repayment low-interest mortgages that were affordable by families of modest incomes; and a baby boom, to produce a sudden surge in demand for family homes where young children could be raised. The first three of these were already in place, though sometimes only in embryonic form, a decade before the boom began. The fourth triggered it.

(Hall, 1988, p.291)

The role of the motor car is related to the first of these 'foundations'. What the motor car did in existing cities was to permit travel in the gaps between the tram and commuter railway arteries radially disposed around the centre (see Figure 3.24). No longer was development confined to narrow strips within walking distance of tramlines or commuter railway stations, but it could spread out to fill the spaces between the tramlines or railways and still allow a reasonable journey to work for inhabitants. Furthermore, individual houses could be less densely spaced than those near access to fixed-rail transport routes, which tended to be close together to maximize the number of accessible residences. The greater volume of land for development lowered land prices, and therefore building plots could be larger. Though the spacing of the houses in motor car-dependent areas was therefore generally less dense than in tram-dependent areas, the effect on the overall configuration of the city was to populate it over a larger area (Jackson, 1985, p.181; Monkkonen, 1988, p.177).

This model of the sequence of suburbanization accounts well for the pattern of growth and change of older established cities of the East and Midwest. And, from the point of view of deconcentrating over-populated late nineteenth-century central cities, it would have pleased such contemporary commentators as Adna Ferrin Weber who advocated suburbanization as a means of tackling problems of urban congestion (see Chapter 2, Extract 2.2). However, in

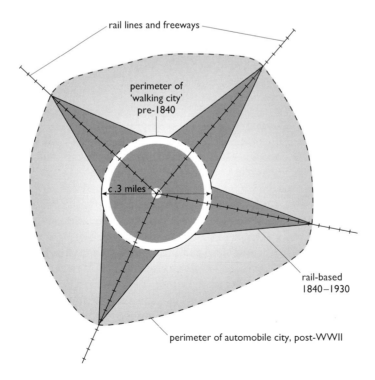

Figure 3.24 The shape of the motor-car city. Compare this diagram with Figures 2.15 and 2.30 in Chapter 2 (reproduced from Monkkonen, 1988, p.178, figure 7)

southern California, a different pattern pertained. Rather than being centralized, its industrial structure was dispersed from the start:

> In this region [southern Los Angeles County and northern Orange County], the discovery of vast amounts of oil in the mid-1890s led to the development of a suburban network. Industrial and residential suburbs appeared at either an oil field or a refinery site; and over the next three decades the discovery of additional oil fields not only led to the development of additional suburbs, but they also were responsible for the industrialization of the greater metropolitan region. By 1930, a suburban industrial network founded by the oil industry surrounded the city of Los Angeles and occupied much of the land throughout the Southland.
>
> (Viehe, 1981, p.3)

A vast, dispersed, politically fragmented urban area consisting of suburbs was the result. The oil industry attracted other industries and cheap, local petrol encouraged car ownership, so that southern California had the highest rate in the US during the 1920s. Over that decade, the number of persons per motor car in California as a whole dropped from 5.2, already well below the national figure of 13, to 2.3 (Flink, 1988, p.140). Some of the deconcentrating suburban development around older cities, such as Pittsburgh, was also the consequence of workers following jobs when some industries relocated out of the city (Tarr, 1978, p.32), but in southern California, the inherently dispersed nature of the oil industry served to initiate fragmented suburban development over a vast area.

3.7 *The motor car and public transport*

Urban public-transport systems eventually declined as motor car ownership in cities spread, but the reasons for this are complex. Paul Barrett (1983) argues that this relationship resulted from the way in which city authorities took responsibility for accommodating the motor car, while treating mass transport as a regulated private business. For example, cities invested public funds in costly infrastructural support for the motor car (see Section 3.4 above), but mass transport was liable for heavy business taxes and had to use its own revenue (from fares) for infrastructural development. These divergent approaches prevented the emergence of integrated transport policies and ultimately worked to the disadvantage of mass transport.

Chicago

During the initial spurt of increase in motor car ownership over the period 1918 to 1927, mass transport ridership increased in Chicago by some 10 per cent. Indeed, such increases were the norm nationally in this period. Per capita ridership increased as well (Foster, 1979, pp.374–5[6]). In Chicago, trams remained self-supporting, possibly engendering complacency on the part of both the city and the tram companies. Chicago tram passengers were very sophisticated, regularly making complex journeys involving transfers in so variable a manner as to defy any economical planning of through-routes. That is, tram riders did not turn to the motor car because of the difficulty of cross-town travel, as is sometimes suggested. Throughout the 1920s, leisure use of the motor car predominated and it was on Sundays, rather than workdays, that competition was first noticed by tram companies (Barrett, 1983, p.162). During the 1930s, the private car began to be used by workers for commuting within

---

[6] An abridged version of this article appears in the companion Reader to this volume: Roberts (1999).

the city. However, the evidence indicates that, for them, the motor car replaced walking rather than mass transport:

> In 1916, 79.2 per cent of workers in a south east side [of Chicago] steel plant had lived within walking distance of their jobs. By 1941, 27.4 percent walked to work, 34.9 percent used streetcars, and 35.2 percent rode in automobiles. At one major stockyards plant, a 1942 survey showed that only 7.8 percent of the workers walked to work, while 28.8 percent drove and 62.5 percent used mass transit. In 1916 the proportion of presumed walkers in the same general area had been between 30 and 47 percent.
>
> (Barrett, 1983, p.210)

In Chicago, the corporate ownership of mass transport put it in the middle of the political arena, while individual ownership of the motor car tended to isolate it from the vagaries and vexations of Chicago politics. After 1913, Chicago's trams came under the regulatory control of the Illinois Commerce Commission. It insisted on low fares but prohibited the abandonment of uneconomical services. Until 1930, the Commission also kept the length of franchises short, thus discouraging companies from investing in improvements. It was not the motor car that undermined mass transport in Chicago, but policy inertia which made self-funding mass transport development increasingly difficult.

Los Angeles

Although we tend to think of Los Angeles as the quintessential motor car city, it was founded in 1781, and began to expand dramatically a century later. Its form, too, was created by developers' deployment of electrically powered mass-transport to open up new areas for building (Jackson, 1985, pp.121–2).[7] However, by the 1910s, there was considerable disillusionment with the electric-traction system in Los Angeles, a mixture of trams and interurban railways: the vehicles were overcrowded and routes seemed planned more to connect up the subdivisions of developers who promoted them than to provide a rational system of transport. Indeed, for the most part, they ran at a loss; their promoters profited from the land deals which the electric lines facilitated rather than from passenger fares. This, in turn, made the lines unattractive as investment propositions, which discouraged improvement. The motor car therefore proved particularly attractive to Los Angeles residents and they showed their reaction to public transport by buying motor cars in large numbers quite early on. By the early 1920s, Los Angeles had more motor cars per person than any other large city (Los Angeles Traffic Commission, 1922, p.6). In 1920, there was a congestion crisis in the central area which led the electric-traction companies to propose a comprehensive parking ban so as to reserve the streets for themselves. As predicted by the business community, the effect on trade was disastrous; motoring customers simply took their business elsewhere. The public outcry was so great that the prohibition remained in effect for only sixteen days before being cut back to apply to rush hours only, with limited waiting during other busy periods (Bottles, 1992).

The crucial decade for the increase in motor-car ownership in Los Angeles, as well as a boom in house-building, was the 1920s. It was during that decade that those responsible for planning Los Angeles shifted toward policies favouring the motor car and a dispersed mode of city development (Foster, 1975; Interrante, 1983, p.93). As in Chicago, although tram ridership increased by 15 per cent during the 1920s, it decreased relative to the population of Los Angeles, which doubled during that decade. In the same decade, the number of motor cars in Los Angeles County increased by 550 per cent. From this time, new roads, rather than new tram routes, became part of the pattern of developing new suburbs.

---

[7] An abridged version of this chapter appears in the companion reader to this volume: Roberts (1999).

Figure 3.25 New roads and new building plots near Los Angeles, 1922. Although it is hard to see, the sign in the background advertises new building plots, while that in the foreground advertises the new boulevard that will serve future residents (photograph: Security Pacific Collection/Los Angeles Public Library)

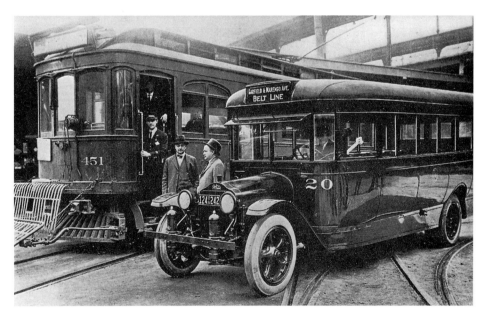

Figure 3.26 A motor bus and a tram car on the line it served, 1923. According to W.J. Showalter (1923, p.399), sixty electric lines around the country were using motor buses to supplement their services in 1923: 'They serve admirably in new territory as feeders to established street-car and interurban traction lines.' Buses provided a much cheaper way of extending less-travelled routes than building additional fixed-rail lines. From the later 1920s, Yellow Truck and Coach, which was 50 per cent owned by General Motors, targeted tram companies as likely customers for buses, and subsequently set up a holding company to finance conversions from tram to bus. In 1936 – the year that it introduced its new diesel buses – General Motors formed National City Lines (NCL) to buy up inefficient tramlines and replace their services with buses. They increased routes, rationalized services and reduced fares. This has sometimes been seen as a motor-industry conspiracy to undermine mass transport. However, the evidence is that NCL's aim was to create a market for their buses while sustaining and improving mass transport by eliminating more expensive and obsolescent tram equipment. In the case of Los Angeles, the electric railway company had already decided to switch to buses some four years before it was bought out by NCL (Adler, 1991; Bianco, 1998; Bottles, 1992, p.194; photograph: Ewing Galloway)

As new tracts of land opened up for development, even the traction companies responded not by extending fixed-rail routes, but by adding much cheaper motorized buses to serve the new areas. With the downtown area facing a traffic crisis as early as 1920, two contradictory plans were drawn up before 1925, one favouring rationalized mass transport, the other favouring the building of a grid-road system linked to a grid-highway system in outlying areas. The latter assumed a new model for a city whose centre would have a limited role as cultural, government and corporate headquarters and whose other business functions would be decentralized to local neighbourhoods with good road access.

It was this second plan that prevailed. According to one planner,

> Instead of the automobile conforming itself to the limitations of the cities, the cities began to conform themselves to the necessities and services of the automobile. That is the BIG thing that has happened here in the Southland.

(Gordon Whitnall, quoted in Foster, 1975, p.470)

Planners consciously questioned the central-city models of the East and chose other options:

> Is it inevitable or basically sound or desirable that larger and larger crowds be brought into the city's center; do we want to stimulate housing congestion along subway lines and develop an intensive rather than an extensive city [and] must all large business, professional and financial operations be conducted in a restricted area … ?

(Clarence Dykstra, quoted in Foster, 1975, p.481)

Figure 3.27 Los Angeles major street-traffic plan, Los Angeles Traffic Commission, 1924 (courtesy of University of California, Los Angeles, Bancroft Library, Department of Special Collections)

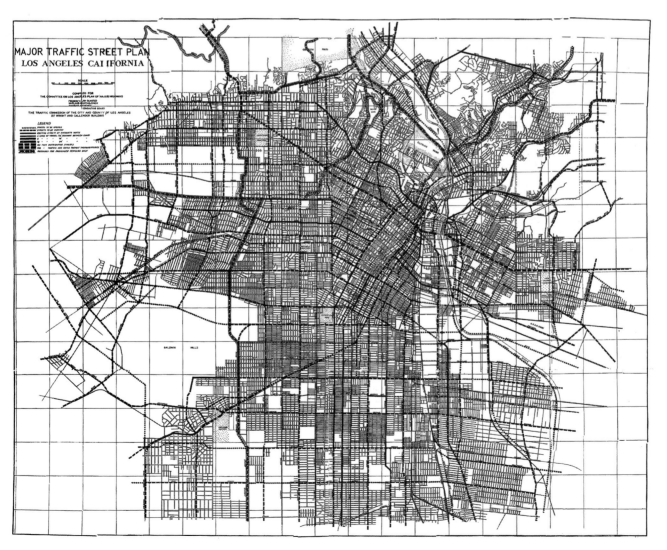

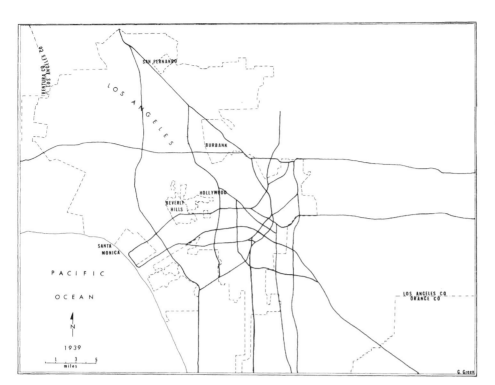

Figure 3.28 Los Angeles Freeway Plan of 1939. This plan followed the broad outlines of a plan put forward in 1937 by the Automobile Club of Southern California (Warner, 1995; reproduced from City of Los Angeles, 1939)

Figure 3.29 The Arroyo Seco Parkway, 1939. Subsequently incorporated as the first leg of the new system, the Arroyo Seco Parkway (later the Pasadena Freeway) was built on the Moses model of grade separation and limited access. It was completed in 1939 (reproduced from Banham, 1971, p.86, figure 32; photograph: Baron Wolman)

And residents supported them by approving the property taxes needed for the various road schemes. The famous Los Angeles freeway system, a giant grid of limited-access roads, was proposed in 1939 after further traffic surveys, and was built mainly after the Second World War (Warner, 1995, pp.138–41; see Figures 3.28 and 3.29).

Reyner Banham points out that, in part, the freeway system paralleled the original electric railway lines which had so dominated Los Angeles transport previously.

> ... the point about this giant city, which has grown almost simultaneously all over, is that all its parts are equal and equally accessible from all other parts at once. Everyday commuting tends less and less to move ... in and out of downtown, more and more to move by an almost random ... motion over the whole area.
>
> (Banham, 1990, p.36)

Planned dispersal also underlay the provision of housing in Los Angeles. And overwhelmingly, that housing was in the form of single-family, detached suburban homes (see Chapter 4), making Los Angeles a suburban metropolis (Fishman, 1987, Chapter 6). In Greg Hise's phrase, in the case of Los Angeles, suburbanization was urbanization (1997, Chapter 1).

3.8 *The lorry*

Surprisingly little has been written about the history of the lorry. According to Clay McShane (1979), some of the earliest arguments for motorized transport were for goods carriage rather than passenger transport. The steam railway had greatly lowered the cost of interurban transport. However, transhipment of goods by horse and wagon or hand-cart was by comparison slow and costly. One rather extreme-sounding estimate published in *Engineering News* in 1896 was that the cost per mile of goods carriage within cities was 2,000 times greater than that between cities (McShane, 1994, p.120). Furthermore, horse-drawn wagons created a double pollution problem: manure from the motive power and dust from the granite blocks favoured by teamsters (wagon drivers) crumbling under wagon wheels (McShane, 1979; Barrett, 1983, p.51). Their lack of manoeuvrability contributed to urban congestion problems.

The word 'lorry' is a corruption of the surname 'Laurie'; the name of the early nineteenth-century inventor of the rail flatcar. 'Truck', the word used in the USA, derives from a verb meaning to trade in a market, especially in garden produce, hence 'truck farm' in the USA for the British term, 'market garden'. The word eventually became applied to the vehicle hauling the goods to market, initially under horse power and later under mechanical power (Lay, 1992, p.168). As John Brinckerhoff Jackson (1994, p.174) points out, colloquial usage includes the implication that the lorry or truck is a money-earning vehicle. As with the motor car, various means of propulsion were tried. Steam power was used by slow, lumbering lorries carrying heavy loads, until well into the twentieth century. Electrically powered delivery vans were also widely used in cities at the turn of the century. Early internal-combustion engine lorries were slow and heavy (see Figure 3.30), especially in relation to the loads they carried, and they were not sufficiently powerful to travel on the poor-quality rural roads of the period. They were urban vehicles, used to ferry goods between railway stations, factories and warehouses and were seen to complement the railway system (Jackson, 1994, p.175). Despite the drawbacks of early lorries, 1914 figures suggest that, over short distances, rail haulage cost 30 per cent more, and horse-drawn haulage 100 per cent more than lorry haulage (Lay, 1992, p.169).

Figure 3.30 The first lorry powered by an internal-combustion engine, 1891. Like early motor cars, early internal-combustion engine lorries were adapted from existing vehicles; the first was put together by Daimler using a standard tray wagon (illustration: courtesy of Daimler-Benz AG Archives)

According to Jackson (1994, pp.176–7), the technical changes that transformed the early, cumbersome vehicle into a practical and flexible piece of mechanical equipment were made not by large manufacturers but by small-time mechanics and engineers through *ad hoc* experimentation and tinkering. Just as rural motor car owners defined how they would use their vehicles, anonymous individuals developed lorry features such as four-wheel drive, the trailer, the tipper lorry, and the low-loader to meet their own requirements.

As Table 3.1 indicates, growth in the number of lorries in the USA paralleled that of motor cars. Before the First World War, about 10 per cent of urban traffic was attributable to motorized goods traffic. A single lorry was capable of doing four times the work of a horse-drawn wagon (Jackson, 1985, p.183). The immediate effect of the use of lorries for deliveries in cities was that the number of horses decreased over the decade 1910 to 1920: from 128,000 to 56,000 in New York; from 68,000 to 30,000 in Chicago; and by similar proportions elsewhere (Showalter, 1923, p.404).

It was the First World War that brought about the transition to long-distance lorry-transport of goods (see Figure 3.31 overleaf). The railway lines were overwhelmed by the transport of war matériel, but there was no restriction on the production of heavy-goods vehicles beyond those needed to meet military demand. Consequently, the number of lorries on US roads increased by some 60 per cent between 1917 and 1918, and goods delivery services emerged. Huge numbers of lorries were used in the war and many soldiers learned to drive them as a consequence. The post-war sale of surplus Liberty trucks put vast numbers of lorries on the road and, in turn, put considerable pressure on road-building. Lorries proved to be quicker than rail transport, especially for less-than-carload[8] loads, because they moved directly from point to point, eliminating the waiting time in terminals at either end for filling or emptying railway wagons and making up or breaking up trains. Companies which ran their own lorries also saved on administration. The lorry became not just an adjunct to rail transport, but a major competitor.

Furthermore, the lorry accelerated a trend towards the relocation of industry away from city centres that was already under way in the nineteenth century. Between 1915 and 1930, especially after the First World War, the flexibility of the lorry made it possible for industries to relocate to less cramped quarters on cheaper land in suburban areas and the proportion of factory employment located in central cities dropped in all major US cities (Moses and Williamson, 1967). Coupled with the substitution of electric power for steam power, the

[8] 'Carload' refers to a boxcar; that is, a railway goods wagon.

Figure 3.31 A fleet of trucks on the Courthouse Plaza in Baltimore, en route from Detroit to France, 1918. Liberty trucks averaged 15 miles (23 kilometres) per hour on this cross-country trip and a third of them did not manage the distance. This highlighted the condition of the nation's roads and gave a military justification to the federal road-funding programme of the 1920s (Lay, 1992, p.169; Showalter, 1923, p.405) (photograph: University of Louisville, Photographic Archives, Caufield & Shook Collection CS33528)

optimization of organization for handling materials on these new sites resulted in a completely new type of factory building (see Chapter 4).

3.9 Automobility

The crucial decade in the technological and cultural shift to 'automobility' in the USA was the 1920s. It was then that the motor car came to be understood as a form of mass transport for cities and that public policies favouring it began to create a climate perpetuating its influence (Foster, 1983; Barrett, 1983). Furthermore, the motor-car-assisted shift to suburbia of the 1920s was qualitatively and quantitatively different from the 'streetcar suburbanization' of earlier decades. Developers took advantage of the independence of the motor car to bring the suburban idyll within the reach of a much wider social range of the population, and it was in this decade that the growth of suburbia came to outpace the growth of many central cities (Interrante, 1983). It was also an important decade in the emergence of the lorry. Motorized transport became thoroughly integrated into the economy and society:

> By the 1920s, a powerful force had evolved, wedding road builders and the motor industry, in which government and business joined happily in promoting as a national policy one mode of travel over all others. The highway-motor complex coalesced automakers, cement, asphalt, and steel producers, and petroleum companies into a common purpose. Along the way, it added such diverse groups as road contractors, insurance companies, banks, and motel operators to name but a few. The highway advocates became so dominant in American life that they were instrumental in changing the character of cities and helping to suburbanize the nation.
>
> (Goddard, 1994, p.ix)

From our perspective at the start of the twenty-first century, when the employment of as much as a sixth of the US work force is related to it, the motor car, especially in its effect on cities, seems to 'represent both the blessings and the curses of technology' (Cowan, 1997, p.225). This will be explored further in Chapter 7.

Extracts

3.1   Showalter, W.J. (1923) 'The automobile industry: an American art that has revolutionized methods in manufacturing and transformed transportation', *The National Geographic Magazine*, vol.44, no.4, pp.337–413.

With thirteen million motor cars and trucks now running on the roads of the United States, and with the annual demand for new ones in excess of three millions, America is both literally and figuratively 'stepping on the gas' in the making of transportation history.

A quarter of a century has brought a development in the automobile industry that has outrun the dreamers, confounded the prophets, and amazed the world.

In 1898 there was one car in operation for every eighteen thousand people … To-day there is one motor vehicle to every eight people … Thirteen million motor cars! Who can visualize them! Five for every freight and passenger car on all the railroads of the United States! Enough to carry half the people of America in a single caravan!

The Lincoln Highway, from the banks of the Hudson to the Golden Gate, is 3,305 miles long. To put them all on that highway, even in traffic-jam formation, would require that it be widened so that fifteen cars could stand abreast!

Round trip to the sun every 21 hours

The service they render is proportionately large. Assuming that the average car is operated only ten months in a year and runs only twenty miles a day, their aggregate travel amounts to seventy-eight billion miles annually …

Three times as many motor-miles on the highways as car-miles on the railways is a marvelous record for so youthful a competitor of rail transportation.

Counts at the New York City ferries and elsewhere indicate that the average car carries $2\frac{1}{2}$ passengers. This means that more than thirty million people take to automotive wheels every day, or more than nine billion annually – eight times as many as are carried by all the railroads.

The transformation in the lives of the people which these figures indicate stands almost, if not quite, unparalleled in any quarter of a century of human existence.

Starting out as a plaything, transformed into a luxury, and then becoming, in turn, a definite element in our standard of living, the motor vehicle has assumed the role of a highly efficient factor in our transportation system, touching the lives and promoting the welfare of America as few developments in the history of any nation have done.

Transportation: the ladder of civilization

… [W]e are, as a nation, according to Moody, the statistician, spending more for our automobile service than is being spent for railroad transportation, shelter, or heat and light – more, indeed, than for any other item in our national budget except clothing and meats …

Cities are spreading out. Long Island is built up for half its length to accommodate those who make New York the metropolis of America: so is New Jersey from Morristown to Long Branch and from Jersey City to the Empire State boundary at Suffern. Even Connecticut, as far as Stamford, Greenwich, and New Canaan, is peopled with those who work in Gotham by day and sleep in the country by night.

Chicago has the same story to tell, with its scores of consequential colonies, its dozens of outlying subdivisions. Philadelphia and San Francisco are but other examples of how men are coming to work in town and live in the country.

Not only in a residential way are cities undergoing a change, but also in a business way. The trek of branch banks far out beyond the business district is but one straw showing the direction of the transportation wind. The lack of parking space down town is making an ever-widening business district and new centers of commercial activity in every major urban community. The era of down-town crowding is forcing the future to change radically our orthodox type of commercial concentration …

Eleven out of every thirteen motor cars in the world registered here

… [E]leven out of every thirteen motor vehicles in the world [are] being operated on American roads, and … twelve out of every thirteen produced in a given period [are] Yankee-made.

Surveying motor-car registration, we find that South Carolina has more cars than Australia or Argentina; that Kansas has more than France or Germany; that Michigan has more than Great Britain and Ireland.

Indeed, New York, Pennsylvania, New Jersey, and Maryland, with a combined population smaller than Poland, and with an aggregate area more limited than New Mexico, have more automobiles in service than the whole world outside of the United States.

… The insatiable demand for new cars, in spite of the tremendous number already in service, is disclosed by the fact that many more will be called into commission this year than were built from the birth of the industry up to the end of 1915.

Available figures indicate that the total car sales for the year will approximate five millions, including two million used vehicles. This means that one family out of every four in the country annually figures in an automobile transaction.

When will the point of saturation be reached?

… All the economists have been predicting its arrival for years. A decade and a half ago it was learnedly urged that the wealth of the country could never support more than two hundred thousand new cars a year. A little later it was being said that when the registration reached the five-million mark it would slow down to the slight annual increase required for the growth in population.

But that mark was passed and the expansion continued, with ten millions as the limit beyond which it seemed impossible to go. To-day that limit has been exceeded and there are once more many considerations which would seem to indicate that the 'point of saturation' is close at hand. Car registration is now up to the point where it is only a million behind the telephone listings of the country, only seven millions behind the total number of families, and even closer than that to the total number of dwellings.

Yet contrary to these considerations, and in spite of the warning from financiers that many people who can't afford them are buying cars, and in the face of the additional fact that 70 per cent of the cars being sold are bought on the deferred-payment plan, the demand goes on unchecked except as affected by seasonal conditions.

… It [the automobile] has sent hundreds of thousands of people into the suburbs, where rents are cheaper and living conditions better, and where the savings in rent offset the car's maintenance costs, leaving the better living conditions as dividends …

Home ownership increases with car registration

… When will the point of saturation be reached, in the light of such direct and indirect returns, and in view of the fact that fewer deferred-payment buyers default on their cars than on household furniture; that definite statistics show home ownership increasing with motor-car registration; that national income increases as automotive transportation outgo swells?

Measured by California's present ratio of car-owners to population, it will not be reached until the present registration of the country is doubled.

Yet even California has not settled down to replacements. Measured by Indiana's existing ratio, the ultimate registration of the country would reach eighteen millions, but Indiana still shows herself far on the sunny side of saturation.

Those whose past predictions have been most nearly justified by the trend of events are making new predictions to-day, and these are that the point of saturation will not be economic, but rather physical. The congestion in the big cities is fast growing so great as to keep thousands of motorists out of the down-town districts.

Big city traffic problems

With all the traffic officers and signal systems, the task of handling the ever-flowing stream of motor cars and trucks grows apace. Some 42,000 motor vehicles pass the crossing at Fifth Avenue and Forty-second Street in New York every twenty-four hours; 4,500 in a single busy hour is not an unusual occurrence.

The block-signal system on Fifth Avenue, with traffic moving in a series of stops and starts, controlled from a central tower, has accomplished much, but even it is destined to prove inadequate. Boulevard traffic regulation, based on the Fifth Avenue practice, has also helped in many cities, but here again inadequacy is only a few years away.

Propositions are now coming from the foremost authorities for the establishment of express streets, where cars will move at rates of from forty to fifty miles an hour, and where gates will be established at intersections, just as at railway crossings. Commissioner Harriss, of New York, says that New York needs three north-and-south highways of this character, with traffic moving on each of them in three parallel lines in both directions. These streets, he says, will have to be four hundred feet wide and elevated in special instances.

Chicago is installing a synchronized traffic-control system similar to that now in operation in New York. This system of towers will extend south on Michigan Boulevard from Randolph Street, with the master tower at Jackson.

So great is the congestion in the famous Loop District in Chicago that proposals are being made to take all pedestrians off of the street level and to provide second-story sidewalks for them. The streets could then be widened to the building lines, almost doubling their present curb-to-curb width, and the sidewalks would be reached by stairways, ramps, and elevators. Vehicular and pedestrian traffic, each out of the way of the other, could move twice as fast as now and many times more safely.

It is pointed out that such a plan would give two display window stories instead of one, and that the thousands of people who now avoid the Loop District because of its congestion would come back to trade there, their reclaimed business being large enough to more than compensate the property owners for the cost of the change.

The day may not be so far in the distance when the horse-drawn vehicle will be legislated off the crowded city thoroughfares, to lessen congestion, just as heavy traffic has been banished from the boulevards to protect the motoring public. Likewise, the day will inevitably come when truck traffic will be separated from passenger-car traffic on the busier highways through the countryside, just as is now the case on the fine Roosevelt Boulevard out of Philadelphia.

But whenever the point of saturation is reached, and by whatever route, it will not come before all manufacturing facilities available to-day will be kept busy making replacements. The average life of a motor car is six years. If 18,000,000 cars shall prove the limit, replacement requirements will call for three millions a year, which represent the present annual production …

3.2   Willey, M.W. and Rice, S.A. (1933) 'The agencies of communication' in Recent Social Trends in the United States: report of the President's Research Committee on social trends, vol.1, New York, McGraw-Hill Book Company, pp.172–9

… The number of motor vehicles: private automobiles

Some form of motor vehicle registration was first required by all states in 1913. Between 1913 and 1931 the increase in registration in the United States was twenty-fold. This phenomenal growth involved a displacement of earlier vehicles, such as the horse carriage and the bicycle. It also involved habituation to the use of the automobile of classes in the population who formerly owned no vehicle of private transportation. Within the space of a few years, for vast numbers motor travel ceased to be a novelty and came to be regarded as a necessity. At the end of 1930 there was one automobile for every 4.63 persons in the population. The ratio varied considerably by states: at the extremes, California contained one automobile for every 2.78 persons, Alabama one for every 9.55 persons. †

… With the acceptance of the automobile the individual citizen in virtually all classes of the population has acquired a vehicle that gives a freedom of control in personal transportation such as never before existed. Potential mobility is increased immeasurably and easy, swift movement over distances formerly traversed but rarely is achieved. The result has been a transformation of social habits. ‡

The motor bus: numbers and control

While all evidence indicates that the private automobile is primarily responsible for traffic losses to steam railroads and electric lines, the motor bus has also assumed importance as a competitor of both. It not merely competes but supplements. In both local and interurban transportation the bus has advantages that indicate for it a lasting function. In 1930 estimates show 48,250 of these vehicles in operation for revenue, and 47,150 for non-revenue purposes. § The non-revenue buses, in consequence of more extensive use for school transportation, have shown the more rapid rate of increase. Of the revenue buses, it is estimated that 13,350 were operating on city routes, and 32,150 in intercity and interstate service. Buses in the latter services in 1930 were approximately two and one-half times as numerous as those in local service. On the other hand, local buses carried 1,350,000,000 revenue passengers, while intercity buses carried but 428,000,000. Passenger-miles of city buses were slightly more than half the passenger-miles of intercity buses in the same year …

There has been steady growth since 1924 in the number of buses operated by electric railway companies and steam railroads. In 1924, *Bus Transportation* estimates, electric railway companies controlled about 3,000 vehicles; in 1930, 11,827, or approximately four-fifths of all city buses. Their hold on local bus operations is still increasing and the problem of relationship between the two types of services seems well on the way toward solution, by a process of unified corporate control combined with co-ordination of functions …

---

† *Facts and Figures of the Automobile Industry*, National Automobile Chamber of Commerce, New York, 1931, p.15.

‡ …The increased production of closed cars has contributed to the general utility of the automobile, since it facilitates wider usage and greater comfort under varied weather conditions. As the automobile becomes generally used, the demand for comfort assumes importance, and increased comfort furthers the use of the automobile. In 1931, 92.9 percent of all cars produced in the United States and Canada were of the closed type, in contrast to 22.1 percent in 1921.

§ These and subsequent data pertaining to buses are selected from annual statistical numbers of *Bus Transportation*, a trade publication.

Highways and highway utilization

What the basic rail network is to railroad passenger traffic, the system of American highways is to motor vehicle travel. Highways and motor vehicles have developed in close relationship, each effecting changes in the other and in the social habits related thereto.

Highway mileage

Although early data are unreliable, the extraordinary development of highways has been apparent even to casual observation. In 1904 the total estimated mileage of 'rural roads' (i.e., excluding streets of municipalities) was 2,151,379, of which 153,645 miles were surfaced; about 144 miles had 'high type surface,' or some form of paving. By 1930 the estimated total had increased more than 40 percent, to 3,009,066. Surfaced roads had grown by 330.5 percent, to 693,559 miles; and high type surfaced roads, almost non-existent in 1904, had grown to 125,708 miles. Whereas surfaced roads in 1904 were 7.1 percent of the total, in 1930 they were 23.0 percent. Of these surfaced roads, the proportion with a high type surface increased in the same period from 0.1 percent to 18.1 percent. ††

These highway extensions, demanded by the automobile, have at the same time facilitated and stimulated its use. With a vehicle at hand over which the user has almost complete control and with highway networks on which it may be freely run, a multiplication of social contacts over wider ranges of territory is all but inevitable. For rural populations the importance is even greater, for enhancement of mobility is accompanied by a decrease in physical isolation as well.

Automobiles and highway engineering

The use of the automobile has introduced entirely new highway engineering problems. Old roadways that served adequately for horse drawn vehicles at once became antiquated. With high-powered cars and high speeds roads must be straightened, curves lengthened, vision increased, shoulders carefully planned, embankments equipped with guards, grade crossings protected and surfaces increased in trueness and durability. These are but typical requirements confronting the engineers who are concerned with the swift and certain flow of traffic.

The extension and improvement of highways brought increased vehicle speeds. Connecticut was first to limit automobile highway speeds by law (15 miles an hour, 1901). By 1923 all states had such statutes and analysis indicates a steady increase in the maximum speed permitted by law. In 1905 the median average for those states where regulations were enacted was 25 miles an hour; in 1919 this had increased to 30; in 1925, to 35; and in 1929 the median average had reached 40. The automobile has been an important contributory influence in increasing the tempo of modern life.

The problem of centralized control

From colonial days onward roads were for the most part a responsibility of local governments and an important reason for the latter's existence. The automobile has made state wide and national highway planning essential. Roads must serve the integrated needs of wide areas throughout which standard construction practices and traffic rules must be formulated and introduced. It is an accepted principle that the poorest unit in any roadway determines the capacity of the entire road. Purely local planning and construction accordingly become anachronistic.

In 1900 only seven states had even rudimentary highway administration; by 1917, highway commissions in some form were found in all. Nevertheless local administration, unrelated to the needs of larger areas, still remains in many respects a troublesome social lag ...

†† Data from United States Department of Agriculture, Bureau of Public Roads, Table D-1, 1929 (unpublished), based on figures compiled by the Bureau as reported to it by state authorities.

Automobile utilization

… In no inconsiderable degree the rapid popular acceptance of the new vehicle centered in the fact that it gave to the owner a control over his movements that the older agencies denied. Close at hand and ready for instant use, it carried its owner from door to destination by routes he himself selected, and on schedules of his own making; baggage inconveniences were minimized and perhaps most important of all, the automobile made possible the movement of an entire family at costs that were relatively small. Convenience augmented utility and accelerated adoption of the vehicle.

… [T]he per capita passenger mileage in passenger automobiles in 1930 was 2,697 miles. In the same year the per capita mileage on all steam railroads was 218.3, a decline of 227.8 from the peak of 446.1 in 1919. Comparison of these figures lends additional support to the conclusion that it is from the competition of the private automobile that the passenger business of the railroads has suffered most. While the comparison is admittedly unfair (since the automobile is used in numerous ways for which the railroad offered no corresponding service) there is here some ground for belief that the lost short haul passenger traffic of the rail carriers has been assumed by the private passenger automobile. Some may have been shifted to commercial buses, but if every passenger carried by bus in 1930 had been carried by the railroads instead, it would have increased the per capita passenger mileage figure of the latter by only 57.5 miles, and brought this to but slightly more than three-fifths of the 1919 figure.

…

The frequency of out of state cars on the highways leads naturally to the tendency to think of the automobile in terms of extended mileage. 'Long' and 'short' are relative terms and long trips of one generation may be short to another. The automobile has done much to revise conceptions of distance, but at the same time it has probably led to misconceptions concerning range of mobility. In the five states covered by the surveys cited, from one-third to one-half of all automobiles were on trips of less than 20 miles, from one-half to two-thirds were on trips of less than 50 miles, and distances of 100 miles were not reached by from three-fifths to nine-tenths of the machines. In Vermont, 42 percent of cars bearing Vermont plates were travelling less than ten miles. Were city data included the average trip mileage would presumably be much reduced. In the western states, where distances in general are greater, 'travel of less than 100 miles a day clearly predominates.' Considering the states as a group, about 38 percent of all local cars were traveling between 20 and 70 miles a day, and about 50 percent, less than 100 miles …

3.3 Lynd, R.S. and Lynd, H.M. (1937) *Middletown in Transition: a study in cultural conflicts*, New York, A Harvest Book, Harcourt, Brace and Company, pp.265–9.

If the word 'auto' was writ large across Middletown's life in 1925, this was even more apparent in 1935, despite six years of depression. One was immediately struck in walking the streets by the fact that filling stations have become in ten years one of the most prominent physical landmarks of the city; even between 1929 and 1933 the filling stations enumerated by the Census of Distribution increased from 41 to 70. Saturday-night parking now extends several blocks out from the main business streets into formerly deserted residential streets; and a traffic officer goes about marking the tires of cars parked on weekdays in the business section to enforce parking ordinances. In 1925 Middletown youngsters, driven from street play to the sidewalks, were protesting. 'Where *can* I play?' but in 1935 they were retreating even from the sidewalks, and an editorial, headed 'Sidewalk Play is Dangerous,' said, 'It is safe to say that children under the age of eight years should not be permitted to play upon sidewalks.' † Many business-class people regard it as a scandal that some people on

† Middletown's county now has about 2,500 automobile accidents a year. One of the most insistent notes in the editorial columns of the afternoon paper in 1935 and 1936 was the editor's repeated protests against fast and reckless driving.

relief still manage to operate their cars. No formal effort has been made by the relief authorities to discourage car ownership and operation, and … people on relief who own cars have been encouraged to use them in various ways to pick up small earnings. Even at the time of the labor-union fervor under NRA, local organizers tell one disgustedly, many Middletown workers were more interested in figuring out how to get a couple of gallons of gas for the car than they were in labor's effort to organize. While some workers lost their cars in the depression, the local sentiment, as heard over and over again, is that 'People give up everything in the world but their car.' According to a local banker, 'The depression hasn't changed materially the value Middletown people set on home ownership, but *that's* not their primary desire, as the automobile always comes first.' More hard-surfaced roads and faster cars mean an increased cruising range, and a local paper estimated in June, 1935, that '10,000 persons leave Middletown for other towns and resorts every fine Sunday.'

In a further very significant sense the automobile was writ large over Middletown in 1935. For the Middletown of today is more dependent upon the automotive industry than was the Middletown of 1925. In 1928 General Motors, in addition to its Middletown plant making transmissions, moved in a large Delco-Remy unit, and a machine shop that was small in 1925 has changed hands and grown to large proportions manufacturing automotive parts. The local press hailed in March, 1934, the national settlement of the labor controversy in the automotive industry in terms of its crucial significance for Middletown: '3,000 Men Will Stay on Job,' proclaimed the headlines. 'The settlement affects directly more than 3,000 [Middletown] workmen and, including their families, perhaps 12,000 residents of the city.' It then went on to list the five leading plants in the city dependent on the automotive industry. A year later, early in 1935, one of these … 'passed its all-time employment peak,' and in November of the same year it broke ground for a plant enlargement and again surpassed its employment record established in the spring; while the Delco-Remy plant also reached its highest production in its seven years in Middletown. To all of which must be added the major increment to the above newspaper list of a sixth plant with the return of the General Motors transmissions unit to the city in the spring of 1935; by June this plant was employing 800 men, with the prospect of early additions that, rumor had it, might run their force up toward 2,000. It is probably conservative to say that, by the close of 1935, half the factory workers in Middletown were producing for the automotive industry. To a considerably greater extent than in 1925, Middletown's life is today derived from the automotive industry – and the city is aware of it to its marrow!

Car ownership in Middletown was one of the most depression-proof elements of the city's life in the years following 1929 – far less vulnerable, apparently, than marriages, divorces, new babies, clothing, jewelry, and most other measurable things both large and small. Separate figures are not available for the city of Middletown, ‡ as distinct from the entire county, but the passenger-car registrations in Middletown's entire county not only registered scarcely any loss in the early years of the depression but, both in numbers and in ratio to population, stood in each of the years 1932–35 above the 1929 level. § Along with this tough resistance of Middletown's habit of car owning to the depression undertow, went a drop of only 4 per cent in the dollar volume of gasoline sales in Middletown between 1929 and 1933, suggesting little curtailment in mileage of cars. That the pressure on Middletown's automobile budget was severe is shown by the fact that purchases of new cars in Middletown's county, after rising from 1,885 in 1928 to 2,401 in 1929, were more than halved, to 1,162, in

---

‡ A careful estimate from all available data suggest that, of the 13,533 passenger cars registered in Middletown's county at the close of 1934, approximately 9,000 were in Middletown. This suggests a rise to about one car for every 5.2 persons in the city from the one car to 6.1 persons at the close of 1923.

§ …Middletown's state experienced a drop in passenger-car registrations in every succeeding year from 1929 through 1933, with a total falling off of 14 per cent over the four years. No explanation is at hand for the fact that Middletown's passenger-car registrations show a 5 per cent gain in 1933 over 1929 (and a 10 per cent gain through 1932) while the state's cars were falling off.

1930; they virtually stood at that figure in 1931, with 1,124 new purchases; they were halved again in 1932, to 556, or only 29 per cent of the 1929 quantity; and in 1933 they began slowly to recover. †† While, therefore, people were riding in progressively older cars as the depression wore on, they manifestly continued to ride.

All of which suggests that, since about 1920, the automobile has come increasingly to occupy a place among Middletown's 'musts' close to food, clothing, and shelter.

No explicit data are at hand as to possible changes in the role of the automobile in the leisure life of Middletown in the depression. It is probable that there has been somewhat less random driving about on summer evenings to cool off and more staying home with the radio. One parent stated that since the whole family had been trying to use the car as little as possible there had been less trouble with the children over the use of the car. But the gasoline-consumption figures cited above suggest that these changes probably were not great, or were counterbalanced by wider use of the car over weekends or for other purposes.

The return of the bicycle to Middletown in the spring of 1935 probably represents little more than a high-school and college-age fad. Ten years ago the bicycles which the research staff used for recreation evoked no end of merriment even from the youngsters, who were wont to call out, 'Aw, why don't you get a car?' But in the spring of 1935, 300 bicycles were sold in Middletown, virtually all of them to people under twenty-one. The presence of the local college undoubtedly affected this fad, since the bicycle craze had struck the State University a hundred miles away the fall before... ‡‡

References

ADLER, S. (1991) 'The transformation of the Pacific Electric Railway: Bradford Snell, Roger Rabbit, and the politics of transportation in Los Angeles', *Urban Affairs Quarterly*, vol.27, pp.51–86.

BANHAM, R. (1990) *Los Angeles: the architecture of four ecologies*, Harmondsworth, Penguin Books (first published 1971).

BARRETT, P. (1983) *The Automobile and Urban Transit: the formation of public policy in Chicago, 1900–1930*, Philadelphia, Temple University Press.

BIANCO, M. (1998) Review of J. Klein and M. Olson, 'Taken for a Ride' (videotape, Hohokus, New Day Films, 55 minutes), published at H-Net in H-Urban (History Listserver, University of Michigan, Ann Arbor), 26 March 1998.

BOTTLES, S. (1992) 'Mass politics and the adoption of the automobile in Los Angeles' in M. Wachs and M. Crawford (eds) *The Car and the City: the automobile, the built environment, and daily life*, Ann Arbor, University of Michigan Press, pp.194–203.

BOWDEN, S. and OFFER, A. (1994) 'Household appliances and the use of time: the United States and Britain since the 1920s', *Economic History Review*, vol.47, pp.725–48.

BURNHAM, D.H. and BENNETT, E.H. (1909) *Plan of Chicago*, Chicago, Commercial Club of Chicago.

CITY OF LOS ANGELES TRANSPORTATION AND ENGINEERING BOARD (1939) *A Transit Program for the Los Angeles Metropolitan Area*, Los Angeles, December.

CONDIT, C.W. (1973) *Chicago, 1910–1929: building, planning and urban technology*, Chicago, University of Chicago Press.

CONDIT, C.W. (1982) *American Building: materials and techniques from colonial times to the present*, Chicago, University of Chicago Press (first published in 1968).

†† In 1923, 32 per cent of Middletown's cars were two or less years old, another 32 per cent three to five years old, and 36 per cent were more than five years old.

‡‡ The bicycle revival was a national phenomenon. In 1935, for the first time since 1899, the national production of bicycles passed the half-million mark. The production total of 639,439 in 1935 was 54 per cent of that in 1899. (See Census of Manufacturers news release of September 5, 1936.)

COWAN, R.S. (1997) *A Social History of American Technology*, Oxford, Oxford University Press.

FISHMAN, R. (1987) *Bourgeois Utopias: the rise and fall of suburbia*, New York, Basic Books (a Division of HarperCollins).

FLINK, J.J. (1988) *The Automobile Age*, Cambridge, Mass., MIT Press.

FOSTER, M.S. (1975) 'The Model T, the hard sell, and Los Angeles' urban growth: the decentralization of Los Angeles during the 1920s', *Pacific Historical Review*, vol.44, pp.459–84.

FOSTER, M.S. (1979) 'City planners and urban transportation: the American response, 1900–1940', *Journal of Urban History*, vol.5, pp.365–96.

FOSTER, M.S. (1983) 'The automobile and the city' in D.L. Lewis and L. Goldstein (eds) *The Automobile and American Culture*, Ann Arbor, University of Michigan Press, pp.24–36.

GIEDION, S. (1967) *Space, Time and Architecture: the growth of a new tradition*, Cambridge, Mass., Harvard University Press (5th edition).

GODDARD, S.B. (1994) *Getting There: the epic struggle between road and rail in the American century*, New York, Basic Books (a Division of HarperCollins).

HALGRIM, R.P. (n.d.) *Thomas Edison, Henry Ford, Winter Estates*, Kansas City, Terrell Publishing Company.

HALL, P. (1988) *Cities of Tomorrow: an intellectual history of urban planning and design in the twentieth century*, Oxford, Blackwells.

HILTON, G.W. (1969) 'Transport technology and the urban pattern', *Journal of Contemporary History*, vol.4, no.3, pp.123–35.

HISE, G. (1997) *Magnetic Los Angeles: planning the twentieth-century metropolis*, Baltimore, The Johns Hopkins University Press.

HOLT, G.E. (1972) 'The changing perception of urban pathology: an essay in the development of mass transit in the United States', in K.T. Jackson and S.K. Schultz (eds) *Cities in American History*, New York, Alfred A. Knopf, pp.324–43.

INTERRANTE, J. (1983) 'The road to Autopia' in D.L. Lewis and L. Goldstein (eds) *The Automobile and American Culture*, Ann Arbor, University of Michigan Press, pp.89–104.

JACKSON, J.B. (1994) *A Sense of Place, a Sense of Time*, New Haven, Yale University Press.

JACKSON, K.T. (1985) *Crabgrass Frontier: the suburbanization of the United States*, Oxford, Oxford University Press.

JOHNSON, D.A. (1988) 'Regional planning for the Great American Metropolis: New York between the World Wars', in D. Schaffer, *Two Centuries of American Planning*, London, Mansell Publishing.

KLINE, R. and PINCH, T. (1996) 'Users as agents of technological change: the social construction of the automobile in the rural United States', *Technology and Culture*, vol.37, pp.763–95.

LAY, M.G. (1992) *Ways of the World: a history of the world's roads and of the vehicles that used them*, New Brunswick, Rutgers University Press.

LOS ANGELES TRAFFIC COMMISSION (1922) *The Los Angeles Plan*, Los Angeles.

LYND, R.S. and LYND, H.M. (1937) *Middletown in Transition: a study in cultural conflicts*, New York, A Harvest Book, Harcourt, Brace and Company, pp.265–9.

MAYER, H.M. and WADE, R.C. (1969) *Chicago: growth of a metropolis*, Chicago, University of Chicago Press.

McSHANE, C. (1979) 'Transforming the use of urban space: a look at the revolution in street pavements, 1880–1924', *Journal of Urban History*, vol.5, pp.279–307.

McSHANE, C. (1994) *Down the Asphalt Path: the automobile and the American city*, New York, Columbia University Press.

McSHANE, C. (1997) *The Automobile: a chronology of its antecedents, development, and impact*, Chicago, Fitzroy Dearborn Publishers.

MONKKONEN, E.H. (1988) *America Becomes Urban: the development of US cities and towns 1780–1980*, Berkeley, University of California Press.

MOSES, L.N. and WILLIAMSON, H.F. (1967) 'The location of economic activity in cities', *American Economic Review*, vol.57, Part 2, pp.211–22.

PURSELL, C. (1995) *The Machine in America: a social history of technology*, Baltimore, The Johns Hopkins University Press.

ROBERTS, G.K. (ed.) (1999) *The American Cities and Technology Reader: wilderness to wired city*, London, Routledge in association with The Open University.

ROWSOME, F., Jr (1956) *Trolley Car Treasury: a century of American streetcars, horsecars, cable cars, interurbans and trolleys*, New York, Bonanza Books.

Scientific American (1895) vol.73, 7 December, pp.357–8.

SENNOTT, R.S. (1993) ' "Forever inadequate to the rising stream": dream cities, automobiles, and urban street mobility in central Chicago', in J. Zukowsky (ed.) *Chicago Architecture and Design, 1923–1993: reconfiguration of an American metropolis*, Munich, Prestel/Chicago, Art Institute.

SHOWALTER, W.J. (1923) 'The automobile industry: an American art that has revolutionized methods in manufacturing and transformed transportation', *The National Geographic Magazine*, vol.44, no.4, pp.337–413.

TARR, J.A. (1978) *Transportation Innovation and Changing Spatial Patterns in Pittsburgh, 1850–1934*, Chicago, Public Works Historical Society.

UNITED STATES DEPARTMENT OF COMMERCE, BUREAU OF THE CENSUS (1975) *Historical Statistics of the United States from Colonial Times to 1970*, Part 1, Washington DC, US Government Printing Office.

UNITED STATES NATIONAL RESOURCES PLANNING BOARD (1937) *Our Cities: their role in the national economy*, Washington DC, US Government Printing Office.

VIEHE, F.W. (1981) 'Black gold suburbs: the influence of the extractive industry on the suburbanization of Los Angeles, 1890–1930', *Journal of Urban History*, vol.8, pp.3–26.

VILLALON, L.J.A and VAUX, J.M. (1979) 'Steaming through New England with Locomobile', *Journal of Transport History*, vol.5, pp.65–82.

WARNER, S.B. (1995) *The Urban Wilderness: a history of the American city*, Berkeley, University of California Press (first published 1972).

WIDMER, E.L. (1990) 'The automobile, rock and roll, and democracy', in J. Jennings (ed.) *Roadside America: the automobile in design and culture*, Ames, Iowa State University Press for the Society for Commercial Archeology, pp.82–91.

WILLEY, M.W. and RICE, S.A. (1933) 'The agencies of communication' in *Recent Social Trends in the United States: report of the President's Research Committee on social trends*, vol.1, New York, McGraw-Hill Book Company, pp.167–217.

Chapter 4: BUILDING TYPES AND CONSTRUCTION

by Gerrylynn K. Roberts

In colonial times, European precedents were followed in building types, architectural styles and construction methods; these were adapted as necessary, or desired, to local materials and geographical and climatic conditions.[1] This was true of both housing and public buildings. During the nineteenth century, though European influences and precedents remained important, the urgency of population growth and urbanization in the individualistic, private-enterprise economy of the USA, led to new, locally derived vernacular forms and recognizably 'American' cities.

4.1   Housing

During the colonial period, in the entrepôt cities of the eastern seaboard, the classic dwelling was the brick-built terraced house (known in the USA as a 'row house'), a building type brought from England. However, by the time of the revolutionary period,[2] inland settlements in New England had begun to adopt wood as the principal domestic construction material and to opt for free-standing houses set back from the street on large plots. This would become the dominant housing form in the United States – the American dream house. Peirce Lewis (1990, p.91)[3] attributes the switch to wood to 'a cultural predisposition for New Englanders to experiment', rather than to any raw-material constraints. However, he also points to the positive influence of the ready availability of wood. Gwendolyn Wright (1981, chapter 5) sees the independent homestead of the largely agrarian citizenry in the early years of the new nation as an important root of the American predilection for detached houses. Other historians suggest that its origins lay in the emerging ideal of the English country gentleman, which could be seen in London's peripheral communities of the eighteenth century as well as in some of the peripheral communities of US cities at the time (Fishman, 1987a, p.239; Jackson, 1985, p.13; 1987, p.302). In the nineteenth century, informed by an English picturesque aesthetic, the detached house would become the characteristic type of suburban house (Fishman, 1987b). It came to symbolize certain virtues seen as 'American': personal independence, family pride and self-sufficiency, freedom of choice and private enterprise (Wright, 1981, p.89).

Balloon-frame construction

Until the 1830s, methods of building US houses also followed European precedents. The building of brick houses by masonry-construction methods required specialized skills and numerous labourers. Wooden houses were of braced-frame design, using medieval beam-and-post construction with mortise-and-tenon joints (Condit, 1982, pp.5–9 and figure 1).[4] This method also required

---

[1]   See Roberts (1999), the Reader associated with this volume.

[2]   See Chapter 1, n.2.

[3]   An abridged version of Lewis's essay appears in Roberts (1999), the Reader associated with this volume.

[4]   An abridged version of Condit's essay appears in Roberts (1999), the Reader associated with this volume.

skilled artisans and numerous labourers to deal with the heavy wooden members. From the early 1830s, a new method of house-building was devised: balloon-frame construction (see Figure 4.1). There is debate about who invented the new construction method, precisely which was the first balloon-frame building and whether it was erected in 1832 or 1833 (Condit, 1982, p.43; Giedion, 1967, pp.351–4; Jackson, 1985, p.126; Sprague, 1981). However, there is scholarly unanimity about its origin in Chicago during a boom in property speculation and its great importance for the rapid growth of US cities and their suburbs. Balloon-frame construction could be adapted readily to the construction of many types of building; it required no specialized skills and few labourers. In contrast to buildings erected by traditional methods, balloon-frame buildings went up very speedily. Even when skilled labour was used, the cost of such houses was some 40 per cent less than those built by traditional methods (George Woodward, quoted in Giedion, 1967, p.349).

The voracious appetite of the rapidly expanding cities for housing meant that the new method was soon adopted (Jackson, 1985, p.126). Technological changes contributed to its spread. The power-driven circular saw had been available since 1814, and the timber industry began to provide wood that was machine-sawn to standard dimensions. This simplified the construction process further, as the builder needed only to cut the timber to length with a hand-saw. The other technological key to the success of the balloon frame was the development over the period 1790 to 1830 of cheap machine-cut nails, which cost 85 per cent less than hand-made wrought-iron nails (Jackson, 1985, p.127). House-building quickly became 'industrial' in the sense of depending on mass-manufactured components, making it easier for anyone to build their own home.

The settlement of the treeless prairies from the 1850s onwards created further huge demands for houses and other buildings that could be assembled quickly (Cronon, 1991, pp.178–80). Soon it was not even necessary for the builder to cut to length. Kits of presawn wood were marketed, as were prefabricated components such as timber frames, walls, roofs, doors and window sashes (see Figure 4.2). Already in the 1850s, precut timbers were being shipped by steamboat up the Missouri River to frontier towns.[5]

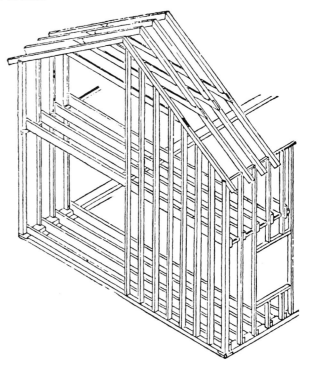

Figure 4.1 Balloon-frame construction. '[The balloon frame] was characterized by the substitution of thin 2-by-4-inch studs, nailed together … for the heavy beams and posts held together by mortise and tenon. Abandoning entirely the bulky members of the New England braced frame, this easy method of permanent construction was like a box. Unlike the timber frame, it required no heavy corner post for stability, and it had lateral as well as vertical integrity, which meant that it could withstand heavy wind loads. The weight rested on the 2-by-4-inch posts 16 inches apart and on the floors, which acted as platforms. By spreading the stress over a large number of light boards of a few sizes, the balloon frame had a strength far beyond the seeming capacity of the wood studs' (Jackson, 1985, p.126). 'Balloon-frame' was apparently bestowed as a term of derision by sceptical artisans who thought the results of the new form of construction would be too insubstantial to survive. During the nineteenth century, it was also referred to as 'Chicago construction' (Giedion, 1967, pp.347, 353; diagram reproduced from Woodward, 1865, p.153)

---

[5]   Exhibition at the Steamboat *Arabia* Museum, Kansas City (MO), 22 October 1998.

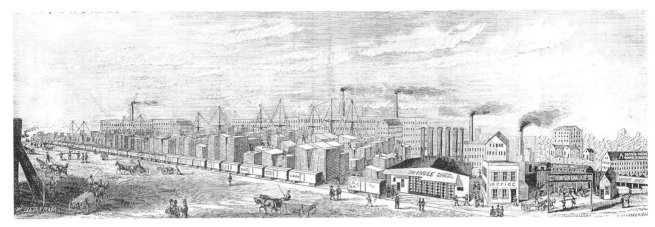

By 1860, prefabrication was big business for firms in Boston, New York and Chicago (Condit, 1982, p.45). An *Atlantic Monthly* writer commented that one Chicago firm in 1867 would

> despatch timber in the form of ready-made houses ... cottages, villas, school-houses, stores, taverns, churches, court-houses, or towns, wholesale and retail, and ... forward them, securely packed, to any part of the country.

(quoted in Cronon, 1991, p.181)

By the end of the century, house-kits (and everything that went into the finished houses) could be purchased from the Chicago mail-order retailers Montgomery Ward and Sears Roebuck. Thus during the second half of the nineteenth century, ownership of the ideal home came within the financial reach of many and the look of the country was transformed. Houses built for rental were also of balloon-frame construction (see Figure 4.3). By the second half of the twentieth century, 90 per cent of US houses would be of this type (see Figure 4.4 overleaf).

Figure 4.2 C.C. Thompson's timber-yard and sawmill, Chicago, *c*.1880. Note the proximity of this yard to both water and rail transport. This particular firm made window frames and other prefabricated components for buildings. Sited at the junction of the prairie and the northern woods, Chicago became the pre-eminent centre of the US timber industry in the second half of the nineteenth century and timber occupied a large area of the riverport. Much of it was shipped westwards by rail (80 per cent in 1860, 95 per cent in 1880), thus providing a return cargo for railway wagons bringing grain eastwards, and so it contributed to railway profits as well (Cronon, 1991, pp.180–83; courtesy of Chicago Historical Society ICHi-03229)

Figure 4.3 Balloon-frame housing in a working-class district of Chicago (photograph: courtesy of Chicago Historical Society ICHi-05803)

Figure 4.4 Balloon-frame housing under construction in suburbia, Palos Verdes, California, 1976 (photograph: courtesy of Professor K.T. Jackson)

Housing types in the nineteenth century

In the years following the Civil War, the romantic cottage ideal became transformed, with the aid of steam-powered jigs and new synthetically coloured paints, into the turreted, multicoloured 'Queen-Anne' houses that are now thought of as characteristic of the late nineteenth-century USA. Their exterior decorative features, such as the elaborate wooden decoration known as 'gingerbread', were factory products, transported by rail and fixed into place by local carpenters. Inside, too, factory-produced woodwork, wall-coverings and window-glass contributed to the decorative effect. Close association with the health-giving properties of nature was also stressed, and it was during this period that the ample porch became a feature of US houses, along with the custom of sitting out in the evening. In this same period, factory-produced plumbing fixtures became common in middle-class houses; heat was provided either by individual room stoves or distributed from basement furnaces, and light was provided by mains gas. House purchases were aided by the new, co-operative providers of mortages: the building and loan associations. These favoured individual suburban properties, seeing them as a good risk, and so further fostered the process of suburbanization (Jackson, 1985, p.130; Wright, 1981, pp.100–7). The process was also aided from the 1840s by pattern-books and magazines that included house plans and detailed specifications for their construction. They helped to create the suburban house as a 'distinct visual example of the attempt to take the city to the country' (Jackson, 1985, p.128), while at the same time creating an American style of house.

The poor in the largest cities could not achieve this ideal, however. For example, in New York City, with its very narrow plot sizes, 25 feet (7.6 metres) in width (see Chapter 6), the cramped, airless, multi-storey tenement was the norm. The earliest tenements were simply large private houses, vacated by the wealthy, which were subdivided for multiple-family occupancy. Occasionally, concerns over public health motivated civic philanthropic groups to build model blocks of improved tenement designs, but in general improvements to tenements were the responsibility of private landlords, the majority of whom were more concerned with maximizing their income. Even the worst properties could always be let in such a rapidly growing city. Legislation of 1867 required fire-escapes to be provided and specified that each water-closet should serve no more than twenty people. The notorious dumb-bell tenement design of the early 1880s is discussed in Chapter 7. Finally, New York legislation of 1901, copied by many other cities, required strict standards of ventilation, fireproofing, sanitary facilities and outdoor courtyards. But the immediate effect of this was to stifle new speculative building and, as the legislation did not apply to existing buildings, to exacerbate the housing problems of the urban poor. In the longer term, however, new buildings did conform. The tenement was characteristic of New York City, but in other cities terraced houses and small balloon-frame houses, often accommodating more than one family, were typical (see Figure 4.3).

Multi-storey living was not, however, characteristic only of the urban poor in the United States. After the Civil War, living in flats (or apartments, as they are known in the USA), became popular among middle- and upper-class urbanites in the larger cities where pressures on land made individual houses prohibitively expensive. By 1860, two-thirds of New York families lived in shared accommodation. Two hundred apartment buildings went up there between 1869 and 1876, and by 1900 'single-family houses of any size were no longer built in Manhattan' (Ford, 1994, p.194). Apartment houses were not affected by New York's tenement regulations. There were 108 apartment houses in Boston by 1878; they became common in Chicago during the post-fire reconstruction of the city from the 1870s and in Washington DC from the 1880s (Wright, 1981, pp.137–8). Purpose-built, luxury apartment buildings, known initially as 'French flats' to give them the cachet of European sophistication, were large, monumental buildings. One strategy to enhance their amenity value was to locate them near city parks, such as Central Park in New York, a prominent recreation area for the city's elite. For developers, the motivation was economic. Expensive land in booming cities achieved greater returns, the greater the number of people who could occupy it. For residents, there was the enticement of a relatively high technical specification, when compared with ordinary homes. The new technology of the lift was an important feature of such multi-storey residences, making every floor equally accessible. Centrally provided steam heat was common. Some buildings, known as 'apartment hotels', included elaborate centralized facilities such as kitchens and laundries (see Figure 4.5). Communal facilities were favoured by 'progressives', who argued for the efficient sharing of expensive technologies, and by feminists, who argued for the rationalization of housework. However, the ideals of independent living and of the home as the private domain of women proved strong and, by the early twentieth century, each apartment was provided with its own, generally tiny, kitchen, attached to mains services through communal lines.

Figure 4.5 Basement kitchen in a turn-of-the-century apartment hotel. 'From these gleaming, busy rooms, food could be hurried to private apartments or to the lobby-level dining rooms, banquet rooms, cafés, and restaurants. Delivery systems for carrying the food included special elevators with metal-lined warming boxes and ammonia-cooled refrigerator cabinets, and subterranean railways with delivery wagons or conveyor belts. Some buildings installed warming boxes at the back entrance of each [apartment] so that food could be left unobtrusively by the apartment staff' (Wright, 1981, p.140; photograph: reproduced from Gilman, 1904, p.141 by courtesy of Professor Gwendolyn Wright)

Despite detractors who saw the congestion implied by apartment-house living as a source of social problems (and suburban living as their solution), it remained the norm in large US cities where, by the 1920s, more apartments than individual homes were being built (Wright, 1981, pp.150–51). For example, in 1900, three times as many apartments as single-family homes were built in Chicago, and the multiple was seven by 1928. These were no longer just prestigious blocks: they were modest apartments in liftless buildings that were constructed on utilitarian lines, and became common for middle- and lower-income living in the biggest cities (Hancock, 1980, p.172).

Twentieth-century housing types

During the early years of the twentieth century, US housing underwent a major change of design. The rambling Victorian house, characteristic of the years after the Civil War, gave way to houses with a much smaller floor plan. In part, this was to do with changing family structures and the decline of servants, but also with the expense of upkeep of large wooden structures, especially after the economic downturn of the early 1890s. In addition, the new domestic technologies of modern plumbing and bathrooms, hot-water heaters, central heating and so on were costly; the *quid pro quo* came to be less space (Wright, 1981, p.171). Americans embraced the changes wholeheartedly. Ellen Richards, a home economist at the Massachusetts Institute of Technology, commented in 1905 on the new level of equipment expected in houses: 'our houses in America are mere extensions of clothes; they are not built for the next generation. Our needs change so rapidly that it is not desirable' (quoted in Keating, 1988, p.54). And since it was so much easier to install the new features in houses as they were being built rather than to install them retrospectively in existing houses, new building was further stimulated. The traditional internal arrangement of a downstairs parlour to serve as a buffer zone between the public and private realms, with a separate dining-room and kitchen, gave way to a much more functional arrangement, with an open plan. There were fewer rooms, and they were multipurpose.

The new form of house in the inter-war years tended to be single-storey with a low-slung roof, a carport or garage, and large windows. From 1895, the *Ladies Home Journal*, an influential magazine with a circulation of two million by the end of the First World War, popularized the bungalow by publishing a series of plans for this type of dwelling. Technical drawings and specifications could be purchased from advertisers for as little as one dollar. Standardized in the USA by the 1920s as a one-storey house with a convertible roof-space, the bungalow had been designed by British engineers for the tropical climate of Bengal. It was low and well-ventilated, and opened out to the garden. The bungalow could be grand or modest; this was the design that made it possible for the working classes to have a 'dream' home. In the USA during the early years of the twentieth century, unlike in Britain, suburbanization became as much a working-class as a middle-class phenomenon, as industry left the centre of cities and workers followed (Harris, 1993). Los Angeles (see Chapter 3), New York, Chicago, Detroit, Milwaukee and Pittsburgh all had extensive working-class suburbs. A 1940 survey identified 174 middle-class, residential suburbs and 149 industrial working-class suburbs (Harris, 1988, p.99). Self-building was very common in such working-class suburbs, and, even at the end of the Depression, ownership rates in these areas were as high as 45 per cent.

Part of the drive for smaller homes was an argument embraced by the progressive movement around the turn of the century. Based on scientific management thinking, it stressed 'efficiency' in daily life, including modernity in the home. The associations were industrial, and the emphasis was on simplicity in order to minimize housework. This was reinforced by general acceptance of the germ theory of disease and a consequent stress on cleanliness and the elimination of possible repositories for germs. Victorian nooks, crannies, and heavy furniture and fabrics disappeared, while high specifications of plumbing and services became common (Tomes, 1990). Kitchens designed along scientific management lines began to include built-in features and numerous mechanical devices. From the 1920s, electricity became increasingly common in US households and appliances proliferated (see Figure 4.6). The growing number of clerical and other jobs at this time meant that fewer domestic servants were available; electrical appliances made it possible for a woman at home to become a 'household engineer'. The amount of sheer drudgery declined, but not the amount of time that women spent in household tasks, owing to ever-increasing standards of housewifery and domestic hygiene (Cowan, 1983).

Figure 4.6 The all-electric home: an illustration from the booklet *Home of a Hundred Comforts*, 1925. At first, electricity was used only for lighting, and only in wealthy homes: in 1907, only 8 per cent of homes had an electricity supply. But as universal supplies became available, it was important to ensure that the electricity generating plant was used to capacity. The domestic market was therefore courted assiduously by the supply industry. Electricity was projected as modern, clean and healthy: 34 per cent of US homes had an electricity supply in 1920, 63 per cent in 1927 and 80 per cent in 1941. An electricity supply was installed in most new housing and, bedazzled by the modernity of New York City's 'White Way' (see p.120), Americans were quick to make their daily lives modern too, by the adoption of electrical appliances. By 1941, 79 per cent of US households owned an iron, 52 per cent had a washing-machine, 52 per cent had a refrigerator and 47 per cent had a vacuum cleaner (Cowan, 1983, pp.93–4; illustration: © Hall of Electrical History, Schenectady Museum)

Pacific Homes are Quickly Constructed

Here is interesting photolog of a typical Pacific Home. Erection completed in 28 days.

Starting

The cement foundation has been completed and carpenters are removing forms and are ready to begin construction work.

3½ Hours Later

A crew of four carpenters has completed and braced the under-pinning, placed the floor joists and is laying sub-floor.

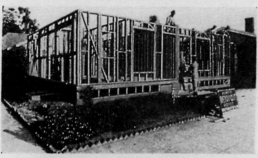

End of the First Day

All walls and partitions are in place, plumbed and braced.

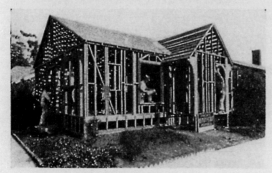

End of the Second Day

By 10:30 A.M. on the second day the ceiling joists and rafters are up. By evening the roof sheathing has been laid and most of the window frames set. The third day, plumbing and wiring are roughed in, valleys and windows flashed.

The Fourth Day

The roof is shingled and exterior lathed, ready for stucco; interior ready for plaster.

The Eighth Day

Plastering has been completed and has thoroughly dried. Exterior has received first two coats ready for color coat. House is ready for cabinet work and other interior finish.

28 Days After Starting

Here is pictured the home complete, ready for occupancy—less than one month's work for average crews working under normal conditions. This is a double house, complete with plumbing, wiring, built-in features, hardwood floors, painting and decorating. It may be inspected on our Los Angeles Exhibition Grounds.

Figure 4.7 Advertisement for Pacific Ready-Cut Homes, 1925; the houses were erected in twenty-eight days (photograph: courtesy of UCLA Special Collections)

During the housing boom of the 1920s, prefabrication provided a cheap and efficient means of building standard houses rapidly (see Figure 4.7). Furthermore, new materials such as precast concrete panels, and even all-steel construction, were widely used in housing experiments. Prefabrication and new materials were not always associated with new designs (see Figure 4.8).

With the collapse of the housing market in the Depression of the 1930s, there was considerable interest in making housing affordable by means of factory-production methods and scientifically determined designs for minimum housing standards (Hise, 1997, chapter 3)[6]. This was reinforced by the policies of the Federal Housing Administration (FHA), which was set up in 1934 to insure lenders against loss on mortgage business. The FHA was a government body created to stimulate the recovery of the hard-hit building industry. It did not establish a public-housing programme, but acted instead as a guarantor of mortgages for lenders in the private sector. The FHA insisted on certain standards for buildings whose mortgages it would insure, but in return would back loans for unprecedentedly high percentages of a house's value and for a longer term than was previously common. One consequence was indeed the regeneration of the US building industry. Another consequence was the further growth of suburbs at the expense of inner cities, since new housing in suburban areas was seen as a better risk for loans; the social effect was to encourage further flight from the cities by those who could afford it, leaving behind the poor (especially the newest immigrants and religious and racial minorities) in deteriorating buildings. Such districts were 'redlined', or encircled in red, on FHA surveyors' maps and could not attract loans for either improvement or purchase (Jackson, 1985, pp.206–16). Thus, as was the case with roads (see Chapter 3), although it had no public-housing policy, the government played a major role in private-sector housing.

During the 1930s, taking automobile-manufacture as an explicit model, the manufacture of prefabricated units was put on an assembly-line basis by Foster Gunnison, who styled himself the 'Henry Ford of housing' (Hounshell, 1984, p.311). His earliest venture exploited non-traditional housing materials of steel framing and asbestos-cement panels. Later versions, picking up on

Figure 4.8 Houses built of precast concrete panels, Forest Hills Gardens, New York, 1918. The precast concrete panels clearly seen in the illustration on the left were used to create traditional 'garden-city' housing, like that shown on the right (photographs: reproduced from Crawford, 1995, p.122)

6   An abridged version of Hise's chapter appears in Roberts (1999), the Reader associated with this volume.

development work by the US Forest Products Laboratory, used waterproof, stressed-skin, plywood panels, all built on a moving conveyor. But the houses proved unattractive to consumers. Potential purchasers could not be persuaded to think of homes as consumer durables. Nor were the houses particularly money-saving, since the shell of a house accounted for a relatively small proportion of its total cost (Hounshell, 1984, p.314).

Not only was factory production of buildings and their components pursued, but also industrial methods of organizing the building of houses on site. During the Second World War, vast quantities of new houses were needed as homes for defence workers in the new industries. The south-west region was a particular beneficiary of federal encouragement to decentralize, and the modern shipbuilding and aircraft industries, already established in the area, led the way. The aircraft industry was attracted to Los Angeles by the presence of a skilled, non-unionized labour-force and the proximity to centres of technical innovation, such as the California Institute of Technology. But the pressures on the infrastructure of the area's cities from the wartime influx were intense. New suburbs for the incoming workers were created near the employing firms. (Thus these suburbs were not the result of a flight from the central city, but were independent, employment-related developments.) Some of the accommodation constructed was of a temporary, barracks-like nature; however, with FHA guarantees, home-ownership rates, especially of the 'minimum house' on a small plot, climbed dramatically during the war. Rather than just selling plots of land which they had subdivided, private developers began to take responsibility for building houses on the plots. Principles of mass building and continuous construction methods were introduced on the new housing estates. For example, the firm of Marlow–Burns organized an area of staging at the edge of their site to which materials would be delivered. There, workers precut and preassembled the materials for distribution throughout the site, while specialized teams of labourers and craft specialists moved sequentially through the site performing their particular tasks (see Figure 4.9). This would become characteristic practice for some of the massive post-war building projects, such as the famous 'Levittown' put up by the Levitt family's building firm on Long Island. In the case of the Levitts, too, their initial experience of 'production-line' building was gained through the necessity of building quickly for war workers.

Furthermore, in some cases, factory-produced components were brought together with factory methods of production on site. For example, planning to make use of wartime production capacity after the war while providing a community service, Henry J. Kaiser, a Californian shipbuilder, established a subsidiary Housing Division in

Figure 4.9 'Well, what kept you? The plumbing contractor has come and gone': a 1947 cartoon by Alan Dunn (by permission of *The Architectural Record*)

Figure 4.10 Kaiser Community Homes under construction, California, 1946. The firm of KCH itself built the development shown here; it was sited very near to a component factory, which can be seen towards the top left of the photograph (photograph: courtesy of The Bancroft Library, University of California Berkeley)

1943. It undertook research and development on construction and household technologies with a view to manufacturing complete housing systems to meet the projected post-war demand. In partnership with the experienced housing developer Fritz Burns, he formed Kaiser Community Homes (KCH) and intended to franchise building operations using components manufactured by KCH and their community designs throughout the country (see Figure 4.10).

During the war, private builders constructed over one million homes according to FHA guidelines, and home ownership increased considerably: by 1947, for the first time since records began to be kept almost fifty years previously, more than 50 per cent of homes were owner-occupied. Post-war demand for inexpensive housing was almost insatiable, fuelled further by cheap government loans for returning soldiers under the 'GI Bill of Rights'. As well as new building methods, new industrial materials were tried to help meet the shortage (see Figure 4.11). However, single-family housing, the American dream house, though very different in type and technology from its forebears, still predominated (Hise, 1997, p.155).

Figure 4.11 'After the second house beyond the polystyrene, you'll come to a two-story phenolformaldehyde. That's it': cartoon by Alan Dunn, 1947 (by permission of *The Architectural Record*)

4.2 The skyscraper

Skyscrapers, as tall buildings came to be known towards the end of the nineteenth century, are perhaps the quintessential American building type. They were developed to accommodate rapidly growing private-enterprise business in the highly congested entrepôt cities of Chicago and New York, where specialized business districts were emerging as spatial components. The early skyscrapers were generally no more than ten to twelve storeys high, so they were no taller than the decorative towers on many railway stations and town halls that were being built at this time in Europe and the USA – or, for that matter, than some Gothic cathedrals. What startled contemporaries was that people were using the full height of the building to carry out their daily tasks.

Though the first phase of skyscraper-building took place before the First World War, there were, by 1914, still only a few office buildings taller than twenty storeys outside New York City. However, in the late 1920s there was an extraordinary building boom, culminating in New York City's Empire State Building of 1931. At 1,250 feet (381 metres), this was the tallest building in the world until the 1970s. Skyscrapers became symbolic of the US city. However, it is important not to overstate the case. The economic reverses of the Depression years brought to a halt expensive private-enterprise building projects such as skyscrapers; and skyscraper-construction was not to be resumed until the 1950s. Although by 1990, almost every large US city had an impressive skyline (Ford, 1994, p.50), even as late as 1960 only three cities other than New York and Chicago had more than two buildings taller than twenty-five storeys: there were eight in Detroit, six in Philadelphia and five in Pittsburgh (Willis, 1995, p.9).

Right from the start, skyscrapers were understood as major technological achievements, involving in creative combination: a number of structural technologies; new service technologies which enabled them to function; and new building methods and organization. To say that the building of skyscrapers involved a range of technologies and that they were technological achievements is not to say that the existence of a particular constellation of technologies is sufficient to explain the building of skyscrapers from the 1880s. Indeed, the technologies had all been available for at least a decade before the first skyscrapers, some since ancient times.

> At least as important, moreover, were new market forces put into motion by the enormous growth in the number of well-paid managerial employees [and less well-paid clerical workers] and their desire to be close to the center of the city, the resulting rise in land values, and a revolution in the way capital could be amassed for building purposes. All these things were necessary to make tall buildings profitable. It was in great part the prospect of profit that finally spurred the architects and engineers to develop and exploit new and existing technologies.
>
> (Bruegmann, 1997, p.68)

New York and Chicago were centres of financial transaction for enormous areas; pressure on land in what were emerging as the cities' business districts was intense as business transactions came to take place in central offices near to other businesses and essential services, such as banking or law firms, instead of at the point of production or transhipment of goods (Ford, 1994, pp.24, 28–9). More office space was desperately needed, but in both cities further outward growth was constrained: the business district of New York City was in Manhattan – an island; Chicago's business district was located on the fork of a river close to where it enters Lake Michigan: railways, which ringed the city centre with stations, depots and rail yards, added to the geographical constraints (see Figures 2.13 and 2.28). In both cases, the only way to create more central office space was to stack it up. It is generally agreed that asking which of these two cities had the 'first' skyscraper is not a particularly useful historical question (nor can it be answered unambiguously). Wherever the first one may have been, it is also generally agreed that, while New York did have some tall buildings dating from the 1870s, the key structural innovations for a new type of building were brought together in Chicago in the 1880s (Bruegmann, 1997; Landau and Condit, 1996; Willis, 1995). The *Chicago Tribune* used the term for the first time in the context of tall buildings in January 1889, in an article entitled 'Chicago's skyscrapers' (Condit, 1982, p.115). However, from the 1890s, New York became the pre-eminent skyscraper city: not only did it have the greatest number of tall buildings, but it also had the tallest tall buildings. This was because of differences between New York and Chicago in standard plot-sizes, the economics of property and the regulatory (political) climate (Willis, 1993, 1995).

In the evocative definition by the architect Cass Gilbert in 1900 (he became famous for his Woolworth Building in New York City some ten years later), a skyscraper is 'a machine that makes the land pay' (quoted in Willis, 1995, p.19). The purpose of most skyscrapers was to supply office space. They were speculative ventures built in the expectation that their owners would turn a profit from letting them. In the pre-Depression phase of skyscraper-building, even many of the skyscrapers bearing famous corporate names were occupied only in part by their corporate sponsors; the rest of the space was intended to be let at a profit. The aim of office-building designers and owners was therefore to fit as many good-quality, high-rent offices as possible on to the plot of land that was available to them (see Figure 4.12).

Figure 4.12 The Monadnock Building, Chicago, as portrayed in the *Lake Geneva Weekly News*, 5 March 1896. The accompanying text reads: 'Occupants of the Monadnock building must not all attempt to leave the building at one and the same time, says the *Chicago Tribune*. A mathematician who had nothing else to do the other day, compiled some figures in regard to the big office structure which are rather startling. He first figured on how the tenants of the building could be placed in the street outside. He then compared the number of occupants in the structure with the population of various Illinois cities. The tenants number 6,000. That is more inhabitants than most Illinois towns have. Yet that is the number of persons who spend the greater part of their lives in the sixteen floors of the Monadnock block. They are so many if they all should attempt to get into the streets around the building at the same time, they would have to stand on each other's heads' (courtesy of Lake Geneva Public Library, Wisconsin)

Some technologies used in the skyscraper

It is somewhat artificial to treat separately the technologies that were deployed in the early skyscrapers, because what was important for creating the new building type was their use in combination to achieve unprecedented results. For example, after the infamous Chicago fire of 1871, which destroyed many buildings constructed with internal metal columns and beams, public trust in this form of construction was regained only when new fireproofing techniques were applied to metal structural members. And the heights to which the new buildings climbed would not have been possible without the technology of the lift to accommodate the buildings' occupants. New foundation technologies were also essential to support the great weights of the large buildings.

Figure 4.13 shows typical urban buildings in Chicago before the skyscraper era, at a time of transition. Balloon-frame buildings can still be seen, but larger, masonry-construction buildings were being built to replace them at an astonishing rate, as land values leapt during a boom period. The masonry buildings were typical of those in many cities. They did not usually exceed five to six storeys, because this was reckoned to be the economic height, above which customers and tenants would not be willing to climb. The buildings shown here were not intended to be used as separate office buildings: most had a mixture of functions, ranging from showrooms, to warehousing, banking, studios and even small-scale manufacturing.

Figure 4.13 Balloon-frame and masonry buildings, 1858. Careful scrutiny of this photograph (one of a series of eleven taken from the highest building in Chicago), roughly one-third of the way down from the top, shows ships' masts on the Chicago River. In fact, they were some of the tallest features on the skyline at the time (photograph: Alexander Hesler; courtesy of Chicago Historical Society ICHi-05724)

Metal-frame construction

What cannot be seen in Figure 4.13 is that a number of the masonry buildings in fact had internal iron frames – cast-iron columns supporting beams, at first of wood or cast iron and, later, of wrought iron. This construction technology was well established in Europe, where numerous engineering treatises on the structural uses of these materials were published (Landau and Condit, 1996, pp.19–23). Cast-iron internal construction was introduced into the United States in Washington DC and Philadelphia early in the nineteenth century. A pioneer building of complete cast-iron framing in the USA was the five-storey factory of James Bogardus in New York City, built in 1848–9. Bogardus manufactured prefabricated iron structural members for buildings. The external walls of his factory had supporting members of hand-bolted cast-iron columns. As the walls themselves therefore had no load-bearing function, they did not need to be of heavy masonry construction. He had devised a technology that would later be developed into the hallmark of the skyscraper –

a fully load-bearing metal skeleton with non-load-bearing external walls. However, Bogardus did not brace his structures against the wind; they therefore lacked rigidity, and his technology was pushed no higher (Condit, 1982, pp.81–4). But cast-iron façades did come into use. The building labelled 'Lithography' in the upper middle ground of Figure 4.13, fairly new at the time of the photograph, was the first building in Chicago to have a metal front. Note how that permitted a much greater percentage of glazing than did its neighbours of masonry construction with their large proportion of brick to glass. In the Chicago fire, all these buildings were destroyed. Their metal frames, though not combustible, had melted or buckled in the great heat.

The period of rebuilding after the fire was initially characterized by modest masonry buildings. Because of the economic depression of the mid-1870s, innovations in the immediate post-fire period were conservative, aimed not surprisingly at fireproofing, and were initiated by the city. For example, construction using wood was prohibited in the central area, and where metal rather than masonry was used for a building's internal structure, it had to be protected from fire, usually by encasing it in plaster or brick. However, by the early 1880s, an increasing demand for office space and a rapidly changing property market made the prospect of higher buildings for offices financially appealing; creative ways of using metal instead of masonry were tried, since the availability of passenger lifts (from the 1860s) made taller buildings practicable. There was a relatively low limit on the height that masonry buildings could reach while still turning an adequate profit. Because the walls were load-bearing, the higher the building was, the thicker the walls had to be, especially at the base, in order to support the weight of the building. But thick walls both restricted necessary daylight and took up valuable space which could not be rented; they therefore lowered a building's earning potential – especially at the all-important street-level, which, even in office buildings, was often dedicated to high-rent retail space requiring large display windows.

This constraint was eventually overcome by eliminating load-bearing masonry walls altogether, using instead a metal skeleton to support the building. The external walls could then just be thin 'curtains' of lightweight cladding. Structurally, this is the important technical departure that defines the skyscraper (though some very tall structures with load-bearing masonry walls continued to be built; see Figure 4.12). But full metal-skeleton construction emerged only gradually over a period of years. William Le Baron Jenney, a Chicago architect trained in France, took the basic idea of the internal metal frame a stage further: in his ten-storey Home Insurance Building of 1884–5, he brought the frame right out to the exterior load-bearing wall and embedded the outermost columns in masonry piers (see Figure 4.14 overleaf). This had already been done in other buildings, but at 159 feet (48.5 metres) the Home Insurance Building was unusually tall: if he were to use a single long metal column in each pier to take the entire weight of the building, he would run the risk of the building's masonry elements being destroyed by the expansion and contraction of the metal in the extremes of Chicago's climate. Instead, Jenney decided to make each column only one storey high. He said afterwards:

> it occurred to me that to overcome this difficulty it was only necessary to extend the lintels under the entire window heads entirely across from column to column and carry the outer walls story by story on the columns as well as the floors.

(quoted in Misa, 1995, p.61)

This principle can be seen in the figures marked 'Fig.3' and 'Fig.4' in Figure 4.16, on p.112 (Jenney, 1885). In the Home Insurance Building, the piers still had some load-bearing function, but, because of the metal columns embedded in them, they could be much thinner than would otherwise have been the case, allowing for larger windows and greater, more usable interior space.

Figure 4.14 The Home Insurance Building, Chicago, designed by William Le Baron Jenney, 1884–5, before additions of 1891; it was demolished in 1931 (photograph: A.J.W. Taylor; courtesy of Chicago Historical Society ICHi-00989)

Holabird and Roche's Tacoma Building of 1889 took the process a stage further. A corner building with two façades fronting the street, it had masonry load-bearing walls at the back and at one side, but the façades were of full metal-skeleton construction. That is, the cast-iron columns, riveted for greater strength rather than bolted as in the Home Insurance Building, were brought right out to the street-front, and only a thin curtain wall of terracotta tiles was fixed to the metal framework as cladding. Its construction astonished passers-by accustomed to masonry methods, because the floors were not clad sequentially – the second, sixth and tenth floors were all done at the same time (Bruegmann, 1997, p.83). The point to be made about both of these buildings is that they were not self-conscious applications of new technologies for the sake of style or innovation, but rather, their architects deployed new technologies in order to meet the requirements of maximizing well-lit, rentable space for their clients. Figure 4.15 shows a further stage in development, with a complete skeleton made of metal columns and beams in William Le Baron Jenney's Fair Store of 1891.

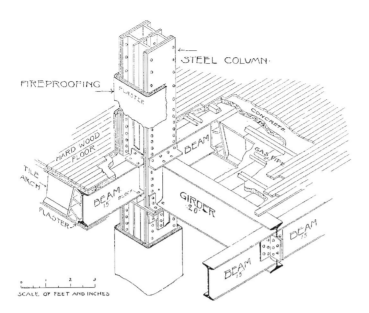

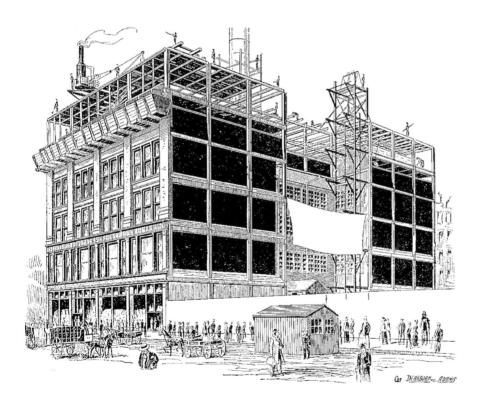

Figure 4.15 *Top* detail of typical column-and-beam joint used in the Fair Store; *bottom* the Fair Store under construction, Chicago, 1891. The Fair Store impressed by its speed of construction; months rather than years were needed to replace a building, thus lowering carrying charges for owners (*Inland Architect and News Record*, 1891, vol.18, plate 8, following p.56; courtesy of Chicago Historical Society ICHi-29568)

By that year, Chicago had six buildings between ten and fourteen storeys high that had been constructed with metal skeletons. In the Fair Store, however, the metal structure was of riveted steel, a material that was beginning to be used in tall buildings. Jenney had already used steel beams in the top three floors of the Home Insurance Building (Misa, 1995, p.60). For its strength, steel was lighter than iron, so structural members could be even less bulky. Ultimately the adoption of steel enabled the very tall buildings that we now think of as skyscrapers to be constructed. However, although steel was recognized to have superior properties for construction, it did not immediately replace iron; precise cost comparisons were involved in the case of each building. Initially, steel was more expensive. Some buildings used a combination of materials because the compressive strength of cast-iron columns was two-thirds greater than that of steel columns.

At the time of transition, the structural cast-iron industry was well developed; the material was relatively cheap and produced in quantity (Misa, 1995, p.50).[7] Steel eventually came to predominate because the production of wrought iron for horizontal members could not be expanded rapidly and cheaply enough to meet the extraordinary demands for structural metals. Puddling, the rate-determining step in the production of wrought iron, was conducted by human labour. It was a heavy and highly skilled job performed under difficult circumstances. Output could be increased only by increasing the number of batches, hence the number of puddling furnaces and the number of highly paid labourers. Attempts to develop mechanical means for puddling proved problematic, and manufacturers switched to steel-production since it could be done in large quantities. Nor should the aggressive marketing and development activity of the steel industry, anxious to find a market other than the declining one for rails (for railway tracks) from the late 1880s, be overlooked. Structural steel had to be of a much higher quality than that needed for rails; the delivery of sufficient high-quality steel was made possible with the industry's change in production method, from the Bessemer process to the basic open-hearth process. In 1889, iron and steel contributed equally to structural shapes (including bridges and ships); thirty years later, one hundred times more structural steel than structural iron was produced (Misa, 1995, p.83).

In addition – depending on which particular mix of interests influenced decision-making in a particular city – the building regulations of different cities favoured different materials. For example, in 1895, for certain aspects of beams, New York's regulations allowed steel a 25 per cent advantage in permitted stresses over wrought iron, while Chicago's allowed 33 per cent. The situation was reversed for columns, where New York allowed steel a 20 per cent advantage over cast iron and Chicago allowed it only 12.5 per cent. So the cost calculations were complex (Misa, 1995, p.67). For related reasons, the construction of buildings using a metal skeleton was initially discouraged in New York, where building regulations required even non-load-bearing curtain walls to be quite thick, thus eliminating some of the space-saving potential of this construction method.

Fireproof construction

Fireproof construction has a history dating to antiquity. However, the 'modern' solution of protecting metal with plaster, brick, concrete or terracotta dates from the eighteenth century. In the mid-nineteenth century, most of the buildings constructed for the growing US government around the country, such as customs houses, followed the European practice of building floors of brick. From the 1860s in France, there were numerous patents for specially designed hollow terracotta blocks which could be used to protect beams and floors. In the USA, this idea was also extended to columns, for which numerous patents were taken out during the 1860s and 1870s. Fire-prevention was a major concern in the new tall buildings, not only for financial reasons (owners had much more to lose) but also because of the difficulty in getting a tall building's unprecedentedly large numbers of occupants to safety. In the Home Insurance Building, Jenney adopted what would become standard practice for flooring: he used hollow terracotta tiles, each in the form of a voussoir (a stone with slightly inclined sides that is used to form a masonry arch) to spring a flat-arched floor, which was then covered with concrete (see 'Fig.5' of Figure 4.16). Figure 4.15 shows the tile system used in the Fair Store to protect structural members.

---

[7]   This discussion of steel-making draws heavily on Misa (1995).

The terracotta tile method of fireproofing the new tall buildings was adopted particularly readily in Chicago because the materials were light. Given the spongy clay that underlay most of the city, anything that would make buildings lighter was welcomed; the higher that buildings went up, and hence the heavier they became, the lighter were the fireproofing and structural materials that were desired. Sara Wermiel (1993, p.20) argues that the high cost of fireproof construction helped to foster the spread of taller buildings as it provided a further incentive for including more rentable space.

Foundations

With bedrock something like 150 feet (45 metres) below the surface in places, Chicago's spongy soil posed particular problems for foundation design, too (see 'Fig.1' of Figure 4.16 overleaf). The traditional method of constructing foundations for large buildings was to drive log piles into the ground in clusters; when set in groups like this, the pilings could support very great weights. However, the maximum depth to which the piles could be driven was governed by the maximum length (and cost) of suitable logs, and so was fixed at about 80 feet (24 metres). Also, because of wood's natural propensity to rot when in prolonged contact with water, the cluster of piles had to be capped off with concrete below ground-water level. Later in the nineteenth century, reinforced-concrete piles as long as 100 feet (30 metres) were developed. They were much stronger than wood and did not have the same problem with water, though, like wooden piles, they could not be driven through soil that contained obstructions.

In Chicago, an alternative solution was to spread the weight of each column over a wider area by building up, as Jenney did for the Home Insurance Building, a pyramidal pier to support it (see 'Fig.2' of Figure 4.16). The amount a building might settle after construction had to be planned for when using such a system, and the trick was to calculate the spread needed to ensure that settlement would be even. Jenney (1885) designed the Home Insurance Building to settle by only 2.5 inches (6.4 centimetres) and was pleased that it stayed within that limit with what he called an 'inequality' of only three-quarters of an inch (1.9 centimetres). A development of this system was to form the piers by setting iron (later steel) rails (later I-beams) in concrete in layers alternating at right angles to each other. This 'floating raft' was a much less bulky system involving less excavation and leaving more basement-space free for lift machinery and services.

In later, taller buildings, caisson foundations became the norm. This system was developed from a bridge-building technology introduced to the USA from Britain in the 1850s. It was used spectacularly in the construction of the Eads Bridge over the Mississippi at St Louis in the late 1860s and early 1870s, and of the Brooklyn Bridge in the early 1880s. For each column, a caisson, or metal cylinder, was driven into the ground in sections and the earth excavated from within it by hand until bedrock was reached. The caisson was then filled, also by hand, with concrete to make a firm foundation for the column. Figure 4.18, p.115, shows a caisson section in the foreground. In Chicago, open caissons were used. In New York, a 'pneumatic' system was pioneered in 1893. In this case, the caissons were closed – that is, the cylinders were capped at the top. Air was forced down with the caisson under pressure, and this cleared out any ground water. Soil was dug by hand from the bottom of the caisson, and passed up to the surface in buckets through an air lock that maintained the pressure. Electricity to provide light was essential for working in pneumatic caissons (Landau and Condit, 1996, p.24; Starrett, 1928, chapter 13).

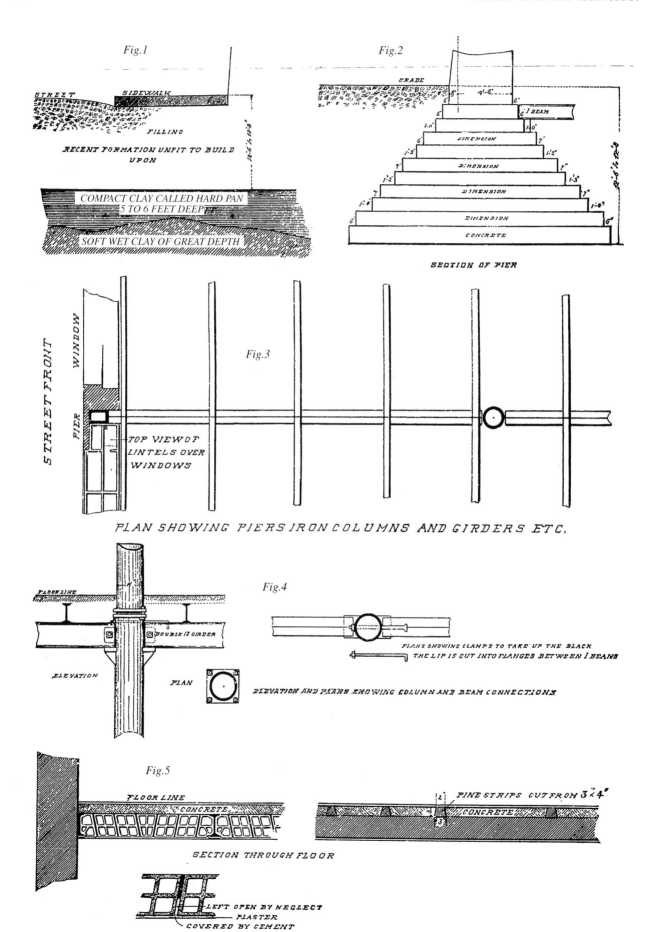

Fig.1

STREET
SIDEWALK
FILLING
RECENT FORMATION UNFIT TO BUILD UPON
12'-6" to 12'-9"

COMPACT CLAY CALLED HARD PAN 5 TO 6 FEET DEEP
SOFT WET CLAY OF GREAT DEPTH

Fig.2

GRADE
4'-0
I BEAM
DIMENSION
DIMENSION
DIMENSION
DIMENSION
CONCRETE

SECTION OF PIER

STREET FRONT

WINDOW
PIER

Fig.3

TOP VIEW OF
LINTELS OVER
WINDOWS

PLAN SHOWING PIERS IRON COLUMNS AND GIRDERS ETC.

Fig.4

FLOOR LINE
DOUBLE 12 GIRDER
ELEVATION
PLAN

PLANS SHOWING CLAMPS TO TAKE UP THE SLACK
THE LIP IS CUT INTO FLANGES BETWEEN I BEAMS

ELEVATION AND PLANS SHOWING COLUMN AND BEAM CONNECTIONS

Fig.5

FLOOR LINE
CONCRETE
PINE STRIPS CUT FROM 3"× 4"
CONCRETE

SECTION THROUGH FLOOR

LEFT OPEN BY NEGLECT
PLASTER
COVERED BY CEMENT

Figure 4.16 (opposite) The construction of foundations in a compressible soil: diagrams illustrating an article by William Le Baron Jenney, 1885. The pyramidal foundation shown in 'Fig.2' was designed to stand on top of the hard pan shown in 'Fig.1'; the vertical measurement marked on the right of each of these diagrams is the same: 12 foot 6 inches to 12 foot 8 inches. The 2-inch difference was the amount Jenney allowed for settling (Jenney, 1885, p.100, figures 1–5)

The lift

The lift, or elevator, is generally held to have first been demonstrated at the New York Crystal Palace Exposition of 1853 by Elisha Otis. His key invention was a safety device that prevented the lift car from falling in the event that its supporting cables broke. Otis installed the first commercial lift in a New York department store in 1857. The early ones were steam-driven. However, the hydraulically powered lift, developed in the 1870s, was capable of travelling much more quickly, up to 600 feet (180 metres) per minute (Otis Elevator Company, 1948, p.12). Not surprisingly, hydraulic lifts were installed in Chicago's new skyscrapers (see Figure 4.17).

For buildings accommodating large numbers of people, calculations of lift capacity were important. Fine offices would lose value were the lift service inadequate. According to the Rand McNally Company's Chicago handbook, put together for the World's Columbian Exposition of 1893, most smaller office blocks had from one to four lifts, while larger ones had eight to twelve (Rand McNally Company, 1893, p.113); the tallest building in the world at that time, Chicago's twenty-one storey Masonic Temple, had sixteen.

Constructing the skyscraper

Organization

Concentrating on the technology of the buildings tends to stress what was new and innovative about skyscrapers. Looking at how they were constructed reveals a mixture of new, and to some extent industrialized, organization and methods, and very traditional practices. On the organizational side, there were important innovations. In the first place, design became very much a matter of engineering and calculation (Condit, 1982). For the early skyscrapers, most Chicago architects drew on the expertise of engineers who had built metal railway bridges, the first large metal structures. Quite apart from the knowledge these engineers possessed, the fact that their metal structures had proved sound was reassuring to clients. Jenney himself had had considerable experience of building railway bridges while in the army during the Civil War, but got a second engineering opinion on the structure of the Home Insurance Building for his anxious client. Architects also worked closely on specifications with those who would be supplying the structural metal, to ensure

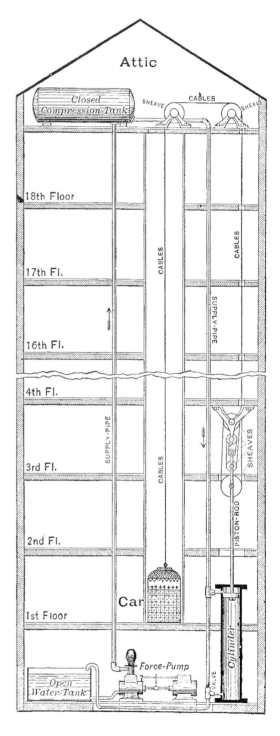

Figure 4.17 A hydraulic lift, 1893. Water was pumped from an open water tank in the basement, which contained 1,200 gallons (4.5 cubic metres) per lift, to a smaller, closed, compression tank at the top of the building. A cylinder with piston placed horizontally in the basement or, as here, vertically in the shaft, was linked by sheaves, cables and pulleys to the cable controlling the lift car. When the car was at the bottom of the shaft, the piston was at the top of its cycle and the cylinder below it was filled with water. To make the lift car ascend, the operator opened a valve in a pipe at the bottom of the cylinder: this allowed the water to flow into the basement tank, while the pressure above the piston forced it down and the cable drew the car upward (Rand McNally Company, 1893, p.115; courtesy of Chicago Historical Society ICHi-29565)

that the innovative shapes and the high quality required were achieved (Misa, 1995, pp.64–6).

Furthermore, an entirely new organizational method of building construction was developed. Traditional practice was for the architect or owner of a building to award separate contracts for each aspect of construction – demolition, masonry, carpentry, plumbing, etc. However, with the construction of the Tacoma Building in Chicago, George A. Fuller, an architect trained in the East, pioneered a new form of constructional organization: he became the first general contractor. His firm took on responsibility for organizing and managing the entire project – in effect, he created a new, specialist area of expertise. Being a general contractor also involved managing financial aspects of the project: to avoid tying up capital too far in advance, for example, the contractor had to arrange for the materials to be delivered to the site as they were needed. The emphasis was on efficiency and speed of construction through good organization, co-ordination and planning (Bonshek, 1988; Bruegmann, 1997, pp.81–3). Reflecting on his experience of building skyscrapers over some forty years in 1928, Colonel William Starrett, who started out in Fuller's office, likened it to a military campaign: 'Building skyscrapers is the nearest peace-time equivalent of war' (Starrett, 1928, p.63).

> Even the organization closely parallels the organization of a combatant army, for the building organization must be led by a fearless leader who knows the fight from the ground up ...
>
> The obtaining of materials near and far and the administration of all those thousands of operations that go to make up the whole are the major functions of the skyscraper builder. Knowledge of transportation and traffic must be brought to bear that the building may be built from trucks standing in the busy thoroughfares, for here is no ample storage space, but only a meagre handful of material needing constant replenishment – hour to hour existence. Yet it all runs smoothly and on time in accordance with a carefully prepared schedule; the service of supply of this peace-time warfare, the logistics of building, and these men are the soldiers of a great creative effort.
>
> (Starrett, 1928, p.66)

A sense of what was involved, more matter-of-factly expressed, can be gleaned from Extract 4.1, 'Speed' by Paul Starrett, William's brother, who also began his career in Fuller's office, and would cap it as the builder of the Empire State Building.

Some construction technologies

Almost by definition, skyscrapers tended to be built in congested areas, which meant that neighbouring properties had to be protected and their services not disrupted; this complicated the supplying of materials. As new technologies were developed, they were deployed in skyscraper construction. As hinted at by William Starrett's comments, the lorry played an increasingly important role for a range of specialist tasks. Its flexibility and load-carrying capacity became critical to the crucial task of delivering supplies to sites (see Figure 4.18).

The application of power to building processes was also important. Initially it was the steam-engine that transformed the labour of handling materials, as it powered diggers, hoists and cranes. The steam-digger was patented as early as the 1830s, and steam-hoists began to be used in building construction in the early 1850s. Some steam-cranes were used in construction in New York City from the 1870s (one can be seen in Figure 4.15, bottom, p.109), but the use of power was not widespread until the 1890s when the economics of tall buildings changed as manual labour costs rose. In a nineteenth-century factory, a single large engine would drive a number of stationary machines, each linked to it by a fixed system of belts and shafts. However, on a construction site, continual spatial change was, by definition, characteristic. Therefore it was necessary to have a small, movable independent engine for each powered device. This was costly in terms of both the

Figure 4.18 Mobile concrete-mixer, 1928. This lorry was delivering concrete to fill an excavated caisson. Note a caisson form in the foreground waiting to be driven down (reproduced from Starrett, 1928, p.160; photograph: courtesy of Thomas Crimmins Contracting Company)

engines and the skilled operatives needed to run them. While manual labour costs remained low, using efficiently organized labourers was cheaper than using powered devices. As labour costs went up and, in New York, building heights did too, manual labour became relatively more costly and power became the norm.

The stress on efficiency also led to new working practices for traditional artisans. Night working became common – under electric arc-lights from the late 1870s and with incandescent light from the 1880s. In an industry that previously had been highly seasonal, winter working was adopted: production sites were covered and heated, and chemicals were added to cement to prevent it freezing (see Figure 4.19) (Condit, 1982). In some cases, new skills were organized in a traditional manner using modern tools (see Figures 4.20 and 4.21, pp.116, 117).

Figure 4.19 Winter working on skyscrapers, 1928. So that work could go on all year round, scaffolds were enclosed in canvas during the winter and solid-fuel stoves provided heat. The mortar was tempered with warm water heated on the stove (reproduced from Starrett, 1928, p.223; photograph: courtesy of Patent Scaffolding Company)

The 'Heater'

The 'Catcher'

The 'Gun-man'

The 'Bucker-up'

Figure 4.20 A gang of riveters. Riveters were organized in gangs or teams of four, each worker having a specialized skill. The riveter's gun, weighing 35 pounds (15.9 kilograms), was powered by compressed air and could hammer in a rivet at the rate of 1,000 blows per minute. Electric welding was considered unsafe for use in tall buildings (photograph: © Arthur Gerlach; courtesy of The British Library)

Figure 4.21 The original title of this photograph was 'The artisans are dead' – though in fact, in terms of organization, riveting was still artisanal. However, skyscraper-construction was undeniably dangerous, with deaths running at between three and eight on sizeable buildings. The insurance premium paid by construction firms for riveters was the highest for any worker (photograph: © Arthur Gerlach; courtesy of The British Library)

Figure 4.22   The Empire State Building, New York City, designed by Shreve, Lamb and Harman, 1932; its mooring mast is not yet in place (photograph: courtesy of The Empire State Building Company)

The Empire State Building

The culmination of the heroic period of skyscraper-building in New York was undoubtedly the construction of the Empire State Building in 1930–31 (see Figure 4.22). From 1923, the total volume of office space in Manhattan had nearly doubled, and more than fifty buildings of thirty-five storeys or more went up (Willis, 1998, p.13). A particular spurt of building occurred in 1929–31, when there was a booming market, easy finance and a climate of speculation. Buildings reached new heights and the Empire State topped them all by 200 feet (60 metres). It was also bigger than the rest in terms of rentable area, with more than twice the capacity of its nearest competitor (Willis, 1998, p.14). That the Empire State Building became the symbolic building of New York City capitalism – and indeed symbolic of skyscrapers in general – was partly the result of an irony of timing. It was not built in the financial district, where there was already a concentration of skyscrapers, but was the first to go up in the Midtown area; when the Depression brought building abruptly to a halt, the Empire State was somewhat isolated from other tall buildings and thus was particularly prominent. Also because of the Depression, the Empire State was not fully occupied, and so did not return an income for its owners until the late 1940s (Willis, 1998, p.30).

The design of the Empire State Building and its contemporaries was in many ways a consequence of the 1916 New York City zoning ordinances (see Figure 4.23) (the origin of these is discussed in Chapter 6). Section 1 of Extract 4.2, by William Starrett, gives a builder's view of their effect. Some of the story of the construction of this mammoth building is described by Paul Starrett, its builder, in Extract 4.3.

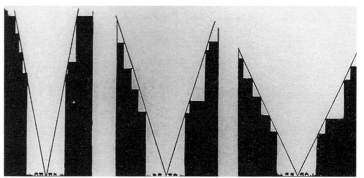

Figure 4.23   Zoning diagrams for New York height districts, 1916. The 1916 zoning ordinance set out 'five formulas based on the width of the street and the angle of the setback. These were described as "districts", which referred to the height of the maximum vertical wall above the street permitted in that area. For example, in a "$1\frac{1}{2}$ times district", where the street was 100 feet wide, the building could rise a sheer 150 feet before the first setback. Above that level, the mass had to step back in a ratio of 1:3, i.e., one foot back for each three feet of additional height' (Willis, 1995, p.71). Unlimited height was permitted on one-quarter of the site. This made large plots particularly attractive, so that the tower section of a building would have room to include sufficient rentable space to offset the great expense of building lifts and services for it (Willis, 1995, p.76). These regulations gave New York its classic skyline (reproduced from Willis, 1995, p.71, figure 75; with the permission of Dr Carol Willis)

4.3 *Electricity and buildings*

Central generation and the 'Great White Way'

Electricity was first introduced into US cities at mid century with the electric telegraph (see Chapter 8). Electric street lighting was introduced from the late 1870s (see Chapter 6) and electric lighting was shortly thereafter adopted for the internal illumination of large city buildings, including the new skyscrapers – though, as mentioned above, its use in housing was not common until much later. Electric lighting in the public arena was soon developed, however, far beyond the utilitarian; it became a characteristic of urbanity (Bouman, 1987, pp.10–14).

To launch his electrical business in an environment where gas lighting was already well established, the American innovator Thomas Edison deliberately sought to capture the gas-lighting market. He aimed to achieve the same standard of illumination achieved by the gas-lighting industry, and he adopted its model of local, central generation and distribution for his pioneering supply of electricity to the public. His first central supply was generated in 1882 at premises in the heart of Manhattan's financial district, a typical location where he could find business customers able to afford the expense of installation. He therefore designed his early generating stations to resemble city office buildings (see Figure 4.24). He was also adept at promoting lighting by means of spectacle, something very appealing to Americans. For example, in 1886, the Statue of Liberty was illuminated dramatically by electricity for the first time (Nye, 1992, p.32).

Figure 4.24 First central station of Chicago Edison Company, built 1887. This generating station, in the heart of the business district, was built in the space to the right of the Home Insurance Building, shown in Figure 4.14 – in comparison to which it was inconspicuous indeed. Once electricity was confidently established and power stations became larger with the introduction of alternating-current generation, they were built on edge-of-town sites and had their own characteristic architectural style (see illustrations in Bouman, 1993[8]) (photograph: courtesy of Commonwealth Edison Company, Chicago)

[8] An abridged version of this essay appears in Roberts (1999), the Reader associated with this volume.

Edison's short-range, direct-current (d.c.) systems were installed in many middle-sized and large cities around the country; they were superseded towards the end of the century by systems based on the long-distance transmission of alternating current (a.c.) at high voltages without loss of energy. The first of such systems was introduced in the USA at Niagara Falls during the early 1890s; the economics were such that a.c. systems made possible a much greater use of electricity. Edison systems were of relatively small capacity and could only serve a small geographical area. Competitor a.c. systems could serve a much larger area more cheaply and also achieve economies of scale by amalgamating the requirements of different types of consumer previously served by separate generating facilities. Cheaper electricity, in turn, encouraged people to use it more (Hughes, 1983, pp.106–39, 201–26). By 1902, there were 2,250 generating plants in the United States and by 1920, 4,000 (Cowan, 1997, p.163). The spread of electricity coincided with a great wave of international exhibitions mounted by city promoters to attract business to their cities. Electrical firms both took part in planning the exhibitions and used the opportunity to be exhibitors. Dazzling lighting displays became symbolic of US industrial and technological achievement (Nye, 1992, pp.33–47). The commercial possibilities of making shops as exciting as the fairs was soon realized, and by the 1890s, electric advertising was common. This was resisted in Europe, but embraced in the USA, where the level of illumination in the streets, and the energy consumed, grew accordingly.

By 1903, Chicago, New York and Boston had five times as many lights per inhabitant as did Paris, London or Berlin (Nye, 1992, p.50). Already in the 1890s, New York City's Broadway was famous as 'the Great White Way' because of the high level of night-time illumination, mostly from advertising. Many smaller cities were tempted to emulate larger ones, and proclaim that they too were modern, by installing 'white ways' of their own, albeit on a more modest scale. Electricity also came to be used decoratively, as a feature of city architecture as well as a means of showing off individual buildings (Bouman, 1993). At night, electric lighting would highlight the more pleasant, or dramatic, features of the cityscape, making a sort of ideal city; the bleak, dreary, or less than tidy was not part of the illuminated landscape (Nye, 1992, pp.60–61).

Electricity and building design

The effect of electricity on building design was gradual. Initially, it was adopted to improve the delivery of existing features. Incandescent electric lights, once available, were adopted instead of gas lights, and electric motors came to be preferred to hydraulic power for lifts. Similarly, central heating and forced-air ventilation were used long before the advent of electricity. However, the small electric motor, developed in the 1890s, made possible much more adaptable arrangements of ventilation systems (see Figures 4.25, 4.26 and 4.27) (Banham, 1969, pp.53–4).

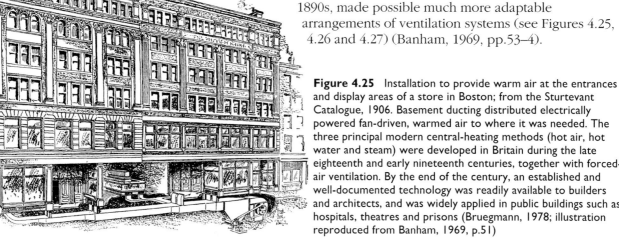

Figure 4.25 Installation to provide warm air at the entrances and display areas of a store in Boston; from the Sturtevant Catalogue, 1906. Basement ducting distributed electrically powered fan-driven, warmed air to where it was needed. The three principal modern central-heating methods (hot air, hot water and steam) were developed in Britain during the late eighteenth and early nineteenth centuries, together with forced-air ventilation. By the end of the century, an established and well-documented technology was readily available to builders and architects, and was widely applied in public buildings such as hospitals, theatres and prisons (Bruegmann, 1978; illustration reproduced from Banham, 1969, p.51)

Figure 4.26 Larkin Building, Buffalo, NY, designed by Frank Lloyd Wright, 1906; exterior view, showing stair towers and duct boxes at the corners. This was one of the first office buildings in the United States to be hermetically sealed and to have a forced-air ventilation system, to keep the office space free of pollution from the neighbouring railway line. Built as a mail-order house, internally it was one vast open space (making mechanical ventilation economic) arranged along scientific-management principles to facilitate clerical work. It had a high level of office equipment for the time (Duffy, 1980, pp.265–9; photograph: courtesy of Buffalo and Erie County Historical Society, Larkin Company Collection)

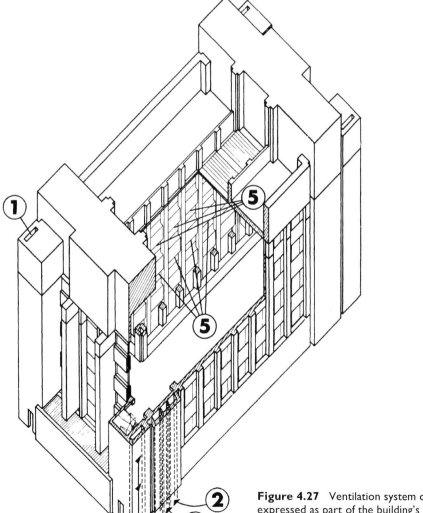

Figure 4.27 Ventilation system of the Larkin Building; note that this was expressed as part of the building's exterior design

Key 1: fresh-air intake; 2: tempered-air distribution; 3: foul air and exhaust; 4: utilities duct; 5: tempered-air outlet grilles under edges of balconies (reproduced from Banham, 1969, p.89)

In terms of their basic design, skyscrapers of the 1930s, for example, were remarkably similar to their predecessors: they were based on the office unit as a cell whose size was constrained by their having to be lit by natural light and ventilated by opening windows. In order to maximize the amount of daylight for office illumination while simultaneously maximizing the number of rentable office units, buildings on very large plots were constructed around light courts. These were open spaces in the centre of buildings which gave them a doughnut-like footprint, with one row of offices on the outer circumference and another row on the inner circumference. High ceilings and large windows also helped to increase the amount of light that could enter a building (Willis, 1993, 1995). However, when skyscraper-building resumed after the Second World War, the use of new, electrically powered technologies, particularly fluorescent lighting and air-conditioning, allowed the removal of these constraints; more varied interior designs were made possible, including much deeper open-plan spaces and movable partitions. What continued to drive change was the perennial imperative of maximizing rentable space. Fluorescent light tubes, commercially available from the late 1930s, gave far more light with much less heat than incandescent bulbs. Heat was still generated by even fluorescent lights, though, and by the building's occupants and, subsequently, by business machines. To achieve the deeper spaces and do away with light courts, it was necessary to remove this heat by artificial cooling.

Air-conditioning had been developed from early in the century as a means of controlling the quality of air, especially its humidity, in industries where sensitive processes were carried out.[9] Engineers came to consider it, more generally, as a means of 'climate control' (incorporating humidity control and temperature control, as well as cleaning and circulating the air). To the public, however, air-conditioning meant cooling, and it was in the new cinemas of the 1920s that many of them first met it. Air-conditioned cinemas advertised this technological feature of their modernity along with the new medium of film.

In office buildings, fluorescent lighting and air-conditioning were complementary technologies which made possible a number of design changes in addition to the deeper spaces. For example, because rooms no longer needed to be illuminated by natural light, the high ceilings that maximized the amount of window-glass were no longer essential; it was thus possible to have more floors in a given height of building (see Figure 4.28). In addition, the introduction of air-conditioning eventually made it possible to do away with opening windows, thus eliminating a major constraint on the design of façades (Willis, 1995, pp.132–43); they were retained in many buildings, though, even when they were not technically necessary, because traditional habits could be slow to change (see Figures 4.29 and 4.30).

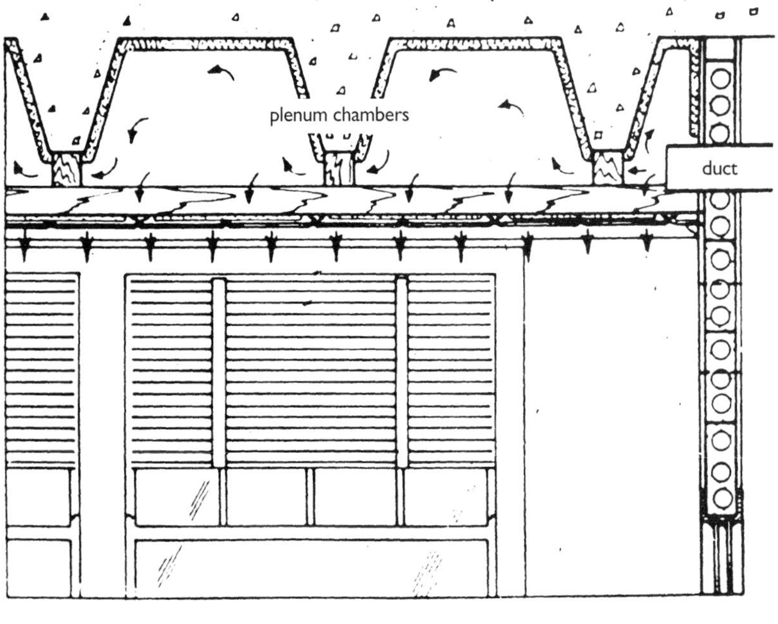

Figure 4.28 Perforated 'Acousti-vent' ceiling system, Burgess Laboratories, 1936. A new ceiling surface, which reduced noise, contributed to the development of the open-plan office. Here, plenum chambers (sealed spaces containing pressurized air), for distributing conditioned air and removing spent air, have been constructed in the space above the artificial ceiling. The space was also used to carry ceiling lights (reproduced from Banham, 1969, p.215)

---

[9]   This discussion of air-conditioning draws heavily on Cooper (1998).

Figure 4.29 Milam Building, San Antonio (TX), designed by George Willis, 1928. Initially, air-conditioning was introduced into new buildings built to a traditional design, like this one. The building was constructed of brick with a reinforced-concrete frame, and was reputedly the country's tallest such building at the time. It is generally said to be the first fully air-conditioned, high-rise office-building. With air-conditioning, opening windows were not technically necessary, but they were provided in the Milam Building. Built in hot, southern Texas, it was the first multi-storey office project taken on by the pioneering air-conditioning firm of Willis Carrier. Though it had not happened here, Carrier predicted that, as long as air-conditioning was incorporated from the start, it would permit dramatic changes in building design – as, in combination with fluorescent lighting and acoustic ceilings, it eventually did (photograph: H.L. Summerville, San Antonio, TX, 1928; courtesy of the San Antonio Conservation Society Foundation)

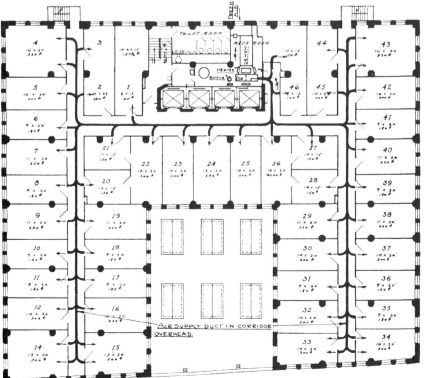

Figure 4.30 Office and ducting plan, Milam Building. The Milam Building's traditional design, with its emphasis on maximizing daylight in the offices by means of high ceilings, meant that the ducting for the air-conditioning had to run along the corridors, with a branch leading off into each office (reproduced from *Building Investment*, 1930, February, p.52; courtesy of the San Antonio Conservation Society Foundation)

The overall effect was to encourage the construction of block-like buildings. Such designs were dependent on custom-designed air-conditioning for their habitability, but to make the installation of air-conditioning affordable, other features were sometimes skimped. In an era of cheap and abundant energy, it was thought feasible to compensate by using inexpensive, and less energy-efficient, materials in the rest of the building's construction. For example, facing materials were often chosen with no thought for their thermal properties; this made many buildings very inefficient in their use of energy (Cooper, 1998, pp.160–61). Modern homes were also energy-inefficient, because their design did not incorporate traditional means of keeping homes cool (careful orientation with respect to the sun, cross-ventilation, deep eaves, high ceilings, external landscaping). The small houses on the huge estates built after the Second World War had ample expanses of glass; these made the tiny interior spaces seem larger, but also admitted the heat of the sun in summer and dissipated artificial heat in winter. Such houses included air-conditioning, or at least space for the system's ducting, ready for installation when the owner's finances allowed.

Whole-house air-conditioning was what we might call the engineer's approach to the technology and was usable only if designed into a house from the start. Taking a different approach, appliance manufacturers developed window units which could be installed in existing homes. These became progressively less expensive to purchase, but were extraordinarily inefficient to run. When air-conditioning was applied to office buildings and houses, it was often less effective than it might have been because it proved difficult to persuade occupants to change their behaviour and accept that they should keep the windows closed. Domestic air-conditioning was slow to be taken up not only because of the expense, but also because of people's reluctance to adapt habits and customs for dealing with high temperatures. Open windows were part of a culture of hot-weather behaviour that changed only gradually: including eating light, cold meals, sitting on the porch in the evening or pursuing outdoor activities. And of course, moving to the suburbs was the solution chosen by some as a way of keeping cool (Cooper, 1998, pp.168–73). Even when air-conditioning was accepted, many users failed to use it optimally, adjusting the thermostat radically depending on the time of day instead of keeping it at an even setting.

After the Second World War there was a boom in the adoption of air-conditioning. In humid, government-led Washington DC, its ability to control humidity was an important feature. However, the use of air-conditioning spread most rapidly in the hot, dry cities of the south-west: Phoenix, Tucson (AZ), Las Vegas (NV), Bakersfield (CA), Dallas, Austin (TX), Fresno (CA). This rapid take-up in areas where humidity control was needed least is somewhat puzzling. However, evaporative coolers (often built by do-it-yourselfers), in which fan-driven air was blown across dampened pads, had been used in these areas from the 1930s; some 90 per cent of Arizona homes had evaporative coolers by the early 1950s, and this may have prepared the ground for the adoption of more sophisticated air-conditioning as a means of cooling (Cooper, 1998, p.177). Subsequently, air-conditioning would become the key to the mushroom-like development of the sunbelt cities of the region (see Chapter 1). In the 1930s, electrical undertakings saw summer air-conditioning as a way of utilizing out of season the large capacity needed for winter; but by the late 1950s, summer peaks were already exceeding winter peaks in some cities. The demand reached crisis proportions during a heatwave in July 1966. Air-conditioning, installed for cost-cutting purposes in new, energy-inefficient buildings, or supplied by inefficient window units in older buildings never intended for it, led ultimately to 'higher consumer costs, overtaxed electrical networks, and a national dependency on imported oil' (Cooper, 1998, p.180).

4.4 Buildings and the motor car

Transport technologies and buildings were clearly closely related. For example, to be financially viable, a skyscraper needed good transport links. At the same time, buildings that concentrated so many people in one place contributed to increasing street congestion. In the case of the motor car, various ways of accommodating it in the city were experimented with during the 1920s, when its use in cities was expanding so rapidly. An alternative approach was to move activities that were not office-based out to the edge of the city where motor vehicles could be more readily accommodated.

Accommodating the motor car in the city

One of the most striking changes made to accommodate the motor car was on the domestic scale: the development of the domestic garage (Jennings, 1990, pp.95–106). Garages were often attached to the house and, together with their driveways, came to dominate street frontages. The front porches that people had once sat out on were no longer built, and families retreated to patios at the rear of their houses, or to their air-conditioned interiors. This major social change in the nature of neighbourliness was exacerbated by the tendency of new suburban estates to have no pavements.

The case of Chicago illustrates a variety of public devices for dealing with the motor car. The city's redevelopment under the Burnham–Wacker Plan occurred as motor-car use was increasing dramatically (see Chapters 3 and 6). One attempt to accommodate the car in a building's design can be seen in the Jewelers Building, which was finished in 1926, the same year as Wacker Drive, the new double-deck riverfront road on which it was located. The building advertised its location on this major, modern thoroughfare to attract occupants. It was named for what was seen as its principal group of clients, jewellers, and security was a major issue. One of the ramps to the lower deck of Wacker Drive came right up beside the building (see Figure 4.31 overleaf). Therefore, in place of a central light court, a twenty-three-storey parking garage was installed at the heart of the building and serviced by an interior lift (see Figure 4.32, p.127). A jeweller could enter the building in the safety of a motor car and then leave the vehicle in the hands of an attendant. The car would ascend to an upper floor in a lift and be rolled into a parking bay by a series of automatic, electrically operated ramps. Effectively, the motor car was seen as an occupant of the building too. Though security was the pitch, having a guaranteed parking space was already in 1926 an important attraction in Chicago. The lift operated successfully until 1940, by which time standard cars were too big for it (Sennott, 1990, pp.164–7). The Jewelers Building was not the only example of an attempt to integrate street and building design, although, as was pointed out in Chapter 3, tearing down city-centre buildings to make car-parks was a more common practice (see also Figure 4.33, p.127).

Figure 4.31   The Jewelers Building, Chicago, 1926. Although it is hard to see, there is a 'garage' sign on the right-hand side of the building, just above ground-floor level (photograph: Raymond Trowbridge; courtesy of Chicago Historical Society, HB-TR 37825-F)

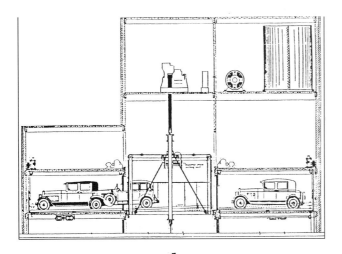

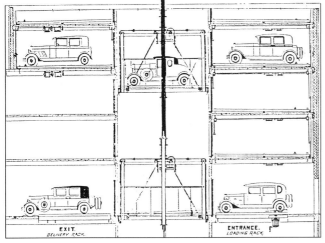

Figure 4.32 Ruth Safety Garage, Jewelers Building, Chicago, 1926 (reproduced from Smith, 1988, p.61)

Figure 4.33 Vertical car-park in the Chicago 'Loop', 1941. In Chicago, in addition to the street-level car-parks, new structures were devised to store the motor car, taking advantage of the technology of the lift. This one was built by Westinghouse in the 1930s, and was a private attempt to provide off-street parking. However, its operation was too slow to be effective. In the 1950s, with the parking problem still acute, the city undertook a programme of building multi-storey car-parks both above and below ground (Condit, 1974, p.332; photograph: courtesy of Chicago Historical Society ICHi-05189)

Railway facilities were also designed to serve the motorist. For example Chicago's Union Station, which opened in 1925, was designed to facilitate passengers' ease of transfer between motor cars – in the form of taxis – and railway carriages (see Figures 4.34 and 4.35, p.128). To avoid congestion in the surrounding streets of this busy district, a system of interior circulation by means of ramps and drives was planned, so that passengers and baggage were delivered or collected from inside the station, rather than outside on the streets (Sennott, 1990, pp.161–3).

Chicago was undertaking an important series of planned developments just as accommodating the motor car was becoming an issue, so some of those developments took the motor car into consideration. Elsewhere there was more antagonism. Section 2 of Extract 4.2 by William Starrett describes the congestion problem in New York City and points to another of the strategies adopted for accommodating ever more cars – the leasing of 'air rights' over railway lines. Following the Grand Central Station development in New York, which Starrett mentions, this was also done very successfully in Chicago in the late 1920s (Rau, 1993, pp.104–6). Using the techniques of skyscraper construction, and considerable structural ingenuity, the space above city-centre railway lines became occupied by large buildings. In return, those buildings paid the railway companies for their air space. This provided a means of ending the visual dominance of the city by railway lines at the same time as providing developers with new, large city-centre sites and the railways with funding for their own developments.

Figure 4.34 Union Station, Chicago, 1920s (photograph: courtesy of Chicago Historical Society ICHi-05319)

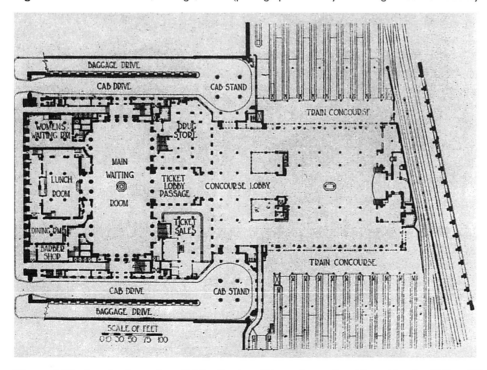

Figure 4.35 Interior floor plan of Union Station, Chicago (reproduced from Condit, 1973, p.273, figure 53)

One of the classic Chicago examples of the period is the Merchandise Mart (built 1927–31), then the largest building in the world in terms of volume. Using a number of catchwords of the period, a Chicagoan commented in 1930 that the exploitation of air rights marked 'a new era in commercial and civic development … typifying an age of increased efficiency through consolidation and concentration' (quoted in Rau, 1993, p.108). For the Merchandise Mart, which was designed to be a department store for wholesalers, managed to integrate road, rail, the El (see Chapter 2) and shipping. Its riverfront site on the edge of The Loop was adjacent to a stop on the El, and the building itself extended over a goods-railway line. It was conceived as an entire shopping street within a single building, and included, in addition to the wholesale displays, a full range of retail shops and such services as an optician, a beautician and a stenographer. Wholesale customers from the surrounding area could complete all their city transactions in one place. It also included the most modern of technological appurtenances – six radio broadcasting studios.

Buildings on the edge of the city

The Merchandise Mart represented the move of one category of activity – wholesaling – out of the heart of the Chicago business district, but in this case not out of the centre of the city. However, already in 1905 there had been a move to find new industrial locations outside the city in order to escape congestion. The Chicago Junction Railway and Union Stockyards Company constructed and promoted the Central Manufacturing District, a purpose-built industrial estate for a community of mixed industries. This step towards industrial dispersal was taken well before the motor lorry became common and was based on existing rail technology. The new Central Manufacturing District was sited beside a railway marshalling yard at the edge of the city, in an area that was otherwise occupied by market gardens. In its first eight years, some 200 industries moved there. Its supporting facilities included: a bank, a club, a telegraph office, a traffic bureau and its own police force and street-cleaning organizations.

In 1923 the estate expanded on to a nearby site, and the company built an elaborate infrastructure for its occupants: the Pershing Road Development included a series of uniform five-storey manufacturing blocks, a central power plant distributing steam and electricity to every building, a central water tower and sprinkler system, a central goods station and handling facilities with linking tunnels to every building, private rail sidings and full sewer and water connections (see Figure 4.36). The new element in 1923 was explicit provision of loading bays for lorries (see Figure 4.37 overleaf) and a fleet of lorries for internal communications. The same company promoted a Central Manufacturing District for Los Angeles in 1925 (Schopp, 1955).

Figure 4.36 Chicago's Central Manufacturing District, Pershing Road Development, 1923 (reproduced from *Central Manufacturing District Magazine*, 1955, vol.39; courtesy of Chicago Historical Society ICHi-29566)

Figure 4.37 New Chicago Produce Market, 1925. Lorry bays became common in new developments of the 1920s. When the Chicago produce market was moved 2 miles (3.2 kilometres) south of the centre as part of the Burnham–Wacker improvements to the riverfront, the lorry featured in its design. Note how the lorries reverse into loading bays that run directly into the market building, much as horse-drawn wagons had done in the old market. This would become a characteristic configuration for later, edge-of-town industrial buildings when transport was lorry-based (photograph: Kaufmann and Fabry; courtesy of Chicago Historical Society ICHi-03241)

During the inter-war years, shopping also began to develop outside city centres. Alternative shopping areas complementary to those of the city centre were developed as high-end retailers began to locate premises in planned suburban communities, where their markets increasingly were, on the assumption that shoppers would use the motor car to get to them. The shopping centres of these peripheral communities served as social foci as well as centres of commercial activity. One of the earliest was the 'Spanish-style' Country Club Plaza, built in the 1920s to the south of Kansas City (MO), to serve the wealthy residents of a property development begun in 1908. A key feature was that, rather like the Central Manufacturing District, the Plaza was owned by a company and run as a business itself; the company could thus control the range and nature of shops provided. Dedicated, free, off-street parking was provided for motorists. Other community shopping centres were started in the period 1927–31, but their growth was halted by the Depression. Shopping centres did not become standard features of the urban landscape until after the Second World War (see Figure 4.38), but the developments of the inter-war years served as precedents (Longstreth, 1997).

4.5   Buildings of tomorrow – or yesterday?

The types of building developed over two centuries of US history were not inevitable, nor due to technology alone. Rather, they were the consequence of numerous small decisions about the deployment and development of technologies in particular social, geographical, environmental and economic contexts. Once certain technological choices were made and rendered material, however, they then provided part of the enduring context of future decisions.

During the twentieth century, the United States became the world's largest consumer of energy. Its energy-use near the century's end far outstripped that even of other highly industrialized nations: the USA consumes per capita about 40 per cent more energy than Germany, 100 per cent more than Sweden and 200 per cent more than Italy or Japan (Nye, 1998, p.6). Buildings – in their own use of energy, in the uses made of energy by their occupants and in transport uses related to their occupancy – are part of this consumption pattern. US buildings have been conspicuously 'high tech' and demanding of energy. However, at the end of the century, as efforts were made to regenerate city centres that had been allowed to languish, with the dispersal of core activities, some historians pointed to the virtues of earlier, less energy-intensive technological models. Skyscrapers ventilated by opening windows and illuminated by natural light provided a model not only for more efficient use of energy, but also for a more human-scale building, whose occupants would be more in touch with nature instead of cocooned by technology.[10] The implications of high-tech communications and the dispersal of city centres are considered in Chapters 7 and 8.

[10]   A point made by C. Willis in an interview on *The Today Programme*, BBC Radio 4, 22 January 1999.

Extracts

4.1 Starrett, P. (1938) *Changing the Skyline: an autobiography,* New York, McGraw-Hill/Whittlesey House, pp.172–7

Time is important in building … We Starretts were noted for speed, as I have told you, and it is worth while to explain some of the factors in that speed.

The first is the builder's organization. On each job we (and this is true of any large modern builder) place a superintendent long in our service. Under him are a production manager; a construction division with foremen on steel, masonry, concrete, carpentry; an expediting department; an inspection department. The quality of the key men in this organization is vital. One of the most necessary requirements in the head of a building company is to be able to size up men …

When we start a building we draw up a schedule, giving the time allotted to each branch of the work, with starting and completion dates, and showing how the time for each trade interlocks with other trades. This schedule is both written and graphic; often it is supplemented with detailed schedules, showing what is required of each trade on each floor in each week. Subcontractors and manufacturers are required to carry on divisions of the work to fit these schedules.

The expediting department has the duty of following the manufacture of every item and seeing that it progresses at a rate which will bring it to the site on time.

The superintendent has an important assistant called a job runner. This man keeps constantly in touch with the architect and his engineering consultant. He secures and distributes all plans and specifications. He checks changes of any sort, notifies everybody affected by them. He carries on all negotiations with subcontractors after the original contract is signed. Through the expediting department he keeps tab on the progress of manufacture of materials.

Daily reports give the previous day's labor production in each important trade, the average daily production to date, so that the head office may know how its actual labor costs correspond with the estimated cost.

Now, consider materials – steel, for example. Steel is usually manufactured in the Pennsylvania mills and, if destined for a New York building, is shipped after fabrication to a point in New Jersey across the river from Manhattan. There it is sorted out and is brought across the river on lighters in the order required. But if 5 per cent of the steel for the first story is missing, the whole building is held up … Often we have to go out and search for missing [railway] cars ourselves. Our man might find that a train composed of a certain number of cars had left Pittsburgh and only a part of the cars had reached Altoona. His job would be to go along the line, riding perhaps in cabooses [guards' vans], searching all the sidings for his missing cars …

Now let us look at the subcontractor as a factor in speed. In my early days, many contractors did most of the work on the building themselves. As time went on and buildings grew larger and more complicated, and subcontractors became more competent, the tendency was to let out a good part of the work on subcontracts. Many of the subcontractors had developed a high degree of specialization and could do better work in their particular fields than could the general contractor …

Subcontractors must work in proper sequence. Delay in metal lath holds up plasterers. Delay in 'roughing,' that is, the plumbing pipes concealed in walls, holds up plasterers, too. Plastering delay holds up carpenters, and so on. To assure proper sequence and stimulate speed, we started the practice of weekly meetings of the subcontractors, presided over by the superintendent or the job runner, and often attended by an executive of the company and by a representative of the architect. Everybody reported on progress. Laggards had to account for themselves.

4.2   Starrett, W.A. (1928) *Skyscrapers and the Men who Build Them*, New York, Charles Scribner's Sons, pp.100–7

[Section 1]

The New York zoning law, passed in 1916, was practically, not aesthetically, intended. Depending upon the width of the abutting streets, the law requires that a building be stepped back at certain heights. These restrictions apply to three-quarters of the ground area of any new building. On the remaining quarter, a tower may be carried to any altitude the owner may desire. The law was intended to protect the rights of lesser buildings and to permit the sunlight to reach the streets a greater part of the day. Its principal, though purely collateral, effect, however, was to give to architectural design in high buildings the greatest impetus it ever has known, and to produce a new and beautiful pyramidal sky-line.

... This New York-born architecture is an adaptation of no other; it is our own, expressing ourselves. It is the sounder for having a reasoned motive rather than individual fancy behind it. Beauty of line and form, rather than beauty of ornamentation distinguishes it. With only ten years' history behind it, the setback is a thing of grace and sweep. There is no reason to suppose that it is a finished form; towers of unimagined beauty should rise in the 1930s, and that is far enough to carry any prophecy these eventful years.

We borrowed the zoning law from Europe, where long ago it became customary to limit any commercial building to a height not exceeding the width of the abutting street, measured from building-line to building-line; and as the purpose was aesthetic, the limitation was absolute. The setback is an American compromise. It is in effect, with local variations, in more than half the American cities of 25,000 population or more.

Washington is our only city, to my knowledge, that follows the European practice of absolute limitation. Boston did so until 1928, when it adopted a zoning plan similar to New York's ...

The workings of the New York zoning law are so complex that a new profession – that of consulting expert in zoning – has arisen there. These experts advise and interpret between the architect and the builder, and the Building Department in the fashion of lawyers.

[Section 2]

The assault on the skyscraper in recent years has been on practical rather than aesthetic grounds ... its enemies charge it with creating outrageous traffic congestion, unsettling land values and putting a disproportionate burden on the municipality for fire protection, water supply and sewage disposal ... Certainly, a building housing 10,000 workers aggravates the traffic problem for blocks around. But the high building is only one factor in a condition practically inescapable in modern life. The motor car is a worse offender than the skyscraper, as is demonstrated every day in such cities of relatively low sky-line as Los Angeles. As well padlock Detroit. London and Paris both have rigidly limited sky-lines and relatively fewer motor cars, yet their traffic problem is similar ...

For twenty years or more men have been drawing dream pictures of a many-decked city, one in which traffic would be segregated by kind and speed on various levels ... It is still a dream, and an impractical one in the light of this day's sun; but it is the best, almost the only vision of a traffic solution we have imagined to date, and it may be that we are seeing its beginnings in the New York Central Building, recently constructed.

Twenty-five years ago, while steam trains still operated into the old Grand Central Station through the original inadequate tunnel, two passenger trains collided in the tunnel with heavy loss of life. The disaster brought a smouldering quarrel between the city and the New York Central Railroad to a crisis, and resulted in an agreement whereby the road was to acquire large areas of valuable contiguous property, replace the narrow tunnel with a wide cut, electrify and build the present terminal.

The expense of acquisition and construction was so great as to threaten the financial stability of that very prosperous railroad, and in casting about for relief, the novel plan of selling air rights to permit the building of commercial structures over the tracks in Park Avenue was conceived …

… Beginning with the Biltmore, a procession of great hotels, apartments, clubs and office buildings rose over the maze of two-level tracks leading into the terminal, and they return a revenue in air rights more than sufficient to pay the interest and sinking fund of the bonded indebtedness of the terminal …

All these buildings are unique in that they have no ground rights, existing by virtue of their leases to the air above the street level. Patently, such superimposed structures have no basements; they take their steam heat from the New York Central's power house, located at one end of the terminal improvement – which heat, by the way, is purely a by-product, inasmuch as the power plant is a necessity to generate electricity for the terminal.

The track space was blasted out of the solid rock of Manhattan – a stupendous undertaking in itself – and as the tracks come in on two levels, it will be seen that the upper must run in part on a floor-like structure corresponding to the basement of the skyscrapers above. Engineers recognized that the vibration of the heavy trains would necessitate a special foundation for the superimposed buildings. Accordingly, the building foundations reach down through the entirely separate structure on which the tracks run, a sort of cage within a cage, with no points of contact, every column footing of each structure insulated from the others by ingenious construction which reduces to a minimum the transmission of the jar and impact of the trains to the footings of the skyscrapers.

4.3 Starrett, P. (1938) *Changing the Skyline: an autobiography,* New York, McGraw-Hill/Whittlesay House, pp.284–303[11]

The story of the Empire State Building … has … all the spirit, the imaginative and technical daring, and even some of the frenzy, that animated the decade of which it was the culmination. A project which began as a speculative venture was transformed into a real and stable undertaking, which will stand as an ultimate glory in New York's skyline. [*]

The site, that of the old Waldorf-Astoria on Fifth Avenue between Thirty-third and Thirty-fourth streets, was one of the most valuable in the world …

… I doubt if there was ever a more harmonious combination than that which existed on the Empire State, between owners, architects, and builder. We were in constant consultation with both of the others; all details of the building were gone over in advance and decided upon before incorporation in the plans.

The wrecking of the old Waldorf-Astoria Hotel was an interesting feat …

The traffic around this busy corner necessitated our interfering with the street as little as possible. Diagonally opposite was Altman's store; all of Thirty-fourth street was a crowded retail center. We built chutes down the center of the building. After providing amply for the protection of pedestrians, with scaffolds around the top of

---

[11] The two photographs in this extract are not from Paul Starrett's autobiography, but are taken from C. Willis's 1998 book *Building the Empire State*. The book reprints an extraordinary notebook record, found in the archives of Starrett's firm, of the construction process of the Empire State Building; Willis reproduced the photographs from the original notebook.

[*] The size and height of the Empire State Building are not, I think, its most interesting or important features … The building covers 197 feet on Fifth Avenue and 425 feet on the side streets, 33rd and 34th; it therefore occupies about 84,000 square feet, approximately two acres. It contains eighty-five office floors and, with the mooring mast above it, rises 1,245 feet above the street, the mooring tower being equivalent to seventeen stories in height. It has about thirteen more stories than any other office building in New York. The building is lighted by 6,400 windows and is served by sixty-seven elevators. It weighs 616,000,000 pounds, including 134,000,000 pounds of steel. When fully occupied, it will have a population of 20,000 persons; this would provide office space for a city of about 60,000. The total investment is approximately $45,000,000.

One question is very often asked of me. 'How is the Empire State Building anchored to the earth?' When we consider the weight of the building as given above, 308,000 tons, it is apparent that the structure requires no anchorage.

the building to catch any falling material, we ripped down the outside walls and dumped them down the inside chutes into trucks. The rubbish we hauled to scows and carried out to sea.

We removed from the old building about 18,000 tons of structural steel, which was sold for the benefit of the owners at a substantial price. We also removed debris amounting to 16,000 truckloads. On this wrecking operation our working force reached a peak of 700 men …

Out of consideration for the adjacent shops and their merchandise, we made special efforts to minimize dust by the free use of water, and when we had finished, we received an expression of gratitude from Colonel Friedsam of Altman's; he declared that they had looked forward to a great deal of discomfort from the wrecking of the Waldorf-Astoria, but to their surprise had experienced none at all.

In four months, we had the hotel down completely and our men were swarming into the excavation. A month later, we began to set steel.

In construction work, there are established records of time which are marks to shoot at; for steel, three and a half stories per week; for brick walls, a story a day; for stonework one to two stories a week. It was clear to us at once that on the Empire State we could never finish the building on time by any such progress. We decided to discard all these plans of operation and determined to erect the Empire State at the rate of *a story a day* [see Figure 4.39]. For the owners and for ourselves we must maintain this schedule.

When I say that we did actually maintain this schedule, I must add that we should have failed, had not the architects and the owners cooperated with us in a really remarkable way.

Never before in the history of building had there been, and probably never again will there be an architectural design so magnificently adapted to speed in construction.

The fundamental fact of the design was simplicity – a straight shaft rising, with a few setbacks, from the sixth to the eighty-fifth floor. An interesting point about that design, by the way, is the matter of its height … When the architects made their preliminary sketches, they found that eighty-five office floors reached about the height which could be constructed with the money available. Studies of elevator equipment showed that eighty-five stories also set about the limit of efficient and economical operation for elevators that could be installed in a building of this ground-floor area. In other words, the height, the beauty of the Empire State Building, rose out of strictly practical considerations.

Given this design, our job was that of repetition – the purchase, preparation, transport to the site, and placing of the same materials in the same relationship, over and over. It was, as Shreve the architect said, like an assembly line – the assembly of standard parts.

But that general simplicity in the design of the whole was underlaid by innumerable simplicities and economies in detail, which the architects worked out in collaboration with us.

For instance, the stonework and its brick backing were carried directly on shelf angles attached to the columns. The result of the combined designing of steel and stone was a reduction of quantity of stone to one cubic foot for every 200 cubic feet of building, whereas most other modern buildings curtained with stone have four times as much. The design also greatly reduced the amount of cutting and fitting of stone on the site. All the stonecutting was done at the yards, before being brought to the site, and

Figure 4.39 The Empire State Building reaches the 85th storey, 1930. 'This overall shot, taken as the main steel framing was topped out, shows the effect of the extensive use of interior hoists. Only a few cranes are visible, unlike most tall building construction sites of the 1920s and '30s. Ordinary construction methods used at the time would have incorporated a set of cranes for the steel, a set for the bulk materials … and a set for the façade panels and windows. The interior hoists allowed for hoisting the bulk and façade materials with little interference from the weather, and more important, with little of the cross interference sometimes created by the different sets of cranes' (Willis, 1998, n.p., opposite facsimile MS p.40; photograph: courtesy of Skyscraper Museum)

because there was a large number of pieces all alike and of simple shapes, we had one-third of the stone ready for setting before delivery to the site began …

Earlier in this narrative, I have explained that in the opening days of steel-frame construction, the outer walls of the building became merely a wrapping, or covering, for the steel; but architects continued to set windows back into the masonry wall, so that the outer walls had an appearance of thickness and carrying strength which they did not really possess. In the Empire State, however, the architects brought the windows out flush with the outer surface, in fact slightly outside it. They not only admitted, but asserted, that the outer walls of a building are mere curtains, supported by the steel structure within. In this sense, the Empire State is the dramatic and complete realization of the skyscraper development, whose beginnings I had seen as a lad in Chicago.

Steel could not be fabricated fast enough at any one plant to meet our schedule. Our subcontractor for steel, Post & McCord, arranged with two of the largest plants in the country to fabricate alternate sections – each section two to eight stories in height.

The building contains 67,000 tons of steel, the largest order ever awarded for structural steel for building use. We planned and executed our steel schedule with such care that the steel went direct from the cars to its position in the frame. We set 10,000 tons a month and finished steel erection fourteen days ahead of schedule.

We handled and distributed all material by industrial cars on narrow-gauge tracks, running completely around the perimeter of the building on each floor, with the tracks extending across the platform of the hoist, so that cars loaded, for instance, in the basement, could readily run on the hoist and off at the proper floor and deliver at almost the spot where the material was to be used [see Figure 4.40]. Brick arriving at the job on the first-floor [ground-floor] level was dumped into large bins in the first basement. These bins, with inclined bottoms, allowed the brick to slide through doors and drop into industrial dump cars after being thoroughly wet down. The cars deposited the brick alongside the bricklayers, without having been handled from the time they came into the building until the bricklayer placed them in the wall. Under the old method of wheelbarrows, we could hoist only two barrows containing 100 brick per trip on a standard platform. With the industrial car and the same hoist, we carried 400 brick per trip.

Mortar was hoisted at the rate of 21 cubic feet per trip in the cars, as compared to 7 cubic feet per trip in barrows; other materials showed similar savings in time and cost. There was, of course, a huge labor saving through decreased handling, fewer hoists and hoistmen.

For the setting of our stonework, we cut out altogether the customary derrick. The stone trucks drove into the building with the stone in crates, which we call skips or slings. Marked for its proper section of the building, each crate was lifted off the truck by a small crane, operating from a monorail on the ceiling, and delivered to the flat-cars of the industrial railway. Taken to the proper floor, it was unloaded at almost the exact location in which it was to be set. Two hoists handled all the stone for the building, not only eliminating a large number of hoisting derricks and engines but, since the hoisting was inside the building, doing away with a grave source of danger to the public.

We not only beat the stone-setting schedule by fourteen days, but for one period of ten consecutive days averaged 1.4 stories a day.

Figure 4.40 Industrial track on the 85th floor of the Empire State Building, 1930. 'As the concrete for each floor set, railway tracks were laid so that materials could be distributed on either flat cars or in rocker dumpcars directly from the hoists to where they were needed with a minimum of manhandling' (Willis, 1998, n.p., opposite facsimile MS p.25; photograph: courtesy of Skyscraper Museum)

As a result of efficiency methods, short cuts, and collaboration with the architect, we built the Empire State Building for something like $2,000,000 under the original estimate.

At peak of construction, we had on the job some 3,500 men in fifty different trades, with a pay roll of $250,000 a week. One of our biggest problems was getting these workmen to the upper floors quickly, to avoid loss of working time. At the beginning, we salvaged four passenger elevators from the old building and installed them in temporary positions in the new framework. Then, as the building went higher, added two high-speed hoists, equipped with every safeguard.

A related problem, and on this job one equally serious, was feeding the men. To bring down the entire gang of 3,500 and send them up again would cost too much in time. We found an experienced restaurateur in the vicinity, and made a deal with him. As the building climbed, we built cafeterias for him at successive floors, the first one at the third floor, then one at the ninth, the twenty-fourth, the forty-seventh, and the sixty-fourth floor. Completely equipped by the restaurant man, these cafeterias remained throughout the job. They served food of the finest quality. The restaurateur made a fair profit and the men bought good food at cheaper prices than they could have found outside the building.

The great height of the Empire State Building (1,245 feet, 225 feet higher than the Eiffel Tower) brought about some interesting adjustments, to take care of the behavior of materials under unusual conditions.

A great many engineering problems confront the builder of a skyscraper. Calculations as to height are made accurately from the drawings, but when steel is erected and the load imposed on it, there is a shrinkage in its length, owing to compression. This shrinkage is small; nevertheless, in the Empire State Building it amounted to more than six inches in the total height. Next, account had to be taken of the swaying of our building in the wind. This is sufficient to crack the stone joints, unless provision be made for it. In every story of the Empire State Building one of the cross joints in the stonework is of corrugated lead of our own invention; it allows a little give, to take care of the movement and at the same time is made watertight. The sway of the Empire State Building has been measured in strong wind-storms and the greatest variation of the mooring mast from the perpendicular was found to be two and a half inches on either side. The amount of the wind pressure under extreme conditions is astonishing; the records show a wind velocity exceeding 102 miles per hour. The anemometer did not register above this.

A steampipe expands an inch in eighty feet of length, at a variation of 100 degrees in temperature. This means a possible expansion of more than fourteen inches in the eighty-five stories of the Empire State Building. Radiators on the upper floors, if attached to the piping, would be raised off the floor when the pipes were hot. To provide for this, there are expansion loops in the piping. These loops are placed at intervals, so as to render the total expansion negligible. Similarly, unless a method were devised to overcome the pressure of a water column from the top of a building, it would amount to nearly 600 pounds per square inch at the lower fixtures, equivalent to the steam pressure in locomotives. At intervals are installed tanks into which the water is piped, and then redistributed downward to lower levels; in this way, and with the occasional use of pressure-reducing valves, the pressure at any faucet is not excessive. Without them, any tenant who washed his hands at an Empire State porcelain basin would have the surprise of his life. He might as well stand under Yosemite Falls.

References

BANHAM, R. (1969) *The Architecture of the Well-tempered Environment*, London, The Architectural Press.

BONSHEK, J. (1988) 'The skyscraper: a catalyst of change in the Chicago construction industry', *Construction History*, vol.4, pp.53–74.

BOUMAN, M.J. (1987) 'Luxury and control: the urbanity of street lighting in nineteenth-century cities', *Journal of Urban History*, vol.14, pp.7–37.

BOUMAN, M.J. (1993) ' "The best lighted city in the world": the construction of a nocturnal landscape in Chicago' in J. Zukowsky (ed.) *Chicago Architecture and Design, 1923–1993: reconfiguration of a metropolis*, Munich, Prestel/Chicago, The Art Institute of Chicago, pp.32–51.

BRUEGMANN, R. (1978) 'Central heating and forced ventilation: origins and effects on architectural design', *Journal of the Society of Architectural Historians*, vol.37, pp.143–60.

BRUEGMANN, R. (1997) *The Architects and the City: Holabird and Roche of Chicago, 1880–1918*, Chicago, University of Chicago Press.

CONDIT, C.W. (1973) *Chicago, 1910–1929: building, planning and urban technology*, Chicago, University of Chicago Press.

CONDIT, C.W. (1974) *Chicago, 1930–1970: building, planning and urban technology*, Chicago, University of Chicago Press.

CONDIT, C.W. (1982, 2nd edn) *American Building Materials and Techniques from the First Colonial Settlements to the Present*, Chicago, University of Chicago Press.

COOPER, G. (1998) *Air-conditioning in America: engineers and the controlled environment, 1900–1960*, Baltimore, The Johns Hopkins University Press.

COWAN, R.S. (1983) *More Work for Mother: the ironies of household technology from the open hearth to the microwave*, New York, Basic Books.

COWAN, R.S. (1997) *A Social History of American Technology*, Oxford, Oxford University Press.

CRAWFORD, M. (1995) *Building the Workingman's Paradise: the design of American company towns*, New York, Verso.

CRONON, W. (1991) *Nature's Metropolis: Chicago and the Great West*, New York, W.W. Norton.

DUFFY, F. (1980) 'Office buildings and organisational change' in A.D. King (ed.) *Buildings and Society: essays on the social development of the built environment*, London, Routledge and Kegan Paul, pp.255–80.

FISHMAN, R.L. (1987a) 'American suburbs/English suburbs: a transatlantic comparison', *Journal of Urban History*, vol.13, 237–51.

FISHMAN, R. (1987b) *Bourgeois Utopias: the rise and fall of suburbia*, New York, Basic Books.

FORD, L.R. (1994) *Cities and Buildings: skyscrapers, skid rows, and suburbs*, Baltimore, The Johns Hopkins University Press.

GIEDION, S. (1967, 5th edn) *Space, Time and Architecture: the growth of a new tradition*, Cambridge, Mass., Harvard University Press.

GILMAN, C.P. (1904) 'The passing of the home in great American cities', *The Cosmopolitan*, vol.38, no.2, December, pp.137–47.

HANCOCK, J. (1980) 'The apartment house in urban America' in A.D. King (ed.) *Buildings and Society: essays on the social development of the built environment*, London, Routledge and Kegan Paul, pp.151–89.

HARRIS, R. (1988) 'American suburbs: a sketch of a new interpretation', *Journal of Urban History*, vol.15, pp.98–103.

HARRIS, R. (1993) 'Industry and residence: the decentralization of New York City, 1900–1940', *Journal of Historical Geography*, vol.19, pp.169–90.

HISE, G. (1997) *Magnetic Los Angeles: planning the twentieth-century metropolis*, Baltimore, The Johns Hopkins University Press.

HOUNSHELL, D.A. (1984) *From the American System to Mass Production, 1800–1932: the development of manufacturing technology in the United States*, Baltimore, The Johns Hopkins University Press.

HUGHES, T.P. (1983) *Networks of Power: electrification in Western society, 1880–1930*, Baltimore, The Johns Hopkins University Press.

JACKSON, K.T. (1985) *Crabgrass Frontier: the suburbanization of the United States*, Oxford, Oxford University Press.

JACKSON, K.T. (1987) 'Suburbanization in England and North America: a response to "a transatlantic comparison" ', *Journal of Urban History*, vol.13, pp.302–6.

JENNEY, W.L.B (1885) 'The construction of a heavy, fireproof building on a compressible soil', *The Inland Architect and Builder*, vol.6, no.6, p.100.

JENNINGS, J. (1990) 'Housing the automobile' in J. Jennings (ed.) *Roadside America: the automobile in design and culture*, Ames, Iowa State University Press, for the Society of Commercial Archeology, pp.95–106.

KEATING, A.D. (1988) *Building Chicago: suburban developers and the creation of a divided metropolis*, Columbus, Ohio State University Press.

LANDAU, S.B. and CONDIT, C.W. (1996) *Rise of the New York Skyscraper, 1865–1913*, New Haven, Yale University Press.

LEWIS, P.F. (1990) 'The northeast and the making of American geographical habits' in M.P. Conzen (ed.) *The Making of the American Landscape*, London, Routledge, Chapter 5.

LONGSTRETH, R. (1997) 'The diffusion of the community shopping center concept during the interwar decades', *Journal of the Society of Architectural Historians*, vol.56, 268–93.

MISA, T.J. (1995) *A Nation of Steel: the making of modern America, 1865–1925*, Baltimore, The Johns Hopkins University Press.

NYE, D.E. (1992) *Electrifying America: social meanings of a new technology, 1880–1940*, Cambridge, Mass., MIT Press.

NYE, D.E. (1998) *Consuming Power: a social history of American energies*, Cambridge, Mass., MIT Press.

OTIS ELEVATOR COMPANY (1948) '50 years of Otis Elevator Company', *The Otis Bulletin*, November, pp.3–36.

RAND MCNALLY COMPANY (1893) *Bird's Eye Views and Guide to Chicago*, Chicago, Rand McNally.

RAU, D.F. (1993) 'The making of the Merchandise Mart, 1927–1931: air rights and the plan of Chicago' in J. Zukowsky (ed.) *Chicago Architecture and Design, 1923–1993: reconfiguration of a metropolis*, Munich, Prestel/Chicago, The Art Institute of Chicago, pp.99–117.

ROBERTS, G.K. (ed.) (1999) *The American Cities and Technology Reader: wilderness to wired city*, London, Routledge, in association with The Open University.

SCHOPP, M.D. (1955) '50 golden years,' *Central Manufacturing District Magazine*, vol.39, no.6, pp.10–19, no.7, pp.10–21.

SENNOTT, R.S. (1990) 'Chicago architects and the automobile, 1906–26' in J. Jennings (ed.) *Roadside America: the automobile in design and culture*, Ames, Iowa State University Press, for the Society of Commercial Archeology, pp.157–69.

SMITH, M.J.P. (1988) 'Parking carcasses', *Inland Architect*, November/December, pp.58–63.

SPRAGUE, P.E. (1981) 'The origin of balloon framing', *Journal of the Society of Architectural Historians*, vol.40, pp.311–19.

STARRETT, W.A. (1928) *Skyscrapers and the Men who Build Them*, New York, Charles Scribner's Sons.

TOMES, N. (1990) 'The private side of public health: sanitary science, domestic hygiene, and the germ theory, 1870–1900', *Bulletin on the History of Medicine*, vol.64, 509–39.

WERMIEL, S. (1993) 'The development of fireproof construction in Great Britain and the United States in the nineteenth century', *Construction History*, vol.9, pp.3–26.

WERMIEL, S. (1996) 'Nothing succeeds like failure: the development of the fireproof building in the United States, 1790–1911', Ph.D. thesis, Massachusetts Institute of Technology.

WILLIS, C. (1993) 'Light, height, and site: the skyscraper in Chicago' in J. Zukowsky (ed.) *Chicago Architecture and Design, 1923–1993: reconfiguration of a metropolis*, Munich, Prestel/Chicago, The Art Institute of Chicago, pp.119–39.

WILLIS, C. (1995) *Form Follows Finance: skyscrapers and skylines in New York and Chicago*, Princeton, NJ, Princeton Architectural Press.

WILLIS, C. (1998) *Building the Empire State*, New York, W.W. Norton.

WOODWARD, G.E. (1865) *Woodward's Country Homes*, New York, G.E. Woodward.

WRIGHT, G. (1981) *Building the Dream: a social history of housing in America*, New York, Pantheon Books.

Chapter 5: TECHNOLOGIES OF WATER, WASTE AND POLLUTION

by Gerrylynn K. Roberts

5.1 Introduction

Urban populations consume vast quantities of land, water and other resources, and create huge amounts of human, domestic and industrial waste. 'Urban metabolism' is the phrase that Joel Tarr has used to describe the processing of such inputs and outputs through urban systems: 'the supply of water and the disposal of wastewater or sewage; the generation and disposal of industrial wastes; and the collection and disposal of solid wastes and garbage from food and consumer products' (1996, p.xxxi). The production of waste long preceded the development of methods for its systematic disposal, for waste was not initially perceived as a problem (Melosi, 1980a, p.18). However, as industrialization and ever more rapid urbanization occurred during the nineteenth century, waste and refuse came to be understood as problems of urban life. Initially, it was a matter of the 'metabolism' of individual cities, regardless of effects elsewhere. Indeed, it is difficult to generalize about water supply, waste and pollution in the USA because their handling was so locationally, politically and temporally specific. In many cases, European methods were adopted, though adapted to the local context of application. One city's solution to the problem of its waste often became a pollution problem for the next city downstream or downwind, a situation persisting well into the twentieth century until some aspects of controlling the infrastructure became federal matters. The story of water supply, waste and pollution is both highly local and 'environmental' in the broadest sense.

5.2 Water supply

In colonial times, water for urban inhabitants was drawn from local wells, springs, streams or cisterns, and this continued to be the case well into the nineteenth century. The central supply of piped water was introduced early only in the largest, most rapidly expanding cities of the new nation. There, its introduction tended to be crisis-driven: owing to a dependence on natural run-off for accommodating waste, the largest cities polluted their own local water sources very quickly and new supplies were sought. Philadelphia, New York City, Richmond (VA), St Louis (MO) and Pittsburgh (PA) had public waterworks by the end of the 1830s. Baltimore and Boston followed in the 1840s. By 1860, the sixteen largest US cities had central water-supply systems. Generally, however, as Table 5.1 shows, it was not until the last two decades of the nineteenth century that most smaller cities gained a central supply, though not necessarily because of problems stemming from the size of their population. Most of the development was funded by private enterprise through a mechanism of municipal franchising – that is, private companies would contract with local governments to provide the service. However, as the century progressed, the provision of pure water came to be seen as a municipal duty, especially in the largest cities. The conflict between profit and essential services inherent in such matters as the provision of adequate fire-hydrants, supplying poor areas and extending supplies as cities expanded, encouraged municipalization: by 1900, more than 80 per cent of the nation's fifty largest cities had public water supplies and, by 1910, more than 70 per cent of cities larger than 30,000 had public supplies (Schultz, 1989, p.164).

Table 5.1   Number of water utilities in the United States, 1800–1897 (Anderson, 1988, p.138*; number of cities from United States Department of Commerce, Bureau of the Census, 1975, A43–56, A57–72)

Year	Number of cities with populations of 2,500–50,000	Number of cities with populations greater than 50,000	Number of water utilities[†]	Percentage publicly owned
1800	32	1	16	6.3
1810	44	2	26	8.7
1820	58	3	30	16.6
1830	86	4	46	20.5
1840	126	5	54	27.8
1850	226	10	83	39.7
1860	376	16	136	41.9
1870	638	25	243	47.7
1880	904	35	598	49.0
1890	1,290	58	1,878	42.9
1900	1,659	78	3,196[‡]	53.2

* *Anderson's sources are: Baker (1889; 1899)*

† *The trend follows, of course, that of the increase in the number of cities (see Chapter 1, Table 1.1); however, some of the larger cities had more than one water utility*

‡ *Figure for 1897*

Seeking safe water: two early systems[1]

Where central water-supply development was undertaken before the Civil War, the motives were usually protection from the dangers of fire (see Chapter 6) on the one hand and disease on the other. The growing requirement of industry for supplies of clean water was also a stimulus (Schultz, 1989, pp.162, 164).

Philadelphia

Philadelphia, the first large US city to construct a central water-supply system (in 1799–1801) included fire control in its objectives. However, what provided the immediate stimulus for the project were the recurrent epidemics of yellow fever in the 1790s. In 1798, a third disastrous epidemic over a five-year period struck the city, claiming 3,500 lives and causing three-quarters of the population to flee (Blake, 1956, pp.5–6). There were conflicting views about the causes of yellow fever, as would be the case with many epidemic diseases. Some believed that it was brought in on foreign ships and called for strict quarantine regulations, which were implemented in many cities. Others argued that the cause was 'the putrid exhalations from the gutters, streets, ponds, and marshy grounds in the neighborhood of the city' (quoted in Blake, 1956, p.6). This was sometimes called the 'miasmatist' or 'anti-contagionist' theory of disease. Civic cleanliness, to be achieved by regular flushing of the streets with water in summer, was the method proposed to eliminate this 'cause' of the disease. Though neither suggested cause was correct – that it is carried by a mosquito was established in the early twentieth century – both strategies of prevention based on them were to some extent effective in that they hobbled the mosquito population.

Benjamin Latrobe, the engineer who designed and constructed the Philadelphia system, believed that the disease was caused by pollution of drinking-water owing to contamination of Philadelphia's public wells by seepage from nearby domestic cesspools. The 300 wells were accessed by

[1]   This section draws heavily on Blake (1956).

public pumps. These were small-scale and localized facilities, and installing them required little technical expertise. By contrast, constructing a central supply involved surveying a considerable area of land, employing engineering experts and using technically sophisticated materials such as hydraulic cement; it also required substantial funds. Expertise was supplied by practical engineers with experience of canal-building. Whether private or public funds should bear the cost was a matter of considerable debate. The level of investment required was large and the returns for private companies uncertain, motivating them to serve only wealthy areas or to promote combined schemes – for example, where canals would provide both transport and water. Philadelphia raised a loan from the state, sought powers to levy taxes for such improvements and established a mechanism for raising further loans through subscriptions from private investors. Thus, the first water utility supplying a major city was a public one.

Its construction was an elaborate engineering project, the particulars of the design being specific to the local topography, as would be the case for all water-supply systems (Anderson, 1988, p.140). Latrobe made innovative (in the USA) use of two locally built, separate-condenser steam-engines (of the type invented by Boulton and Watt) for pumping water collected from a nearby source, the Schuylkill River, which at that time defined Philadelphia's western boundary. A series of basins, tunnels and wells conducted the water to the Centre Square Engine House in the middle of the city where it was pumped up into wooden reservoir tanks that held some 20,000 gallons (72 cubic metres); it was then distributed by gravity throughout the city in pipes of wood, which were cheaper and more readily available than iron pipes. The elegant classical design of the Centre Square Engine House turned the new water system into a tourist attraction. However, the system was a financial disaster; few customers signed on to pay for pure water from the new supply because water, though suspect, was still available without charge from the old town pumps. And it was possible for everyone to tap the new supply for free at the system's hydrants, owing to a regulation stipulating that the new water should be free to the poor. The system was also costly to run for technical reasons: wooden pipes were not robust, and were eventually replaced by iron; the two steam-engines had numerous wooden parts which needed frequent repair, leading to irregularity of supply, and they consumed huge quantities of coal. The quantity of water supplied would be seen as inadequate for the city's expanding population as early as 1811. Extensions and alternative supplies were soon being constructed, but Philadelphia was generally recognized as a pioneer of municipal water supply.

New York City

In New York City too, it was a yellow fever epidemic, in 1798, that finally focused more than a decade of desultory debate among city authorities about a central water supply. State sanction was sought for the establishment of a municipal water-supply system at about the same time as Philadelphia was making its similar move. The plan was to service every street with water for fire-fighting and street flushing in return for a small tax on property fronting the street; residents would have the option of paying more for individual domestic connections. Ostensibly on the grounds of objections to the level of debt the city would incur, the move was blocked at state level by two nationally famous politicians, Aaron Burr and Alexander Hamilton. Not long afterwards, Burr's own proposal for the establishment of a private water company was passed (see Figure 5.1 overleaf). Extraordinarily, this Manhattan Company was empowered to use surplus capital to establish and run a bank, New York's third. This was in fact the nub of the proposal; it was Burr's aim to establish a bank under the control of his political party, in contrast to the other two controlled by an opposing party, in New York City's highly political financial environment (Blake, 1956, pp.44–51).

Figure 5.1   Reservoir of the Manhattan Company Waterworks, Chambers Street, New York City, 1825; from a painting by G.P. Hall. Built in 1799 of flagstone, clay, sand and tar, this imposing structure in fact accommodated only 132,600 gallons (477 cubic metres) instead of the proposed 1,000,000 gallons (3,600 cubic metres). The reservoir was filled by a horse-powered pump, which was cheaper than steam, from an old well, instead of by expensive aqueduct from the more distant Bronx River, which had been proposed by the city. Here, too, wooden water-pipes were chosen over the more costly alternative of iron, and were soon to become problematic. When the company finally substituted iron for wood, it failed to replace fire-hydrants, resulting in poor supplies for fire-fighting. To compensate, over the period 1817–29 the city built forty public rainwater-cisterns, though the Fire Department wanted 100 (Blake, 1956, p.123; illustration courtesy of Museum of the City of New York)

The banking side of the Manhattan Company flourished, but as early as 1804 there were serious concerns about the adequacy of water supply. By 1830, New York City's population had more than trebled and the company could not keep pace, despite increasing its supplies by adopting the European technology of deep-well drilling. In 1832, private water-carriers were still so necessary that their combined revenues were thirty times as great as those of the Manhattan Company. However, it was the devastating first visitation of cholera in that year that spurred the city to take action. It was made known that residents of Philadelphia had fared much better. This was widely attributed, erroneously, to Philadelphia's ample supplies of water for flushing streets, enabling it to eliminate filth and associated smells, which were thought to be the cause of the disease. (In fact, Latrobe's hypothesis about the cause of yellow fever was nearer the mark for cholera, and accounted for Philadelphia's easier passage in 1832; this would not be understood, however, until the 1850s.)

Having no local supply of fresh water, since all its rivers were tidal, New York City's authorities proposed to take the matter in hand by building an aqueduct to bring fresh water by gravity from a remote site at Croton Water to the north-east. Calculations indicated that even with an expensive long-distance aqueduct, New Yorkers would profit from the annual savings on fire losses and insurance premiums, let alone the savings on medical costs and loss of business arising from epidemics. After much politicking, the plan was put to the electorate, who voted overwhelmingly in its favour. Authorized in 1835, in which year there was a devastating fire that concentrated minds further, it would take seven politically fraught years before water flowed through the Croton Aqueduct (see Figure 5.2). This was partly because of the sheer magnitude of the engineering operation and partly because of opposition and compensation claims from affected landowners and speculators. And funding for the project was severely disrupted by a national financial collapse in 1837; this was alleviated only by selling water stock in Europe. The level of civic debt incurred was unprecedented in the USA.

Figure 5.2 A distant view of High Bridge, New York, which carried the Croton Aqueduct across the Harlem River on to Manhattan, 1849; from a painting by Fanny Palmer. The chief engineer, John B. Jervis, had canal-building experience. Built by some 3,000 navvies, the aqueduct was, effectively a canal of brick and masonry, 32 miles (51.5 kilometres) long. The water was sometimes carried underground, and sometimes carried along embankments or over bridges, from behind a dam on the Croton River, six miles above its confluence with the Hudson, to the Harlem River at the north end of Manhattan. It crossed the Harlem in an iron pipe supported by the famous High Bridge (completed some seven years after the aqueduct, which initially made the crossing on a temporary structure). From there, it was conducted by pipes to a receiving reservoir that could hold 150 million gallons (540,000 cubic metres) and then to a distributing reservoir that could hold 20 million gallons (72,000 cubic metres). The receiving reservoir was located at Yorkville, between 79th and 86th Streets in what would later become Central Park; the distributing reservoir was further south, at Murray Hill (now the site of the New York City Public Library), between 40th and 42nd Streets. Considerable savings were effected by using pipes cast directly at the blast-furnaces, rather than from pig-iron re-smelted at local foundries. The problem of rust in iron pipes was dealt with in the 1840s by coating their interior with tar. Iron became the principal material for making urban water-pipes from then on. Late in the nineteenth century, pipes consisting of a thin sheet of iron surrounded by cement came into use (Armstrong *et al.*, 1976, p.233; illustration courtesy of The Eno Collection, Miriam and Ira D. Wallach Division of Art, Prints and Photographs, The New York Public Library. Astor, Lenox and Tilden Foundations)

However, it did not take very long for consumption to outstrip supply. This was the result not only of rapidly increasing population (see Chapter 1, Table 1.2), but also higher use of water per capita, which stemmed from its ready availability. The comforts of daily life increased as bathtubs and water-closets were installed to achieve standards of cleanliness that were seen as morally desirable. Numerous additions and extensions were made to the Croton system up to the end of the century. Supply problems were further compounded in 1898 when New York City's boundaries were redrawn; the city greatly increased in size, and much of the annexed territory had inadequate water systems. In 1913, again after a great deal of political wrangling, a completely different watershed was tapped to the north-west of the city, 120 miles (193 kilometres) away in the Catskill Mountains. But even the vast additional quantities thus made available were all too soon predicted to be inadequate. During the inter-war years, ways were sought to tap the headwaters of the Delaware River, which rose in New York State, north-west of the city. However, the Delaware was also an asset of New Jersey and Pennsylvania, states whose boundary it formed as it flowed down to the Atlantic. Those two states involved the US Supreme Court to control the amount of 'their' water that was being taken by New York City. Ultimately, the regional planning of water supplies became the way forward.

Water rights and the growth of Los Angeles[2]

Establishment of the right to take water from distant sources was a critical issue in the transformation of Los Angeles from pueblo to metropolis, as water was aggressively sought from the hinterland, using the might of state and federal authorities. Shortly after California achieved statehood in 1850, the city of Los Angeles reaffirmed the principle of public responsibility for water supplies which had operated under previous Spanish and then Mexican rule. With Amerindian labour it ran the extensive, pre-existing irrigation system. For some thirty years, from the 1860s, it ceded responsibility to a private water company. However, the Los Angeles River was the principal source of water, and the city faced growing competition from users further upstream. As the city grew and commerce succeeded agriculture as the basis of the economy, Los Angeles called on the courts to adjudicate on its community rights to the river's waters under previous Mexican law. The city wanted a guarantee of its own supplies and the right to sell surpluses to areas outside its limits. In 1895, the courts finally supported the city's position on its rights – but only to the water it needed, not to the surplus. This resolution of the water rights issue had the effect of encouraging annexation of neighbouring districts, and the spread of Los Angeles, to increase the permitted market for water. In turn, annexation was attractive to those districts precisely because it meant acquiring amenities such as water supply. Furthermore, as the population in the annexed areas increased, the demand for water also increased, feeding back again into the process of growth.

By the turn of the century, city authorities reckoned that the Los Angeles River would soon be inadequate to supply further expansion. To foster growth, Los Angeles cast its eye on the Owens River, some 235 miles (378 kilometres) away; this required the negotiation of construction work on federal lands as well as the purchasing of private water rights. To fund an aqueduct, the necessary works and purchases, Los Angeles – with voter approval – floated bonds of some 245 million dollars. It also promoted, in conjunction with the aqueduct, a hydroelectric project (which would compete with private generating firms) to meet the city's increasing demands for electricity, and eventually established a joint municipal utility. Water and electricity started to flow in 1913. To eliminate conflicts with Owens Valley farmers, the city became a major land-holder in the area, doing away with the agricultural economy. Aqueduct water was not affected by the constraint on selling of surpluses; however, the city developed a strategy of selling them only to territories that were ripe for annexation, again stimulating its own growth. So while water was needed to expand the city and to make the desert attractive for population settlement, by transforming it into green suburbs (a policy leading to considerable stress on water resources in the late twentieth century), the city expanded even further because of it. Transport (see Chapters 3 and 7) was not the only technology involved in shaping Los Angeles.

---

[2] This section draws heavily on Hundley (1992).

5.3   Drainage and sewerage

The increase in water supplies brought a concomitant increase in the amount of waste for disposal; the used water had to go somewhere. Where local wells or springs supplied residents, the classic solution was the cesspool/privy vault, which relied on the principle of natural seepage and could result in the contamination of drinking-water (see Figure 5.3). Though this was not an issue where central supplies were drawn from remote sources, the dramatically higher per capita consumption, which occurred universally, meant that cesspools were overwhelmed by the increased amount of effluent. Natural seepage could not deal with the extra load. If householders did not empty their cesspools more frequently, at considerable extra cost, there were overflows. Another recourse was to direct waste to storm drains or open gutters (Tarr, 1996, p.115[3]). These options were not just unpleasant: the prevailing anti-contagionist theory, developed by English sanitarians in the 1840s, that diseases were caused by foul air arising from putrefying matter meant that they also gave rise to serious health worries.

Figure 5.3 Privies and pump, Pittsburgh, 1909. The privy vault was a hole in the ground, sometimes lined with stone, located in the backyard or cellar. Liquid waste drained away into the soil, while solid waste had to be removed from time to time by scavengers who were either private or under contract to town authorities. In some areas, farmers worked as scavengers to secure fertilizer (Kellogg, 1914; photograph: courtesy of Carnegie Library at Pittsburgh)

Ogle (1996, chapter 2) points out that many Americans had rigged up ingenious private water-supply arrangements involving local streams, private wells or roof-top rainwater cisterns, so that they might have the personal convenience of piped water to kitchens, bathtubs and water-closets long before the widespread adoption of central supplies in the late nineteenth century. However, the change in scale as central water became available, and the

---

[3]   An abridged version of chapter 4 of Tarr's book appears in Roberts (1999), the Reader associated with this volume.

concerns of sanitarians became prominent, was marked, especially in the use of water-closets – although there was some resistance by those who feared that the system would admit unhealthy foul air to their homes. This tendency was reinforced by a major expansion in the sanitary-appliance industry: the adoption of factory methods of production made it relatively cheap to install facilities in new buildings where piped water was laid on from the start. Often urged on by the concerns of sanitarians, many towns saw water itself as a possible new technological solution to the problem of the unprecedented volume of waste. Disposal by 'water-carriage' – that is, by the construction of sewerage systems – as had been done from the 1840s in England and in Hamburg, was the technology adopted. Like water supply systems, sewerage systems were vast enterprises requiring large-scale capital investment and engineering expertise. Unlike them, however, sewerage systems were not undertaken by private enterprise. Although they were built on contract by private firms, centralized sewerage systems, benefiting the health of the whole of a city's population and requiring city-wide organization, were run as municipal enterprises from the start.

The case of Chicago[4]

The case of Chicago will serve to illustrate some of the issues involved. Chicago grew extraordinarily rapidly from little more than a trading post in 1830 to a booming Midwestern entrepôt city by the 1850s; by 1870, it was the fifth-largest city in the nation and had a thriving industrial base; by the turn of the twentieth century it was the second-largest city in the USA. Initially, water was drawn from a well, but this soon proved inadequate. However, because of its location on the south-western edge of Lake Michigan, the city was very well situated for access to fresh water; the lake was a virtually limitless resource (see Figure 1.2) and water-carriers soon set up business. However, Chicago was low-lying and its land was marshy, making it difficult for cesspools to drain naturally or for rainwater to run off. In wet weather, streets rapidly became impassable quagmires; this was a major problem for an aspiring centre of commerce and development.

Chicago promoters worked on the problem of rainwater run-off by digging roadside ditches from almost the moment the city was laid out in the 1830s. But these tended to back up during heavy rainfall. They were also used indiscriminately for the disposal of refuse, which further compromised their effectiveness. The installation of the first waterworks by a private company in 1842 exacerbated the situation. In 1847, the state of Illinois gave the city power to establish a municipal sewerage and drainage system to be paid for by those property owners who benefited from it. At the same time, Chicago was authorized to surface streets with planks of wood in an effort to combat the mud. In 1850, certain principal streets were laid with wooden sewers that were triangular in section and conducted street run-off into the river. In the face of inadequate and polluted water supplies, the city took over the waterworks from 1852, but increased water supply meant increased quantities of contaminated ground water, so that as many as 5.5 per cent of the city's population died in the cholera and typhoid epidemics of 1854 (Cain, 1978, p.23). So extreme were Chicago's circumstances that the state authorized the establishment of a Chicago Board of Sewerage Commissioners in 1855. Ellis Chesbrough, then city engineer of Boston, became the board's chief engineer.

Chesbrough recommended the installation of a comprehensive sewerage system that would accommodate both street-water run-off and domestic sewage, discharging either into the river or directly into the lake. Adopting

[4]   This section draws heavily on Condit (1973) and Cain (1978).

English technological precedents, Chicago's was the first such system in the USA and, financed by special assessments and bond issues, was the cheapest of the various solutions Chesbrough investigated. It was necessary to raise the ground-level of the city so that the brick sewers, which varied from 2 to 6 feet (0.6 to 1.8 metres) in diameter, could be installed at a gradient which allowed them to discharge into the river. In practice, the sewers were laid on the surface at the necessary height and the streets were filled in around them (see Figures 5.4 and 5.5). This process of 'raising the grade' took some twenty years to complete. The river was dredged to deepen it, making it a more satisfactory channel for handling the load anticipated from the new combined sewerage and drainage system, and the dredged material was used to raise the grade. These two complementary processes eventually solved the problem of quagmire streets and made possible a proper paving system (see Figure 2.16).

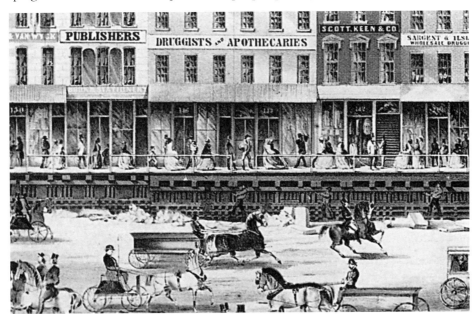

Figure 5.4 'Raising the grade', north side of Lake Street, Chicago, 1857. So that they would not be below the new ground-level, entire buildings were raised on screw-jacks by as much as 10 feet (3 metres). In this illustration you can see men at street-level turning the screw-jacks below the pavement to raise the buildings; the resulting space beneath them was filled by material dredged from the river supplemented by the city's refuse (Colton, 1994, p.130; photograph: Charles R. Clark; courtesy of Chicago Historical Society ICHi-26605)

Figure 5.5 Installing Chicago's main drain, 1858. The access cover in the foreground (a light-grey circle in the road, a little to the left of centre) shows where the main drain lies. By the date of this photograph, some 30 miles (48 kilometres) of sewers had been constructed. They were flushed periodically by means of specially built horse-drawn wheeled 'flush-tanks' that contained 60 barrels (6.8 cubic metres) of water. One tank-load emptied down each access point periodically was a very effective means of flushing (Cain, 1978, p.32). On the plank pavement beside the access cover are iron water-pipes awaiting installation. Look carefully at the pavement: just to the left of the pipes, it steps down and then up again. This shows that the street level has been raised along the entire length, but some of the buildings are still below it (photograph: Alexander Hesler; courtesy of Chicago Historical Society ICHi-05736; detail)

A less beneficial consequence of Chesbrough's design was that the Chicago River became an open sewer that polluted the lake; increasing amounts of industrial effluent flowed into it as well. This in turn stimulated the city fathers to seek further technological fixes. In 1861, the city's sewerage and water authorities were integrated in a single Board of Public Works so that the two sides of the issue would subsequently be considered together. Figures 5.6, 5.7 and 5.8 show the essential features of the mature Chicago water-supply system, built just after the Civil War.

Figure 5.6 Water-tower and pumping station, Chicago, c.1905. Because the land was so flat and the water source on a level with it, a tall standpipe (inside the tower) was needed to provide pressure for the water's distribution by gravity to large tank reservoirs in other parts of the city. The steam-engines in the pumping station lifted the water from the level of the lake and supplied the standpipe. This pumping station, which survived the famous fire of 1871, was Chicago's third. The first was built by the private water company in 1842 to serve the central area; water was pumped into two elevated wooden tanks from which it was distributed by gravity through wooden pipes. When the city took over the water supply a decade later, it substituted a new municipal plant. Water was drawn through a pipe by gravity from some 600 feet (183 metres) off shore; by this means it was hoped to take water from beyond the polluted zone of the lake. The water was pumped to three iron reservoirs in different districts of the city. The capacity was extended from an initial 2 million gallons (7,200 cubic metres) per day to 6.4 million gallons (23,000 cubic metres) per day in 1863. Towards the end of the decade, the second pumping station was replaced by this now-famous landmark, with its neighbouring Gothic water-tower. The new waterworks had a capacity of 35 million gallons (126,000 cubic metres) per day (photograph: Barnes–Crosby; courtesy of Chicago Historical Society ICHi-19124)

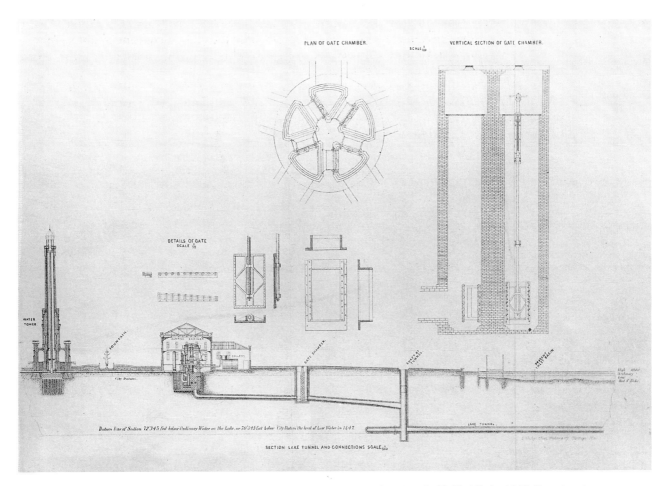

Figure 5.7 The Chicago waterworks: engraving from *8th Annual Report of the Board of Public Works*, 1869. To solve the problem of the water supply's being polluted by the outflow from the new sewerage system it was decided to take water from much further out in the lake. The new pumping station was linked by a brick-lined, masonry and timber tunnel 2 miles (3 kilometres) long, which was constructed 69 feet (21 metres) below the lake-bed. The excavation was done by hand, but rails were run into the tunnel to expedite removal of the dug clay by mule-drawn wagons. Water was drawn by gravity from the lake into the tunnel through a vertical pipe. The pipe was topped by a structure on the lake surface called an 'intake crib' (courtesy of Chicago Historical Society)

Figure 5.8 Water intake crib, Lake Michigan at Chicago; this is the original 2-mile crib, built in 1867 (courtesy of Chicago Historical Society ICHi-29571)

Among the possibilities contemplated by Chesbrough in 1855 was that of directing Chicago's outflow away from the lake in a south-westerly direction, towards the Mississippi drainage basin via the Des Plaines and Illinois Rivers. The Illinois and Michigan Canal had linked the Chicago River with the Des Plaines since 1848. From the 1860s, the pumps supplying river water for the canal were also used to flush the contents of the river through the canal and away to the south; the canal was deepened in 1865 to create a greater flow. However, the capacity of the canal was inadequate to keep the river and lake clear of pollution by this means. A disastrous wet summer of 1879 resulted in the whole system backing up into the lake for thirty days. More pumps were installed. In 1885, however, a 6-inch (15 centimetre) downpour so overwhelmed the system that resulting epidemics killed an estimated 12 per cent of Chicago's population (Miller, 1996, p.426). After prompting by the Civic Association – a group formed by some anxious Chicago businessmen two years after the 1871 fire, to lobby for better fire protection and water supplies – the state established the Sanitary District of Chicago. The Sanitary District was an early example of a regional authority; it was charged with building and running a new canal purpose-built to a standard that would permanently reverse the flow of the Chicago River and solve the pollution problem. In the event, a joint 'Sanitary and Ship Canal' was constructed, which also served as an artery to link Great Lakes shipping with the Mississippi. It was the most ambitious civil engineering project of the time; more earth was shifted in its construction than was shifted in the building of the Panama Canal a few years later. It finally opened as a 'Sanitary Canal' on 2 January 1900, but it would be some years before it could take large ships.

Figure 5.9 'Mrs Partington of St Louis vs the drainage canal'. Published in the *Chicago Tribune* on 4 January 1900, two days after the opening of the canal, this cartoon showed Chicagoans poking fun at the sensibilities of St Louis' citizens. St Louis, though, was indeed downstream of Chicago's sewage. The rumour that the state of Missouri was about to take out an injunction to prevent pollution coming its way hastened the opening of the canal in great secrecy, beginning at dawn. When the Sanitary District of Chicago was expanded after 1903, it drew so much water that the lake was quickly lowered by 4 inches (10 centimetres), which brought such protests from Wisconsin, Michigan and Canadian cities that international negotiation was necessary (courtesy of Chicago Historical Society)

The installation of sewerage systems in turn brought about further problems, to which technological solutions were eventually sought. Typically, a city deposited its effluent into rivers at a point downstream of the city itself (see Figure 5.9), or else into the open sea. When the anti-contagionist theory of disease was still prevalent, it was argued that this practice had no consequences for cities further downstream or that were affected by the prevailing ocean currents because the sewage was rendered harmless (defined, at the time, as unable to give off noxious odours) by oxygenation and dilution in the large volumes of water. Surprisingly to contemporaries, deaths from epidemic disease (typhoid being a particularly sensitive indicator), did not fall in a number of cities that installed sewerage systems. This was primarily because they were located downstream of other major cities with sewerage systems. Atlanta (GA), Pittsburgh, Toledo (OH) and Trenton (NJ) are particular examples. In many other cities, though downward, the trend was not dramatic (Tarr, 1996, p.190).

The 'germ' theory of disease gradually came to prevail over the anti-contagionist theory from the late nineteenth century. Based on new scientific research techniques, it held that diseases were the result of specific causes, or germs, which were transmitted in various ways. Though not incompatible with the earlier theory (Tomes, 1990), it tended to refocus medical activity away from broad public health measures and towards scientific searches for specific causes, so that these could then be eradicated. Under germ theory, however dilute the sewage, the germ, if a water-borne one, would still exist and be a possible source of disease. Even after the germ theory was accepted, however, and bacteriological techniques were developed to evaluate water quality, the idea of a city paying to process its sewage in some way before disposing of it was not favoured, as the cost would bring no benefit to the city incurring it. Most cities had, like Chicago, built combined street-drainage and sewerage systems. Sewage treatment was particularly costly for such cities because of the large volume of water to be treated; cities that had built separate sewers for run-off and sewage, or sewers that dealt only with sewage, were in a better position (Tarr, 1996, p.188). Eventually, in the early years of the twentieth century, filtration techniques and chemical treatments were applied to water supplies coming into cities, to guarantee their purity (see Figure 5.10). The disease rates improved, and thus the cost of these improvements directly benefited the city that incurred them (Tarr, 1996, pp.122–8).

Figure 5.10 Proposed filtration plant with recreation features, Chicago, 1925. The lakefront was the logical place for locating a filtration plant. In this period, however there was an emphasis on beautifying the area, as part of the implementation of the Burnham–Wacker Plan (see Chapters 3 and 6), and so a combined filtration and amenity facility was proposed (Ericson, 1925, n.p.). Chicago also experimented with the chlorination of its water supply from 1912, and made the practice general by 1916. The city's typhoid rate dropped dramatically: by 1919 it had the lowest rate in the nation (Cain, 1978, p.127). Similarly, the city built an experimental filtration plant in 1928. Though the building of a full-scale plant was approved in 1930, the onset of the Depression prevented its construction until the city received a federal grant from the Public Works Administration. But the Second World War again delayed it, and a filtration plant did not come into operation until 1947 – almost twenty years after the early experiments (Armstrong et al., 1976, p.228; illustration courtesy of Chicago Historical Society ICHi-29567)

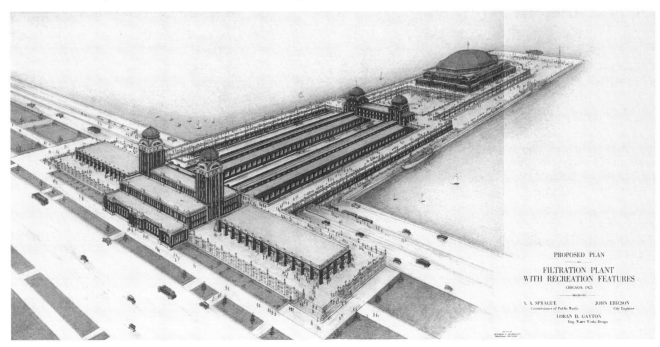

PROPOSED PLAN
for
FILTRATION PLANT
WITH RECREATION FEATURES
CHICAGO, 1925

In Chicago, it was recognized by 1909 that the dilution method on its own would soon be inadequate to purify all Chicago's sewage and effluent, which was increasing to well beyond the permitted capacity of the system. The chief engineer of the Sanitary District approached the problem by means of experimentation, funded in part by the meat-packers, the pollution from whose firms alone was calculated to be equivalent to that of one million additional inhabitants. A number of European treatment methods were investigated for their effectiveness in different circumstances – trickle filters,[5] Imhoff tanks,[6] activated sludge[7] (Armstrong *et al.*, 1976, pp.407–9) – with a view to reducing by 50 per cent the load of pollutants requiring dilution that would enter the main Sanitary Canal. Several facilities were built in the 1920s. Some of these developments were planned in the context of lawsuits brought by other Great Lakes states to limit the amount of water that Chicago could draw from the lake to put through the canal, hence affecting its capacity for dilution. A Supreme Court decision of 1930 did indeed impose limits which reinforced the move to sewage treatment (Cain, 1978, chapter 6). A prominent result of this decision was the building in the 1930s of lock gates at the mouth of the Chicago River, in the heart of the city.

5.4 Refuse disposal and street cleaning[8]

Alongside the installation of central water supplies and municipal sewerage systems in the last two decades of the nineteenth century went a transformation in how refuse was perceived – from a mere nuisance to a health hazard requiring a solution (Melosi, 1980b, p.105[9]). The movement for dealing with refuse also drew on the anti-contagionist theory, that foul odours from filth caused disease, and was similarly beneficial despite the mistaken attribution of cause. Extract 5.1, at the end of this chapter, indicates just how inadequate, haphazard and varied the situation was across the large cities of the USA, while suggesting something of the magnitude of the problem. According to contemporary surveys, Americans seemed to produce more rubbish than their European counterparts, which reflected the higher average income in the USA. In 1905, fourteen US cities produced on average 860 pounds (391 kilograms) of mixed rubbish per person every year, while the average over eight English cities was 450 pounds (204.5 kilograms), and over seventy-seven German cities, 319 pounds (145 kilograms) (Melosi, 1981, p.23).

Rubbish was commonly piled up in streets and alleys, or tipped on to vacant land or into waterways. And the problems of disposing of the dung and carcasses of horses consequent on a horse-drawn urban transport system should not be underestimated (see Chapter 2). By the 1890s, the 'nuisances' became impossible to ignore (Figure 5.12, p.156, shows how bad they could be). In large cities it was simply no longer practicable to regard refuse disposal as the responsibility of the individual, and there was a gradual change over the next three decades. Refuse collection became municipalized in many cities, replacing unsatisfactory and often openly corrupt

[5]   This method, first introduced into the USA in Reading (PA), could cope with high volumes of wastewater. It involved spraying untreated effluent over a bed of coarse rocks covered with biological growths. Oxygen absorbed from the air, or during passage through the rocky bed, increased the biological oxidation and mineralization of organic matter.

[6]   First introduced into the USA in Madison (NJ), these tanks worked on the principle of sedimentation. Solid matter dropped through slots in the bottom of a sedimentation chamber to a sludge compartment, where the organic matter was allowed to digest until it stabilized and was then either dumped or sold as fertilizer. More than half of the treatment works in the USA used this system by the end of the 1930s.

[7]   This method, which was developed just before the First World War, relied on combining sewage with large quantities of aerobic bacteria and then pumping air through the mass to stimulate bacterial reduction of the pollutants. The need to dispose of the resulting sludge stimulated further developments.

[8]   This section draws heavily on Melosi (1981).

[9]   An abridged version of Melosi's chapter appears in Roberts (1999), the Reader associated with this volume.

arrangements with private contracting firms. Municipalization of street cleaning occurred even more rapidly. By the early 1920s, no major city used contract cleaning (Melosi, 1981, p.137). Rather as had occurred in Britain from the 1840s, civic action by the concerned amateur sanitarians who first raised the problem gave way to professional engineers, and to solutions devised by them. Various new technologies came to be applied; but, as always, cost was a prime consideration.

Women took an active role in cleaning up their cities. In New York City, the Ladies' Health Protective Association (LHPA), founded in 1884, became involved in the campaign for cleaner streets. Typical of the 'municipal housekeeping' movement, the LHPA argued from the point of view of aesthetic and domestic concerns about the standards of cleanliness that a civilized populace should share. Their motives reinforced the sanitary movement, but were distinct from it:

> In their street-cleaning appeal … they had said that 'even if dirt were not the unsanitary and dangerous thing we know that it is, its unsightliness and repulsiveness are so great, that no other reason than the superior beauty of cleanliness' should compel New Yorkers to do what was necessary to achieve a level of comfort and self-respect.
>
> (quoted in Hoy, 1995, p.75)

Figure 5.11 shows one of the results of the zeal of Chicago's 'municipal housekeepers'. Looking at Chicago women's organizations at the turn of the twentieth century, Flanagan (1996, pp.173–6) suggests that women's achievement in the environmental sphere was even wider than that of municipal housekeeping. In stressing the importance of making cities 'liveable', they broadened the idea of the role of municipal government and communal responsibility for city welfare away from a narrower (male) focus on making the city 'profitable'.

Figure 5.11 Travelling refuse-burner, Chicago, 1893. In Chicago, where refuse was collected by private contractors, the expectation of a vast visiting population for the World's Columbian Exposition led to some efforts to make improvements, at least in the areas that tourists were expected to visit. A women's organization, the Municipal Order League, led by the capable civil servant Ada Sweet, was a very effective lobbying body (Hoy, 1995, pp.76–7). Incineration, adapted from English practice, was one of the new technologies. Though fixed plants were more likely than mobile burners, Chicago had both and they were rivals (Melosi, 1981, p.115; illustration reproduced from *La Nature*, 1894; photograph: Mary Evans Picture Library)

New York City

Perhaps the most famous example of the sanitary transformation of a city was that achieved by Colonel George E. Waring as commissioner for the New York City Department of Street Cleaning from 1895 to 1898. New York City had long had ordinances mandating clean streets and forbidding the tipping of refuse, but they were seldom observed by inhabitants. At the same time, handled through private contracting, street cleaning and refuse collection and disposal were notorious sectors of graft and corruption in a city whose politics was run on patronage (Corey, 1994, pp.2–5). After the health education and sanitary

Figure 5.12 Varick Place, New York, 17 March 1893 (photograph: courtesy of The Science, Industry and Business Library, The New York Public Library. Astor, Lenox and Tilden Foundations)

Figure 5.13 The same street, 29 May 1895 (photograph: courtesy of The Science, Industry and Business Library, The New York Public Library. Astor, Lenox and Tilden Foundations)

measures undertaken during the Civil War, which drew on English precedents, there was a movement nationally in favour of preventative public-health activities (Hoy, 1995, chapter 2; Tarr, 1996, p.11). New York City established a Board of Health in 1866 as part of a new law specifying sanitary standards for the city and its buildings (see Chapter 4). The board had wide powers to investigate sanitary matters, including street cleaning and refuse collection and disposal, but little power to enforce standards (Duffy, 1974, pp.19–21). It was the task of local ward aldermen to appoint the contractors – the board could only censure. The situation reached crisis proportions in 1880: a private, citizens' body, the Sanitary Reform Society, drew up a bill to establish a Department of Street Cleaning to be run by a commissioner, a direct appointee of the mayor. The department was established in 1881, but had little success in the early years. This was the organization that Waring used to transform the city.

A committed sanitarian and experienced sanitary engineer, Waring brought military organization and discipline to the department. Extract 5.2, at the end of this chapter, is Waring's own explanation of how he ran the department and what he achieved. The keys to his success were good organization and good public relations, along with the engendering of civic pride through community action: for example, schoolchildren were encouraged to form street-cleaning leagues (Hoy, 1995, pp.78–80). And the results were visible (see Figures 5.12 and 5.13). His methods were adopted, at least in part, by many other cities. According to Melosi (1980b, pp.115–18), Waring perceived street cleaning and refuse disposal in a wide urban environmental context. He was not concerned narrowly with health, though that was an important motive, but also with improving the quality of life generally, including aesthetically and morally.

Technologies of refuse disposal

Disposal methods had varying effects on the built environment. The most common method was straightforward tipping. Before collection and controls, refuse was tipped on to any convenient patch of vacant land or into nearby bodies of water. Once systematic collection began, New York City chose, notoriously, to dispose of refuse at sea – with unpleasant but unsurprising effects on neighbouring shorelines (see Figure 5.14). Until 1914, Richmond, like several other cities, disposed of refuse in open tips (see Figure 5.15 overleaf), which were offensive to nearby residents.

Figure 5.14 Loading boats for disposal of refuse at sea; this photograph is taken from Waring's *Report of the Department of Street Cleaning of the City of New York for 1895–96–97* (Waring, 1898; photograph: courtesy of The Science, Industry and Business Library, The New York Public Library. Astor, Lenox and Tilden Foundations)

Figure 5.15 Open tip in Richmond, before the city's cleanup ordinance of 1914 (photograph: courtesy of *American City* magazine)

However, in some areas, tipping was literally the basis of the built environment. The reclamation of Boston's Back Bay area during the second half of the nineteenth century is a well-known example. To take another case, 20 per cent of the land area of Manhattan, with Brooklyn, the Bronx and Queens (boroughs of greater New York City), is built on landfill, a practice started in the seventeenth century by Dutch colonists.[10] It was not only the swampy areas of Manhattan that received refuse. The harbour, too, was a tipping area. From the eighteenth century, the disposal of refuse along the harbourfront was organized, to extend the island by landfill. 'Water lots' (plots of land that were under water) were sold to merchants for development, with the understanding that they would be built on after reclamation. The waterfront was the centre of business activity at the time, and property there was highly desirable. The relationship of the harbour with the city's refuse was symbiotic. The refuse had to go somewhere, and tipping along the harbourfront was the least costly alternative for carters who did the collecting. Also, to some extent, landfill in the harbour area was necessary, to create essential sea walls and other structures. The practice continued into the mid-nineteenth century, when so much landfill had taken place in the harbour area that it was beginning to encroach on navigation in the rivers; it was prohibited by legislation in 1871 (Corey, 1994, pp.72–6, 81). Landfill was also used to build streets in residential areas further north on the island, a practice that was increasingly opposed by nineteenth-century sanitary reformers such as the city health inspector, Dr Augustus Viele:

> It is a lamentable fact to record that 104th Street, from the East River to Fifth Avenue, 86th and 89th Streets, from the East River to Second Avenue, have been graded and filled with ashes and garbage. No more pernicious material for the filling of streets can be found, especially where the streets are so often opened for sewers, water and gas connections and repairs.
>
> Experience has taught us that the exhalations from the never-ceasing decomposition of this material will continue to percolate and gain the surface for *years* after the original deposit has been made, particularly when made in low, sunken and damp localities.
>
> (quoted in Corey, 1994, pp.80–81)

[10] 'Landfill' is a term that can mean the waste material used to reclaim an area of ground, often under water, for development, including building; it can also be used to mean the process of disposing of rubbish in this way, or the area filled in by this process.

In Chicago, too, from the time that organized refuse collection began in 1849, rubbish was used constructively to fill in marshy areas and to reconstruct the lakefront, which was in danger of eroding. Throughout the second half of the nineteenth century, the city operated a deliberate policy of using its refuse, including the rubble from the fire of 1871, in land-reclamation projects. Among other major projects, this policy eventually resulted in the lakefront amenity area that would become Grant Park under the Burnham–Wacker Plan. Former quarries and clay-pits within the limits of the expanding city were also used as tips, the reclaimed land subsequently being used for parks or school grounds (Colton, 1994). The 'sanitary landfill' was introduced in the 1930s from Britain, first to New York City and Fresno (CA), as an improvement on the open tip, with its foul smells, and a planned way of reclaiming land for development. In this method, refuse is placed in trenches in layers, and sometimes treated with chemicals; once a certain depth is reached, the refuse is covered with a layer of earth. The process is repeated as often as is necessary until the site is filled. After the Second World War, this became the preferred technology for the disposal of refuse (Tarr, 1996, p.22–3). Well into the twentieth century, urban or suburban piggeries, too, were used for the disposal of waste material; this was a favoured method in New England, but it was also used quite widely around the country, for example in Los Angeles (Armstrong *et al.*, 1976, p.448; Melosi, 1981).

From the 1890s, two new technologies for the treatment of waste came into use, both of which required the construction of a special plant: the technologies were incineration and reduction. (Extract 5.2 gives brief explanations of these methods.) In addition, the long-established practice of separating useful material from rubbish – especially rags and paper for the paper industry – which had previously been done informally, by scavengers, was formalized: rubbish was separated in specially contructed buildings, and any profit accrued to the city (see Figure 5.16). As discussed in Extract 5.2, Waring deployed all three methods in New York City, with the result that disposing of refuse in the sea was largely avoided between the late 1890s and 1917.

Figure 5.16 Refuse-disposal plant on the East Side, New York City, 1898. The first sorting plant was built in Manhattan in 1898. Energy from two incinerators was used to power a conveyor belt which moved salvageable materials past workers, who sorted them (photograph: courtesy of The Science, Industry and Business Library, The New York Public Library. Astor, Lenox and Tilden Foundations)

The technology of incineration was developed in England, where it was found that once a certain, very high, temperature was reached, no further fuel was necessary for the process: the material being incinerated itself effectively became the fuel and was completely consumed. Following the introduction of the technology in England, the first successful incinerator in the USA was built in 1885 on Governors Island in New York City. The incinerator was sited on an island, well away from urban areas; such locations were chosen because the economics of incineration were very different in the USA compared with those in Europe. Americans threw out more rubbish than did Europeans, and their rubbish was wetter, because it consisted of more organic matter – an early sign of a more affluent society; they also tended to burn only garbage in incinerators, rather than mixed refuse (in US usage, strictly, 'garbage' is organic matter, such as kitchen waste; 'mixed refuse' includes paper, glass, metal, etc.). As a consequence, the incineration process could not reach the high temperature needed for the European self-burning technology to be employed; much more fuel had to be used in US incineration plants, which made them far smokier and less suitable for locating in urban areas. In the early twentieth century, experiments were done to combine incineration with power-generation; a number of these were reasonably successful, providing enough electricity to power the plants themselves and to supply power to neighbouring structures (Corey, 1994, p.102). However, there was little incentive for the widespread development of such schemes in the USA, where power was available at low cost, as, generally, was land for tipping (Armstrong *et al.*, 1976, p.449; Melosi, 1981).

New York reduction plants, too, had island locations. Reduction promised to be the ideal way to dispose of organic waste: the process involved steaming the material and then pressing it to extract grease, oils and other useful products. The major drawbacks of reduction were air pollution of the surrounding area, as a result of the huge quantities of putrefying waste, and the fact that only a very small percentage of waste material could be treated in this manner. There was also a run-off of foul-smelling effluent. No one wanted a reduction plant in their locality. The great attraction was the revenue to be gained from recycling. The grease and oil were sold for manufacturing purposes and the dried residues were sold for fertilizers. Unsaleable residues were burnt as fuel in the plants themselves.

New York City's respite from tipping at sea was relatively short-lived. The by-product value of waste reclamation collapsed in the 1920s, but the volume of refuse continued to grow. Residents began to become more vocal in their objections to having incinerators as neighbours. The practice of tipping at sea, which had resumed after 1917, was finally prohibited by the Supreme Court in 1934, at the behest of the neighbouring state of New Jersey.

5.5 *Industrial pollution*

The concentration of industries in and around cities was a major contributor to urban pollution in the late nineteenth and early twentieth centuries. Among the earliest offenders, dating from the pre-industrial period, were the malodorous organic-based trades that served the needs of the cities, such as slaughterhouses and the associated animal-processing businesses of rendering and tanning. In addition, in certain cities, the nineteenth-century industries powered by steam and fuelled by coal caused major smoke pollution. Bituminous (soft) coal, which burned very inefficiently and smokily, was the worst source of such pollution. Cities that used bituminous coal as their principal fuel, because it was a local, and therefore cheap, resource, suffered greatly; those that used anthracite (New York, Boston and Philadelphia) or natural gas (San Francisco) were more fortunate. The topography of some cities made them likely to experience a temperature inversion (a very stable situation where lighter warm air lays on top of heavier cold air, thus acting as a lid on any vertical movement of the lower layer), and this served to trap the smoke: the bituminous-coal-using cities of Pittsburgh, Cincinnati, St Louis and Chicago had particular problems (Grinder, 1980, p.84). Smoke pollution was compounded by the use of steam for transport – by locomotives and by tugs in busy harbours – as well as by the use of coal for domestic heating.

Organic industries in New York City

The sanitary movement, galvanized by the Civil War, consciously followed English precedents. It gave rise in New York City to a civic organization, the Council for Hygiene and Public Health, which reported on environmental conditions in the city in 1865 and noted in particular the existence of 173 slaughterhouses. These were more than just nuisances: because of the smells from putrefying matter, they were thought to be deleterious to public health. Having started out on the edge of settlement, the traditional location for such activities, the slaughterhouses were by the 1860s surrounded by housing: many were located in crowded tenement areas (see Figure 5.17). It was not only the poor who were troubled by noxious neighbours. As such businesses could afford to, or were forced to by changing regulations, they leap-frogged the settled area to re-establish themselves on the new edge of the city – resettling in precisely the sort of semi-rural area in which the wealthy, seeking to evade the congestion and conditions of the centre, also resettled.

Figure 5.17   Slaughterhouse and fat, hide and tallow business in a crowded tenement district of the Fifteenth Sanitary District, Manhattan: 'Because they [stenches] were such an annoying, discomforting, sometimes even frightening part of people's lives, these odors played an important role in the geographical evolution of the city and led to civil litigation, environmental regulation, and technological innovation' (Rosen, 1997, p.77; illustration: courtesy of General Research Division, The New York Public Library. Astor, Lenox and Tilden Foundations)

The council's investigation prepared the way for the establishment of the Metropolitan Board of Health in 1866, which acted very quickly to forbid the slaughtering of cattle below 40th Street; this put further pressure on slaughterhouses to move north. The hog-pens and slaughterhouses of the meat-packers, which employed large numbers of the poor, were not affected (Duffy, 1974, pp.128–9). Eventually, slaughterhouses were forced to concentrate in particular areas and combination into large, well-equipped abattoirs was encouraged. The rendering industries were also tackled, many of them being forced to shut down or at least to substitute closed steam boilers for open vats.

Smoke

Smoke pollution, though recognized, was more difficult to deal with than water pollution. In the first place, until well into the twentieth century, smoke was seen by many to be a sign of thriving industry, and therefore of progress and prosperity. No one doubted that smoke was dirty and therefore costly, but its effect on health was rather less well understood than that of water pollution; it was not implicated directly in epidemics, for example. In addition, smoke could always be seen to be somebody else's problem; individual producers argued that they could not justify the cost of amelioration unless the majority of producers joined them in taking action. Finally, the technology for dealing with the problem was not so clear-cut (Tarr, 1996, pp.14–15). The eventual solutions were technological, but their implementation was influenced by changes in values, both private and public, that were ultimately reflected in legislation. The cases of Chicago and Pittsburgh illustrate some of the issues.

Chicago

The first ever smoke-abatement ordinance in the USA was passed in Chicago in 1881, stimulated by the Civic Association – the same organization that had started the process resulting in the Sanitary and Ship Canal (see Section 5.3). The ordinance was, however, ineffective. In 1892, another group of Chicago businessmen organized the Society for the Prevention of Smoke with the explicit objective of cleaning up the city in time for the World's Columbian Exposition, which was to take place in 1893. With the exception of one individual, the principal people involved in the society were all financial backers of the Exposition, and they funded the society personally. Their original plan was to work by education and persuasion, but with only 40 per cent compliance from the smoke-producers approached by the society's engineers, they turned to the courts. They were not overly successful there. However, the society belived that it had succeeded none the less in cleaning up the city centre in time for the Exposition. The Depression of 1893 undid its successes, though, and the society disbanded (Rosen, 1995).

Women's groups were also active in smoke-abatement issues. In 1908, the Anti-Smoke League mounted a concerted campaign seeking the electrification of the smoke-belching, steam-driven Illinois Central Railroad, which ran along the lakefront (see Figure 2.28), despoiling an area of middle-class housing and reclaimed parkland. Like campaigners for an unpolluted water supply, these progressive women claimed a natural, legal and moral right to clean air; they readily assimilated the germ theory of disease to sanitarian ideas to promote the health of the city (Platt, 1995, pp.71–2; Tomes, 1990). However, the more general effect of the germ theory on public-health issues was to refocus concern on the treatment of specific diseases and away from a broad social environmentalism. At the behest of Chicago businessmen, the campaign was effectively buried in an array of expert investigations and then completely forgotten in the more pressing concerns of the First World War – though the Illinois Central was electrified within the city in 1926 (Platt, 1995).

Pittsburgh

Pittsburgh made various attempts at smoke control in the nineteenth century. However, it was a coalition of private business interests and constructive public intervention that made smoke abatement succeed in the twentieth century. It was clear from the 1930s at least that the city was dying. Crucial individual business leaders targeted the regeneration of the city's central business district as the key to revitalizing the regional economy generally. The first strand was the enforcement after the Second World War of a smoke control law; this had been passed in 1941, following the successful precedent of St Louis a year earlier. Importantly, this affected the production of smoke in the domestic situation, as well as in industry, commerce and transport. By then, it was strongly suspected that smoke was implicated in respiratory ailments. Cleaning up the city was seen as the first stage in its regeneration (see Figures 5.18 and 5.19) (Lubove, 1995, pp.106–19).

Figure 5.18 Central Pittsburgh before smoke control, 1936 (courtesy of Carnegie Library of Pittsburgh)

Figure 5.19 Central Pittsburgh after smoke control, 1947; although taken from a different angle from the photograph in Figure 5.18, this shows roughly the same area (courtesy of Carnegie Library of Pittsburgh)

What finally made smoke-reduction possible was the development of new technologies that enabled bituminous coal to be replaced by cleaner fuels – natural gas, diesel fuel and electricity – both domestically and by industry (Grinder, 1980, p.100; Tarr, 1996, chapter 8). It also required political and financial will, as well as a change in perception on the part of the public. In the case of Pittsburgh, for example, the impetus to eliminate smoke pollution and clean up the city predated the widespread shift to natural gas by domestic consumers, but it was the switch that made the policy a success (Tarr, 1996, p.257).

Other air pollutants

Visible smoke, of course, was only one form of air pollution, and it was understood primarily in local terms. In the 1950s and 1960s, it was recognized that other constituents of fuels were also pollution hazards. Chimneys were static sources of air pollution; the technological fix adopted, which would dissipate the pollutants well above the level at which a local impact would be felt, was to build taller chimneys. However, as was the case with sewerage, one area's solution became a problem for other areas – in this case, for areas downwind of the emission (Tarr, 1996, pp.18–19). Mobile sources of air pollution required other solutions. Urban air pollution resulting from railway engines that burned bituminous coal was eventually eliminated by technological substitution – of steam-engines by diesel-electric engines. Cities, however, had little influence over powerful railway companies in bringing this about (Tarr, 1996, p.280). Pollution by photochemical smog, which resulted from the use of the internal-combustion engine in motor cars and lorries, is discussed in Chapter 7.

5.6 Water, waste and pollution in the development of the city

For a variety of reasons, urban politicians and developers turned to technologies of water supply and waste management during their struggles to manage rapidly growing cities driven by the values of private enterprise. In much the same way as they became dependent on the existence of transport networks, inhabitants of US cities in the late nineteenth century became dependent on the provision of water supply and waste infrastructures as well (Tarr and Dupuy, 1988). A growing community of engineers, one of the products of the expansion of US education after the Civil War, was instrumental in promoting elaborate technological solutions and also provided the expertise to implement them. It was not only essential to install such systems, but also to maintain and extend them as cities grew. Engineers took an increasing role in planning this aspect of city services during the nineteenth century; consequently, they were influential in planning fundamental features of cities well before there was a formal planning profession, or a recognition of planning as an issue of city development (Peterson, 1979; Schultz, 1989; Schultz and McShane, 1978). Crucial to the process of infrastructural development was a growing perception, fostered by coalitions of businesspeople, industrialists and sanitary reformers, that local government authorities did indeed have responsibilities for seeing that services were provided for their citizenry (Monkkonen, 1988, pp.218–22).

By the end of the nineteenth century, good water supply and sanitary infrastructures had come to be as important in promoting cities as good transport. And by the 1920s, the principal technological features of those systems had been developed. However, in the twentieth century, organizing water supply and sewage disposal could no longer be solely a local matter in many areas, but raised regional issues and stimulated federal legislation and adjudication. Furthermore, during the Depression, federal funding programmes, developed as part of the 'New Deal' to create employment, became the principal providers of new water and sewerage infrastructures, accounting for 50 per cent of expenditure on waterworks in the 1930s (Armstrong *et al.*, 1976, p.231). Federal involvement was extended after the Second World War, particularly in large-scale projects that considered water usage as a whole, including sewerage: for irrigation, transport, recreation and power supply as well as for supplying the needs of new and existing cities. In the second half of the twentieth century, such perspectives began to inform the understanding of another of the consequences of city development: pollution.

Extracts

5.1 'Disposal of refuse in American cities', *Scientific American*, 29 August 1891, vol.65, p.136

The disposal of the refuse in cities, while it has been a problem in the sanitation of our larger towns, is yet to be solved. There is probably not a city of any size in the United States where the disposal of wastes is satisfactory or conducted in such a manner as to meet the demands of cleanliness and hygiene. If there is a perfect plan adopted, there are to be found defects in its execution which render ineffective the methods used. The report of Mr Walter V. Hayt, General Sanitary Officer of the Chicago Board of Health, recently published, gives a summary of different cities as to their methods of 'collection and disposal of garbage and refuse'. From this we learn that in New York the garbage is collected by the city teams, loaded upon flat boats, removed to sea and dumped. The garbage is removed from Philadelphia by small contractors in a very unsatisfactory manner; some feed to hogs, others sell to farmers, and at times it is buried or accumulated on the ground awaiting slow decomposition. Cremation is recommended. In Brooklyn refuse is moved to the sea. Collections are made daily from hotels, from dwellings twice a week in winter and three times a week in summer. St Louis collects its garbage and discharges it at certain dumps. It is not satisfactory, and cremation is also here recommended. The following is said regarding the disposal of garbage in Boston:

'There are now about three hundred and fifty thousand loads of garbage, ashes, street sweepings, and other miscellaneous *debris* gathered up by the city teams annually, and carted away to different places and for different uses. The annual cost to the city in hauling this large amount of material is about $500,000, about $100,000 of which is spent in collecting garbage. With the growth of the city and the gradually increased distance to which such matter must be carried for disposal comes a corresponding increase in the expense. The question occurs to us, cannot this growing expense be lessened, and much of the offense now attending the necessary storing of garbage in the houses and yards, and the hauling and carting of it in the streets, be avoided? It is found that with trifling cost in arranging the kitchen stove, each family can easily burn all its refuse as it is made, and before it becomes offensive, and thus save all subsequent expenses and nuisance incidental to its being kept on the premises for several days and then carried through the streets by the city teams.'

Baltimore says that 'the mode of disposing of the garbage, night soil, and street dirt of the city, as at present carried on, though far from perfect, is, with our present facilities, the best our circumstances will allow. The offal and filth of cities *must* be deposited *somewhere* when gathered from every house and locality, and is from the very nature of things a nuisance to those who reside in the neighborhood of the dumps. Much, however, could be done in mitigation of this evil by more stringent regulations in the manner of removing both garbage, ashes, and night soil. The garbage carts are very poorly adapted, in their present construction, for the transportation of garbage and ashes through the streets of the city, and are a source of constant complaint by citizens whose olfactories are greeted throughout the summer months by the odor of decaying vegetable and animal matter, and at all seasons saluted by a shower of dust from uncovered ash carts.'

In 1883 Cincinnati contracted with a private company for two years to remove, for the sum of $2,500 annually, all 'vegetable garbage, dead animals, and slaughter house offal within the limits of the city.' The company must remove all animal matters from dwellings and hotels at least three times a week, and daily from all slaughter houses. Ashes and other refuse must not be mixed with vegetable garbage. It is said that the city has the best end of the contract.

The Cleveland method is as follows, as taken from the report: *'Something Necessary?'* 'A long delayed sanitary necessity remains unprovided in this city, to wit: a satisfactory gathering of house garbage. No city of this size can be considered well taken care of which makes no public provision for this purpose. Private methods are entirely inadequate and mainly inefficient for the prompt and cleanly removal of such material. It is an old topic in these reports, but must continue to be urged until some relief is obtained. Whatever disposition is finally made of garbage, a systematic gathering should be inaugurated as speedily as possible. At present we are without funds for such service, but could money be better employed for the real welfare of our citizens?'

From the report it is gathered that the total cost of the scavenger service in Chicago for the year 1889 was $253,140.72. The area covered aggregated 174 square miles, including in this 2,047 miles of streets and about 3,000 miles of alleys. It is admitted that this service is only 'fair'. In one ward only is there a daily service. Outside of this it was tri-weekly. About 225 teams have been regularly employed by the contractor to whom this collecting is let, and by the city, and an average of 2,000 cubic yards per day is removed and deposited in clay holes, removed by train beyond the city limits or from February 19 to October 15, 1890, burned by the Chicago Garbage Reduction Company. This company burned 7,208 tons of garbage in this period. For the year 1890 a total of 325 teams was demanded, and Mr Hayt recommends radical improvements in the contract system of collecting, 'if that is to be continued'. These recommendations cover the employment of a minimum number of teams, a wagon box of standard size, with a proper canvas cover, wagons to be plainly marked with ward and number.

In all the cities cited the methods of disposal are not satisfactory in one. The method adopted or the service rendered is at fault. The lack of funds is responsible in some instances for the deficient work in the disposal of garbage. In the first place there is too much garbage produced, and in the next that which must necessarily be produced is not properly cared for, and the problem how best to dispose of the refuse of cities is still unsolved. The fact is some garbage must be produced, and we do not believe that any method which does not look to the destruction of this garbage will be a success.

5.2 Waring, G.E. (1897) *Street-cleaning and the Disposal of a City's Wastes: methods and results and the effect upon public health, public morals, and municipal prosperity*, **New York, Doubleday and McClure, pp.37–52, 187–91**

Street-sweeping

Naturally the most obvious, as well as the most important, part of the work of street-cleaning is that which is done in removing accumulations from the surface of the streets. In New York forty per cent of the entire disbursement of the department is for sweeping, and sixty per cent of the laboring-force is employed in this part of the work, which here is done entirely by hand.

Machine-sweeping was formerly almost universal, especially when work was done by contract; and, as a rule, contract street-cleaning throughout the country is executed in this way. At the beginning of operations under the present administration there was still a considerable amount of work done by machines, which were employed almost universally at night. The dust raised by them, even with preliminary sprinkling, constituted such a nuisance as to make it improper to sweep by machine during the day. After very careful comparisons of cost and of the character of the work done, it was determined that there was little, if any, economy in using machines if they were made to do the best work of which they are capable, and that it was not possible, under any circumstances, to do such uniformly good work by machinery as by hand. In the summer of 1895 the use of machines was entirely abandoned. Two years' experience with hand-work has satisfied me that it is incomparably more advantageous than machine-work, and it is not likely that the latter will again be resorted to in this city.

We have four hundred and thirty-three miles of paved streets (which alone receive our attention); and we have actually at work, at this writing, about fourteen hundred and fifty sweepers – broom-men. This gives a little less than one third of a mile, on an average, to each sweeper. There are naturally great deviations in this respect, the actual number used in different parts of the city varying about from one to a mile to seven to a mile, according to the character of the pavement, the character and density of the population, the character of the district, whether manufacturing, resident, tenement, etc., and the character and amount of traffic …

… The sweepers are dressed entirely in white duck. The coat is a sort of Norfolk jacket, with a leather belt and metal clasp, with metal buttons; trousers which are rather loose; and they wear helmets … Each wears on his left breast an oval metal badge bearing his number … Each suit costs one dollar and twenty-five cents …

Each sweeper is supplied with the following implements: a two-wheeled bag carrier and a sufficient number of jute bags for his day's work; a broom of African bass [fibre from a certain type of palm tree] with a steel scraper at its back, a shovel, and a short broom. In summer he carries also a watering-can and a key for opening hydrants. If he has any considerable amount of asphalt in his beat, he uses for this a steel scraper about three feet broad, which is very effective for taking up fresh droppings and other accumulations.

If the section is traversed by one or more avenues of heavy traffic, a number – and sometimes all – of the men of the section are worked in gangs early in the morning for the first thorough cleaning of these. After that they disperse and go each to his own route …

The sprinkler must be used always in dry weather, during the season when it is allowed to open the hydrants – from April to November. Fines are imposed for raising a dust. The accumulations on the streets, of whatever character, are, where necessary, loosened by the scraper, and are then swept into little piles within a short radius. These are then, with the aid of the broom and shovel, transferred to the bag, which is held open by the carrier. When the bag is filled it is stood on the edge of the sidewalk. In wet weather, when the sweepings are in a state of solution, they are allowed to stand in piles until the free water has drained away; but even then the material is wet and heavy, the bags are much less easy to handle, and the cart-horses are apt to be overloaded because of this…

At present the work is divided about as follows:

$63\frac{1}{2}$ miles are swept once a day;
$283\frac{1}{2}$ "   "   "   twice a day;
$50\frac{1}{2}$ "   "   "   three times a day;
$35\frac{1}{2}$ "   "   "   four or more times a day.

This makes a total average sweeping of 924. This is not perfunctory work. The streets are really clean, and except for the littering, which the police have not yet succeeded in preventing, they always look clean. Mud is unknown, and dust is vastly diminished in comparison with former conditions.

Carting

Next in importance to the sweeping of the streets is the work of removing not only the product of the sweeping, but all domestic and some trade wastes, such as ashes, garbage, paper, and rubbish. In the New York department thirty-two per cent of the disbursement is for 'carting', and twenty-five per cent of the laboring-force is employed in this part of the work. This includes about six hundred drivers, with horses and carts. Most of the stable force is charged to carting.

The carts start out from the various stables at an early hour, and go to the sections to which they are assigned, the same men generally working on the same routes year in and year out. They first remove a load of ashes. After this they devote themselves to the carting of garbage until this is all removed. The rest of the day is occupied in collecting the remaining ashes and the sweepings that may have been gathered during the day ...

The garbage hauls are very long, as there are only six garbage-dumps for the whole city. These dumps are supplied with scows or other vessels by the Utilization Company. The department loads the garbage upon these vessels, and its connection with this portion of the work is then at an end. For street-sweepings and ashes we have seventeen dumps at different points, so located as to be within convenient reach, except in the case of the district lying west of Central Park ...

The street-sweepings are collected in bags, as described in the previous [section]. The bags are loaded on the carts without being untied, and are emptied at the dump, where they are cleaned and dried for the next day's use.

Thus far ashes are almost entirely collected from metal cans and other receptacles which are set on the sidewalk inside of the stoop-line or in the areas in front of the houses. It is in contemplation soon to extend throughout the city an improved system which has been in successful operation for more than a year. Under this system, each house is supplied with a can supported on a tripod six or eight inches above the floor. It has a hinged cover, and the bottom is closed by two flap-doors. The cartman takes a bag into the house or area or back yard – for it is only required that the can be kept out of sight of the street and protected from the rain – passes the bag under and around the can, and attaches it to the top frame of the tripod. He then closes the cover to prevent the flying of dust, and operates the mechanism which opens the doors at the bottom. The ashes run out into the bag, which is tied and set on the sidewalk to be removed with the sweepings. These bags also are emptied only at the dump.

Formerly the practice prevailed of removing in the same cart the sweepings shovelled from little piles in the streets and the entire waste of the house, which was put indiscriminately into the receptacles – garbage, ashes, paper, rubbish, and everything, save only such large objects as furniture, mattresses, etc. In 1896 the separation of these materials was taken in hand, and has now been completely effected.

The treatment of all other material than garbage, sweepings, and ashes remains to be described. The removal of this constitutes what we call the 'paper and rubbish' service. It is ordered that all wastes of this class be kept in the house, or at least under cover and out of sight of the street. A cart of special construction is used for the removal of this material. It is a very large low-hung box on two wheels, and is drawn by a lighter and more active horse than is required for the heavy loads of the sweepings and ash service. This material is called for only on the exposure – as in the basement window – of a special 'call' card. This is red and of diamond shape, with the letters 'P.R.' in conspicuous form in white on the front. Printed instructions are given on the reverse side.

This card relates to the following articles: paper, general rubbish, bottles, rags, tin cans, excelsior [shredded straw or paper used for packing], pasteboard boxes, old shoes, leather and rubber scrap, carpets, broken glass, barrels, boxes, discarded furniture, wood, and all metals.

Thus far the carts carrying these wastes dump their loads upon the scows which also receive the sweepings and ashes, but measures are now being taken to deliver them at 'Picking-yards', where a thorough sorting will be done and everything of saleable value culled out and made ready for the market ...

Final disposition of garbage

A time-honored custom of the city of New York has been to send its garbage to sea with all of its other wastes, save only the fat and bones collected by the scow-trimming Italians at the dumps. By far the largest proportion of garbage consists of vegetable refuse, much of which floats in sea-water. As a result of this method of disposal, the bathing beaches of New Jersey and Long Island have often been made unfit for use by the immense amount of offensive material washed ashore, especially during storms …

The outcry for years against this fouling of the beaches has been loud and strenuous. Efforts have been made on the part of the authorities of the State of New York and of the United States to seek a practicable remedy. This remedy has at last been found in the separation of garbage from all other material, and its delivery to a company which is charged with its care. My expert assistants have been actively engaged in the consideration of this subject since the very beginning of this administration. The result of their investigations is well set forth in the following report of Mr Macdonough Craven, in chief charge of the investigation, written in December, 1895, as follows:

Report of Macdonough Craven on the preliminary investigations made for the Department as to garbage and its treatment

When it was decided, early in the year, to dispose of the city's garbage by some better method than the old process of dumping at sea, an effort was made to learn what system would be best suited to the city of New York, with its limited space and its large amount of material to be cared for daily, and what the economies of such a system might be …

… The average of all bids from companies which proposed to cremate or destroy by fire was ninety cents per ton of garbage delivered, to be paid by the city; and from companies which proposed to utilize the garbage, or convert its available parts into grease and fertilizer, the average of all bids was fifty-five cents per ton …

… a circular letter was prepared and sent to each of the companies, proposing an examination of its plant and system by two competent men from this department …

… The tests were of necessity summer tests, when garbage becomes most quickly offensive, when any odors arising from the treatment would surely be noticeable, and when also garbage contains most water and is least valuable for utilization purposes.

More than three thousand tons of garbage in the cities of Buffalo, St Louis, Philadelphia, Brooklyn, and New York were treated by different methods, under the supervision of your inspectors.

One point made clear by the investigation is that when garbage is collected daily from each house, from clean cans, and conveyed at once to a properly equipped reduction plant, it has not time to ferment, even in summer, before it is safely stowed away within the steam-tight cooking-tanks of the reduction plant; and that under these conditions, and under experienced management, the operations of such a factory can be carried on with little more offense than arises from a large kitchen…

Kitchen refuse consists of animal and vegetable scrap, containing and mixed with a large amount of water. The animal scrap is of value for utilization purposes, because it furnishes the principal part of the grease and ammonia which are the saleable products of garbage; and since the cost of treating such waste is approximately the same, be it rich or poor, it is plain that the commercial value of garbage varies almost directly as its proportion of animal matter. If the amount of grease and ammonia recovered are sufficient to defray the expense of treatment, the people of any city may have their garbage disposed of without cost …

One can scarcely conceive of a crematory which destroys garbage by fire becoming a self-supporting concern, since considerable fuel is necessary and the only residue is ashes; but the fact that there are garbage 'utilization' plants at once suggests that under certain conditions the utilizable material may pay for its own extraction. It is perhaps needless to say that the word 'garbage' – which is so loosely used in this and a few other cities to denote any kind of waste, or a mixture of them all, including ashes and street-sweepings – is for the purpose of this investigation limited to animal and vegetable refuse from markets and kitchens. Only this is desirable in a utilization plant. A small admixture of cans, bottles, and berry-boxes entails extra expense for separation, but is not prohibitory of the process, while any such mixture as we have in New York to-day, of ashes, garbage, and a little of everything, *is* prohibitory. Garbage must be separated from everything else to be effectively and properly treated, and the other things must be separated from garbage to find, in their turn, any useful outlet …

Conclusion

... Few realize the many minor ways in which the work of the department has benefited the people at large. For example, there is far less injury from dust to clothing, to furniture, and to goods in shops; mud is not tracked from the streets on to the sidewalks, and thence into the houses; boots require far less cleaning; the wearing of overshoes has been largely abandoned; wet feet and bedraggled skirts are mainly things of the past; and children now make free use as a playground of streets which were formerly impossible to them. 'Scratches', a skin disease of horses due to mud and slush, used to entail very serious cost on truckmen and liverymen. It is now almost unknown. Horses used to 'pick up a nail' with alarming frequency, and this caused great loss of service, and, like scratches, made the bill of the veterinary surgeon a serious matter. There are practically no nails now to be found in the streets.

The great, the almost inestimable, beneficial effect of the work of the department is shown in the large reduction of the death-rate and in the less keenly realized but still more important reduction in the sick-rate. As compared with the average death-rate of 26.78 of 1882–94, that of 1895 was 23.10, that of 1896 was 21.52, and that of the first half of 1897 was 19.63. If this latter figure is maintained throughout the year, there will have been fifteen thousand fewer deaths than there would have been had the average rate of the thirteen previous years prevailed ...

It is not maintained, of course, that this great saving of life and health is due to street-cleaning work alone. Much is to be ascribed to improvements of the methods of the Board of Health, and not a little to the condemnation and destruction of rear tenements; but the Board of Health itself credits a great share of the gain to this department ...

Furthermore, during this administration the employment of private ash-carts and private sweepers has greatly decreased as people have found that the department service could be relied on.

... And, after all, how much did it cost all the people of this city for all that was done in 1896, including the removal of snow and the renewal of 'stock and plant'? The total sum is $3,283,853.90. And how much is that?

It is almost exactly three cents per week for each one of us!

The progress thus far made is satisfactory. An inefficient and ill-equipped working-force, long held under the heel of the spoilsman, has been emancipated, organized, and brought to its best. It now constitutes a brigade three thousand strong, made up of well-trained and disciplined men, the representative soldiers of cleanliness and health, soldiers of the public, self-respecting and life-saving. These men are fighting daily battles with dirt, and are defending the health of the whole people. The trophies of their victories are all about us – in clean pavements, clean feet, uncontaminated air, a look of health on the faces of the people, and streets full of healthy children at play.

This is the outcome of two and a half years of strenuous effort – at first against official opposition and much public criticism ...

I venture to predict a recovery, from the sale of refuse material, of at least one half the cost of the whole work.

These diagrams set forth the actual relation between the work of former years and that of Mayor Strong's administration:

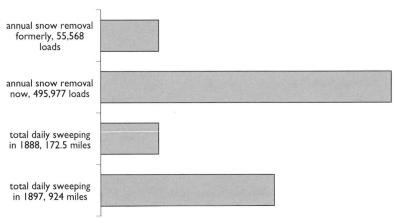

annual snow removal formerly, 55,568 loads

annual snow removal now, 495,977 loads

total daily sweeping in 1888, 172.5 miles

total daily sweeping in 1897, 924 miles

References

ANDERSON, L. (1988) 'Fire and disease: the development of water supply systems in New England, 1870–1900' in J. Tarr and G. Dupuy (eds) *Technology and the Rise of the Networked City in Europe and America*, Philadelphia, Temple University Press, pp.137–56.

ARMSTRONG, E.L., HOY, S.M. and ROBINSON, M.C. (eds) (1976) *History of Public Works in the United States, 1776–1976*, Chicago, American Public Works Association.

BAKER, M.N. (ed.) (1889) *The Manual of American Water Works*, New York, Engineering News.

BAKER, M.N. (1899) 'Waterworks' in E.W. Bemis (ed.) *Municipal Monopolies*, New York, T.Y. Crowell.

BLAKE, N.M. (1956) *Water for the Cities: a history of the urban water supply problem in the United States*, Syracuse, NY, Syracuse University Press.

CAIN, L.P. (1978) *Sanitary Strategy for a Lakefront Metropolis: the case of Chicago*, DeKalb, Ill., Northern Illinois University Press.

COLTON, C.E. (1994) 'Chicago's wastelands: refuse disposal and urban growth, 1840–1990', *Journal of Historical Geography*, vol.20, pp.124–42.

CONDIT, C.W. (1973) *Chicago, 1910–1929: building, planning and urban technology*, Chicago, The University of Chicago Press.

COREY, S.H. (1994) 'King Garbage: a history of solid waste management in New York City, 1881–1970', Ph.D. dissertation, New York University, Ann Arbor, Mich., UMI, Order no. 9514375.

DUFFY, J. (1974) *A History of Public Health in New York City, 1866–1966*, New York, Russell Sage Foundation.

ERICSON, J. (1925) *The Quality Problem in Relation to Chicago's Water Supply: official report to Col. A.A. Sprague,* Chicago, City of Chicago, Bureau of Engineering, Department of Public Works.

FLANAGAN, M. (1996) 'The city profitable, the city livable: environmental policy, gender and power in Chicago in the 1910s', *Journal of Urban History*, vol.22, pp.163–90.

GRINDER, R.D. (1980) 'The battle for clean air: the smoke problem in post-Civil War America' in M.V. Melosi (ed.) *Pollution and Reform in American Cities, 1870–1930,* Austin, The University of Texas Press, pp.83–103.

HOY, S. (1995) *Chasing Dirt: the American pursuit of cleanliness*, Oxford, Oxford University Press.

HUNDLEY, N. Jr (1992) *The Great Thirst: Californians and water, 1770s–1990s*, Berkeley, Cal., University of California Press.

KELLOGG, P.A. (ed.) (1914) *The Pittsburgh District: civic frontage, the Pittsburgh survey*, New York, Survey Associates.

LUBOVE, R. (1995) *Twentieth-century Pittsburgh*, Pittsburgh, University of Pittsburgh Press, vol.1.

MELOSI, M.V. (1980a) 'Environmental crisis in the city: the relationship between industrialization and urban pollution' in M.V. Melosi (ed.) *Pollution and Reform in American Cities, 1870–1930,* Austin, The University of Texas Press, pp.3–31.

MELOSI, M.V. (1980b) 'Refuse pollution and municipal reform: the waste problem in America, 1880–1917' in M.V. Melosi (ed.) *Pollution and Reform in American Cities, 1870–1930,* Austin, The University of Texas Press, pp.115–33.

MELOSI, M.V. (1981) *Garbage in the Cities: refuse, reform, and the environment, 1880–1980*, College Station, Texas A&M University Press.

MILLER, D.L. (1996) *City of the Century: the epic of Chicago and the making of America*, New York, Simon and Schuster.

MONKKONEN, E.H. (1988) *America Becomes Urban: the development of US cities and towns 1780–1980*, Berkeley, Cal., University of California Press.

OGLE, M. (1996) *All the Modern Conveniences: American household plumbing, 1840–1890*, Baltimore, The Johns Hopkins University Press.

PETERSON, J.A. (1979) 'The impact of sanitary reform upon American urban planning, 1840–1890', *Journal of Social History*, vol.13, pp.83–103.

PLATT, H.L. (1995) 'Invisible gases: smoke, gender, and the redefinition of environmental policy in Chicago, 1900–1920', *Planning Perspectives*, vol.10, pp.67–97.

ROBERTS, G.K. (ed.) (1999) *The American Cities and Technology Reader: wilderness to wired city,* London, Routledge, in association with The Open University.

ROSEN, C.M. (1995) 'Businessmen against pollution in late nineteenth century Chicago', *Business History Review,* vol.71, pp.351–397.

ROSEN, C.M. (1997) 'Noisome, noxious and offensive vapors, fumes and stenches in American towns and cities, 1840–1865', *Historical Geography,* vol.25, pp.49–82.

SCHULTZ, S.K. (1989) *Constructing Urban Culture: American cities and city planning, 1800–1920,* Philadelphia, Temple University Press.

SCHULTZ, S.K. and McSHANE, C. (1978) 'To engineer the metropolis: sewers, sanitation, and city planning in late-nineteenth-century America', *Journal of American History,* vol.65, pp.389–411.

TARR, J.A. (1996) *The Search for the Ultimate Sink: pollution in historical perspective,* Akron, Ohio, The University of Akron Press.

TARR, J.A. and DUPUY, G. (eds) (1988) *Technology and the Rise of the Networked City in Europe and America,* Philadelphia, Temple University Press.

TOMES, N. (1990) 'The private side of public health: sanitary science, domestic hygiene and the germ theory, 1870–1900', *Bulletin of the History of Medicine,* vol.64, pp.509–39.

UNITED STATES DEPARTMENT OF COMMERCE, BUREAU OF THE CENSUS (1975) *Historical Statistics of the United States from Colonial Times to 1970,* Part 1, Washington, DC, US Government Printing Office.

WARING, G. (1897) *Street-cleaning and the Disposal of a City's Wastes: methods and results and the effect upon public health, public morals, and municipal prosperity,* New York, Doubleday and McClure.

WARING, G. (1898) *Report of the Department of Street Cleaning of the City of New York for 1895–96–97,* New York, Reform Club Committee on Municipal Administration.

Chapter 6: TECHNOLOGY AND THE GOVERNANCE OF CITIES

by Gerrylynn K. Roberts

6.1   Introduction

The main focus of this series has been the social relations of technology as exhibited in the physical form and fabric of towns and cities. In this chapter we will look at a particular dimension of those relations – the interaction of technology with some aspects of the governance of cities. Both the nature of urban government and the way that technology was involved in it changed considerably from the late nineteenth century onwards. The technological 'networking' of cities during the nineteenth century raised issues about whether and how city authorities should be involved with technologies. At the same time, US cities adopted a new attitude towards government, becoming more active providers of services instead of passive regulators (Monkkonen, 1988, chapter 9; Tarr, 1989). In some cases, technologies were deployed as part of this process; in others, the process affected technology, or technology to some extent stimulated the process. For example, the technologies of street lighting and fire-fighting became essential tools of city governance, while the development of large-scale, networked transport and utilities infrastructures in the nineteenth century involved city authorities in making numerous, incremental decisions that amounted to what might be seen as piecemeal planning. The emergence of the modern planning movement in the twentieth century was to some extent stimulated by these earlier approaches. Modern planning then became part of the sphere of government, and part of the context within which decisions about urban technologies were made. Planners, in turn, were strongly influenced by the technologies that were available to them.

6.2   Street lighting

Public safety was often the justification for installing lighting on city streets. Rushlights or oil lamps were used in the eighteenth century, as urban nightlife – both for commerce and for entertainment – became more common. In the nineteenth century, manufactured gas made it possible for factories and shops to stay open later; this resulted in more people being on the streets after dark – traditionally a threatening time. At the same time, city authorities organized new, networked systems of street lighting using the more reliable manufactured gas, generally through franchises to private contractors (Bouman, 1987; Tarr, 1989, p.222). It made good business sense too; good light became one of the features that city promoters advertised and, concomitantly, it became a characteristic of urbanity (Bouman, 1987, pp.10–14). In that context, intercity competitiveness encouraged the spread of lighting (Platt, 1991, p.11).

A workable system for manufactured street lighting was first installed in 1812 in London, from where the technology, based on the carbonization of coal gas, quickly diffused. The gas was produced in centrally located works, stored on site in gas-holders (conspicuous new features of the cityscape), and piped to individual points of use. The first street-lighting system in the USA to use manufactured gas was installed by Baltimore entrepreneurs by 1816. By

1828, New York City's principal streets were lit by gas, and by the 1840s, most medium-sized cities had gas-works, although they did not necessarily use coal. The number of urban gas enterprises increased from 30 to 221 in the decade from 1850 to 1860; by 1870, there were 390 (Tarr, forthcoming).

Street lighting was not profitable for the companies that provided the service, but it did serve to publicize the illuminating power of gas – as did the bright lights in public buildings, where the new illuminant was quickly adopted. Companies were permitted to decide which areas to serve, and generally chose only the more prosperous districts. In Chicago, for example, where residents had to pay for their own poles and fixtures for street lights, the city provided naphtha lamps instead of gas in poor or outlying districts. It was only after the advent of competition from electricity, from the 1880s, that wider markets were sought, including that of domestic lighting, cooking and heating. US gas companies also met the competition from electricity by supplying electric lighting as well. By 1889, 25 per cent of US gas companies were dual utilities, and by 1899, almost 40 per cent were (Tarr, forthcoming).

Unlike water companies, gas utilities tended to remain in private hands. However, since they had to disrupt the public streets to install and service their delivery systems, gas companies needed public franchises and were subject to regulation. Street lighting, too, was operated through municipal franchises, with concomitant regulations. If they had to be regulated at all, the companies apparently preferred state regulation, which became common during the early twentieth century, to municipal regulation, which was widely seen as subject to the vagaries of local political influence. Once natural gas began to become generally available, from the late 1920s, state and occasionally even federal regulation became the norm, since gas was no longer locally produced, but crossed state boundaries.

Figure 6.1 Servicing the carbon rods in an arc-light, c.1895. Arc-lighting is produced by sparking across a gap between two rods of carbon that functioned as electrodes. They needed frequent replacement as the carbon was consumed. By 1886, some 30 miles (48 kilometres) of New York City's streets were illuminated by arc-lights spaced at roughly 250-foot (75-metre) intervals (Nye, 1992, p.31; photograph: © Hall of Electrical History, Schenectady Museum)

Competition from electricity began in the 1880s, when electric arc-lamps were introduced for lighting the city streets and some large public buildings, such as theatres and department stores. The first permanent arc-light system for lighting a public space in the USA was installed by Charles Brush in Cleveland, in 1879. Brush sold sets of dynamos (generators), steam-engines and arc-lamps as one-off systems to towns, firms or whomever required lighting. New York City experimented with Brush's arc-light system for street lighting by 1880 and soon began to move away from gas (see Figure 6.1). However, it was the introduction by Edison of his system for the generation and distribution of electricity for incandescent street lighting from the early 1880s that brought real competition to gas. For Edison directly targeted the gas-lighting market and patterned his electricity system on the central generation and distribution of gas (see Chapter 4). The disruption of city streets by the installation and servicing of utilities – whether by gas or electricity companies, or by water or sewerage utilities – posed major problems for city authorities in the nineteenth century, and raised important issues of municipal control (Rosen, 1986, p.38) (see Figure 6.2).

Figure 6.2 Chicago Edison Company laying the first network of underground 'tubes' (cables) in The Loop, 1887. In many cities, by the 1890s, after countless incremental developments, the area below ground had to be reorganized before any further development could take place. Water, sewerage, gas, electricity and transport services all involved digging, in some instances by competing companies delivering the same service. In 1891 alone, Manhattan utility companies made 59,000 transverse excavations on 391.5 miles (626 kilometres) of paved streets. If longitudinal excavations are added, we can say that on average this amounted to at least some digging in every 35 feet (10.5 metres) of street (Rosen, 1986, p.38; photograph: courtesy of Commonwealth Edison Company, Chicago)

6.3 Coping with fire

Fire and changes in the built environment

Fire was a fearsome feature of city life throughout the nineteenth and early twentieth centuries. Few major cities escaped devastating fires, with the resulting loss of both property and life. Dealing with fire was a major issue for city authorities. J.K. Freitag, from whose 1912 handbook *Fire Prevention and Fire Protection as Applied to Building Construction*, Extract 6.1 is taken, vividly described the consequences of fire for the built environment:

> it has been estimated that we burn up during every 'normal' week of the year, 3 theaters, 3 public halls, 12 churches, 10 schools, 2 hospitals, 2 asylums, 2 colleges, 6 apartment houses, 3 department stores, 2 jails, 26 hotels, 140 flats and stores, and 1600 homes.
>
> (Freitag, 1912, p.5)

While not discounting the terrible personal, social and financial losses entailed in such figures, it should also be noted that to some extent major fires provided an opportunity for the regeneration of cities. Major city fires were rather like forest fires, which clear the land and allow new growth: they did away with structures that were obsolete but that, because they retained some financial viability, were not likely to be eliminated in the normal course of events – and so could act as a brake on badly needed new development. The trick – not always performed successfully – was to harness that opportunity.

Rosen (1986) analysed the aftermaths of the major fires in Chicago (in 1871), Boston (1872) and Baltimore (1904): she shows that, although the need for infrastructural reorganization and the opportunity for post-fire redevelopment was clearly recognized by various groups in all three cities, they were not all equally successful in achieving infrastructural changes: local political, economic and technical factors affected the extent to which individual cities would grasp the apparent opportunity. Chicago was the least successful at achieving change, and Baltimore the most successful. Chicago was the only city of the three to lose extensive residential areas as a result of fire; these were in working-class districts, so large numbers of people needed to be rehoused urgently – there was no time for long deliberations on possible changes to the urban structure. A municipal election was imminent, too, which made the matter political – it was important to contain any possible unrest. In Boston, already taxed to a very high level for improvements before the fire, there was resistance to a higher level of expenditure than was needed to provide any more than a minimal solution. Another disincentive to change was the view that, if property lines were redrawn to straighten and widen Boston's warren of narrow streets, there were bound to be losers as well as winners. More generally, technological complexities, and the fact that many improvement goals were mutually exclusive, also tended to inhibit development. One thing stands out: success was partly a matter of timing. The Baltimore fire was the latest of the three. The city's reconstruction therefore took place in the political climate of the progressive movement. In addition, the timing of the fire put Baltimore in a position to exploit new urban technologies developed in the late nineteenth century. Another factor was that the three cities had different mechanisms for decision-making, and different groups, with conflicting interests, had varying degrees of access to those mechanisms. Change could be stymied by decision-making gridlock. It was not until the reforms of municipal government in the early twentieth century, when the progressive movement was most influential – the Progressive Era embraced the deployment by government of trained experts, often technical, to solve society's problems –

that the pace of infrastructural improvement quickened, through measures such as compulsory purchase and the regulation of private development (Rosen, 1986, pp.335–6).

Fire-prevention[1]

Regulations

Regulations to control various aspects of the built environment in order to prevent fires or to mitigate their effects were an important means of dealing with the ever-present threat. Monkkonen (1988, p.93) argues that such regulations were a service rather than a constraint because they were aimed at the common good. Most cities, including pre-1871 Chicago, had local ordinances defining fire limits – areas within which the construction of wooden buildings was forbidden. In 1871, New York and Boston promulgated building regulations that exercised more comprehensive control. New York's regulations required that buildings taller than 70 feet (21 metres) be of fireproof construction (Wermiel, 1996, p.288). The law also allowed for the strength of iron structural members to be spot checked by inspectors. A building department with a team of eighteen inspectors was set up, but enforcement was lax and corruption rife. Boston created an eight-strong building inspection department which, instead of checking buildings that had already been constructed, checked the building plans, against which it issued permits before building work could begin. Wooden roofs and exterior wooden decoration were prohibited (Wermiel, 1996, p.171). Though a height limit was discussed in Boston from the time of the fire of 1872, there was no enforceable limit until 1892 when a 125-foot (38-metre) cap was imposed – a much lower level than was technically feasible at the time. The ability to fight fire was among the reasons for deciding on the cap; also important were aesthetic and health considerations about blocked air and light, and property developers' concerns about the rate of return on tall buildings and the skewing of land values (Holleran, 1996).

Fireproof construction

Some new technologies of fireproof construction in the nineteenth century were discussed in Chapter 4. Although they were becoming prominent during the late 1870s and 1880s for the tall buildings of New York and Chicago, fireproof construction methods based on metal and terracotta were not adopted universally. In New England, in the final years of the century, 'slow-burning construction', adapted from textile-mill construction techniques, was used, especially in factories (Wermiel, 1996, chapter 5). Because of the comparative expense, the USA rarely adopted British iron and brick construction for factories, with the exception of those running highly inflammable processes such as sugar-refining works or gas-works. Slow-burning construction consisted of floors formed of thick wooden timbers, tightly jointed to prevent air gaps, and roofs of similar design. This was supplemented by fire-fighting equipment: buckets of water, piped water from roof-top cisterns topped up by pumps from nearby water sources, sprinklers and hydrants. The strategy was that slow-burning construction would allow time for the fire-fighting devices to be deployed, so that fires could be contained and quickly extinguished. Good watch systems and alarms were essential to this strategy. External stair towers, which could be isolated from the main building by fireproof doors, were built as fire-escapes. The towers also gave protection to water mains and hydrants and access for fire-fighters.

---

[1] The sections on fire-prevention and safety draw heavily on Wermiel (1996).

The slow-burning system was still used for mills in the early twentieth century, but, as electric motors came to be used in factories, the workspaces needed to be reorganized to accommodate them; reinforced-concrete construction methods made it possible to build the new, horizontal plant designs. Such designs required more space, so firms began to set up at the edge of cities – land was less expensive there, and also there was more room for expansion and for manoeuvring lorries and parking motor cars. At the same time, supplies of the requisite timber were becoming rarer and more expensive. The cost advantage of slow-burning construction diminished rapidly.

However, certain elements of the system remained. Automatic sprinklers (where the water from overhead pipes was released by the melting of solder plugs set at intervals) linked to alarms were first devised in Britain during the 1860s, but not much used there because of insurers' concerns about water damage. By contrast, American insurers of mills favoured them because they tackled the source of most blazes – the building's contents – quickly, before fire could spread to the structure, which was far more expensive to replace. They were in widespread use by the 1880s, not only in factories, but also in department stores, theatres and hotels. Indeed, from 1885, when Boston and New York introduced stringent regulations for public buildings, they were a requirement in theatres.

As skyscrapers became more common, many cities began to require fireproof construction. In the first place, there were concerns about fighting fires on upper storeys. Ordinary water pressures meant that, from the ground, it was impossible to fight a fire at the top of an 80-foot (24-metre) building – even a 60-foot (18-metre) building posed a challenge. In London, the approach to this problem was generally to impose height restrictions; in the USA, it was to require the fireproof construction of entire buildings if they exceeded a certain height (Wermiel, 1996, pp.287–8). Quite apart from regulations, skyscraper owners had every incentive to use fireproof construction to protect their large investments. At the same time, they had every incentive to build higher, to increase the amount of rentable space and the returns on those investments (see Chapter 4).

Safety

Following the public outcry after a serious fire disaster, New York began to require fireproof fire-escapes in multi-storey, multi-occupancy dwellings from the 1860s, and in factories from the 1870s. A feature of the Boston regulations of 1871 was a requirement that every building have a way of escaping from a fire in an emergency, if only by means of a permanent ladder leading to the roof. The Boston regulations drawn up after the fire of 1872 required that public buildings of a specified capacity have non-combustible, enclosed staircases and outward-opening exit doors: such features were unusual in the nineteenth century. In general, protecting property was considered before protecting life. According to Sara Wermiel (1996, p.346), matters of egress began to receive attention in the fire-protection field only after effective methods of fireproof construction had been developed. Because lives were still being lost, a need for separate measures was recognized. However, the provision of adequate means of escape was costly and the investment was non-profit-earning, so it occurred only after legislation. Not until a law was passed in Massachusetts in 1877 did owners anywhere in the USA have legal responsibility for the safety of occupants of their buildings. Theatres were a particular focus of attention.

The legislative precedents were set in Britain from the 1840s, where walled-in, fireproof stairways were preferred; a ladder brought to a fire was also considered to be a fire-escape – feasible in London because of restrictions on building-height. In the USA, with so much retrospective provision necessary because of its later start, fire-escapes took a different form: external metal platforms and ladders fixed to the sides of buildings became typical city features. Inside public buildings, the posting of floor plans showing exits and the illumination of exit signs by electricity became common from the mid-1880s. The simple provision of escape facilities was not the only issue. Means had to be devised to enforce their correct maintenance, and the public had to be educated to make proper use of them, rather than heading automatically for the usual exits.

The development of legislation and building codes relating to fire-prevention and fire-protection was generally conservative technologically – that is, it was based on existing technologies. However, such regulations did create a market for and stimulate innovations in construction materials and other devices, which subsequently featured in new regulations. So effective were the changes over the nineteenth century that the earthquake-induced San Francisco fire of 1906 was the last urban conflagration in the USA.

Fire-fighting

The most visible manifestations of the requirements of fire-fighting in the built environment were: station houses, where equipment was kept and fire-fighters were trained; watch-towers, which, until the advent of telegraphic alarms in the middle of the nineteenth century (see Chapter 8), were crucial in the chain of early detection and audible spreading of the alarm; and street accoutrements, such as hydrants and cisterns, that were specially constructed to serve the needs of fire-fighters (see Chapter 5).

New York instituted the first fire-watch in 1648 (when it was still called New Amsterdam). Volunteer fire companies, often supported by insurers, began to be formed in the eighteenth century and, by the early nineteenth century, every US city had several, often rival, companies. In some cities, the fire stations which served as their base of operations were municipally funded. Hand-operated pumps were the principal equipment of volunteer companies (see Figure 6.3). They were dragged through the streets by firemen, and required some fifteen to thirty people for efficient use (Greenberg, 1998, pp.10–12, 139).

Figure 6.3 Detail from an engraved notice of a meeting of the Hand-in-hand Fire Company, New York, c.1753. From the seventeenth century, all citizens were responsible for fire-fighting; most large colonial towns required that homes have two buckets ready in case of fire, and all males were expected to turn out to form a 'bucket brigade' when called upon. Here, a bucket brigade conveys water from a public pump to a machine, from where it is pumped by hand to form a jet to direct at the fire. Ladders and hooks on the ends of tall poles for manoeuvring timbers are the other equipment visible (courtesy of The I.N. Phelps Stokes Collection, Miriam and Ira D. Wallach Division of Art, Prints and Photographs, The New York Public Library. Astor, Lenox and Tilden Foundations)

Figure 6.4 stresses the role of horses in fire-fighting, but it was the steam-engines they pulled that constituted the important technological transition in the middle of the nineteenth century. The first viable steam-engines for fire-fighting in the USA were manufactured in Cincinnati (OH) in 1852. Over the following two decades, volunteer companies were replaced by municipally paid fire services in most major cities as the steam-engine was adopted. The new technology of the steam fire-engine afforded cities the means to take charge of the streets, displacing the (sometimes rowdy) volunteer companies. Cities replaced men by machines, often against considerable opposition. Because a steam-engine could pump about three times as much water as a hand-operated pump, and in a shorter space of time, fewer of them were needed. Furthermore, with no pumping by hand involved, fewer fire-fighters were needed on each engine. Steam-engines also eliminated another manpower requirement: they were too heavy to be dragged by even large numbers of men, so horses (and facilities for them) became necessary. Cincinnati reckoned that the numbers of men in fire companies could be reduced by between 60 and 90 per cent. In addition, running and maintaining steam-engines required new specialist skills. Encouraged by insurers, this transition marked a shift in nineteenth-century US society: more faith was put in machines, and there was increased dependence on municipally provided expertise (see Figure 6.5) (Greenberg, 1998, chapter 5; Monkkonen, 1988, pp.105–8). Similarly, the use of the telegraph for sounding the alarm eventually led to centralized control of fire-fighting at the expense of local volunteer companies – again with the support of insurers. The alarms were often linked to police stations as well, enabling greater centralized control of cities by formal authorities.

Figure 6.4 The last run of the fire-horses of Detroit, 1922. Though the introduction of horse-drawn steam-engines had been bitterly contested in many cities in the middle of the nineteenth century, they were treated with great romanticism and nostalgia when replaced by motorized equipment in the early 1920s. Mirroring the closure of many a volunteer company at mid-century, the final run of the last of Chicago's fire-horses was conducted with great ceremony: an alarm was pulled to bring out the horse-drawn engine for a final display while motorized equipment was driven in to the station to replace it (Showalter, 1923, p.406; photograph: Mark W. Stevens; courtesy of *The National Geographic Magazine*, vol.44, p.406)

Figure 6.5 Fighting a fire in a chemical warehouse with motorized apparatus, New York, 1923. By 1923, most major cities had made the transition to lorries for fire-fighting. They were argued to be more efficient and convenient for reaching fires quickly, where rapid response was crucial to prevent a fire's taking hold; also, they were not subject to disease (Showalter, 1923, p.407; photograph: Paul Thompson; courtesy of *The National Geographic Magazine*, vol.44, p.560)

6.4 Planning

After the USA became a nation, the layout of most cities was planned by survey in advance of settlement, to define parcels of land that could be sold. Federal surveyors implementing the Land Ordinance of 1785 (see Chapter 2) set the basic grid co-ordinates, and many city plans simply subdivided those grid-squares further – a process known in the USA as 'platting' (dividing the land into plots) – which resulted in grid-plan cities. Beyond that, however, except for the location of a few key public buildings, there was seldom stipulation of how land should be used; as preceding chapters have shown, US cities grew in a rather chaotic, unplanned way as their populations expanded dramatically to occupy the land platted by private developers for financial gain – often with little regard for the pressures that such development put on the environment. To be sure, there was a certain level of incremental, piecemeal planning in the nineteenth century, especially in the areas of landscape-architecture – with the development of parks – and of engineering, including sanitary reform (see Chapter 5) (Peterson, 1979; Schultz, 1989; Schultz and McShane, 1978). Indeed, Hammack has argued further that, though not comprehensive, the commercially motivated piecemeal planning of the nineteenth century was remarkably wide in scope. It implies, too, at least some level of forethought, intent and co-ordination, as decisions were shaped by the economic and political environment (Hammack, 1988, p.143); and in many US cities there were some visionary individuals, such as the landscape-architect F.L. Olmsted, who worked on comprehensive plans. However, it was not planning in the sense of public deliberation and governmental regulation (Krueckeberg, 1983, p.8).

The modern profession of planner was not articulated in the USA until the Progressive Era, during the early decades of the twentieth century, when civic action to make use of technology to affect the form and fabric of cities for mutual improvement became politically feasible. In the history of planning, the World's Columbian Exposition of 1893, held in Chicago, is often seen as a milestone, demonstrating to US citizens how their dreary, haphazard cities could be beautified and improved by means of comprehensive planning. The original impulse was aesthetic, stimulating the 'City Beautiful' movement of the late nineteenth and early twentieth centuries with its stress on 'Renaissance classical' styles (classical styles as filtered through the Renaissance) and generous street layouts. It is argued that the pre-war progressive movement helped to transform the City Beautiful idea of the early twentieth century into the City Practical (or Scientific) idea of the inter-war years, in which professional planners were heavily involved (Reps, 1965, pp.497–502, 524–5). However, while the Exposition and the City Beautiful movement to which it gave rise were certainly influential, the rise of modern, professional planning was no sudden phenomenon: it derived from many incremental, parallel activities that had taken place much earlier in the nineteenth century. European thinking was one stimulus but, in contrast to Europe, the emergence of modern planning in the USA was stimulated by private interest groups, not by muncipal government (Sutcliffe, 1981, chapter 4).

The development of planning up to the Second World War moved from designing the basic layout of cities, to piecemeal improvements of specific civic areas, to planning infrastructural development on a regional scale. Chapters 7 and 8 consider more general relations between planning and the technologies of transport and communications in more recent periods.

Washington DC[2]

Unlike most new cities in the new nation, Washington DC was planned: it was conceived in 1791, not as a speculative, commercial venture, but as a monumental city with vistas and architecture thought to be suitable for the capital of a vibrant new country. George Washington, the first president, agreed a greenfield (or in the event, green-swamp) site for the new city, on the Potomac River, along the border between Maryland and Virginia. The transfer of the seat of government was scheduled to take place at the start of the new century, in 1800. A survey of the proposed site was soon undertaken by Andrew Ellicott, a professional surveyor and member of the American Philosophical Society, and Major Pierre Charles L'Enfant, son of a court artist at Versailles. L'Enfant had trained in art in Paris before volunteering for the

---

2   This section draws heavily on Reps (1965, chapters 9 and 18) and Scott (1969, pp.47–57).

Revolutionary Army of the American colonies. He served under Washington as a military engineer and was later responsible for remodelling New York's City Hall for use as the venue for the US Houses of Congress while New York City was the nation's capital. Once the survey was accomplished, L'Enfant was commissioned to prepare a plan of the new city (see Figure 6.6). The design vocabulary of L'Enfant's Parisian roots was obvious; it was imperial rather than republican, and looked backward rather than forward, but it would eventually serve the desired symbolic function and influence the design of other US cities, especially state capitals.

Washington DC grew slowly, in part because the federal government did not expand until after the Civil War brought new emphasis on the Union and an increasing bureaucracy. The original plan was followed in broad outline, but with departures that considerably lessened its effectiveness. The Capitol Building and the White House were built much as planned. The dome of the former (completed in 1864 following reconstruction of the building after a fire in 1814) made innovative use of cast iron. The obelisk of the Washington Monument was erected over the period 1848–84, as donations permitted. It

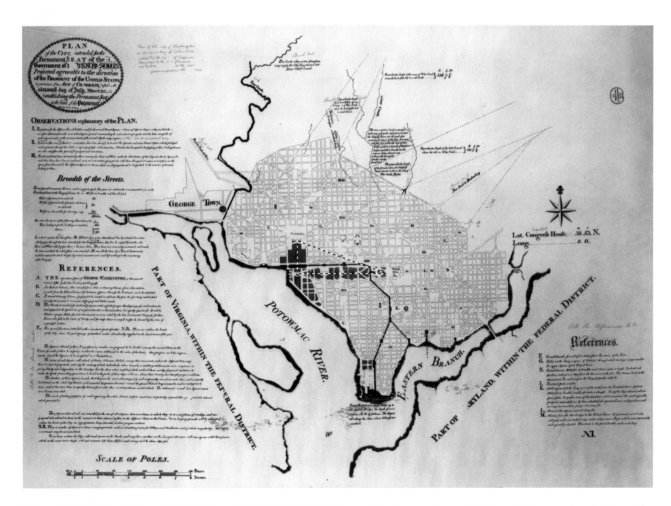

Figure 6.6 L'Enfant's Plan of Washington DC, 1791. L'Enfant's Plan contained many elements that would become key features of Washington DC – in particular, the Capitol Building, the White House (the two separate buildings expressing architecturally the constitutional separation of legislative and executive powers) and a monument to George Washington, to be linked to the Capitol by a grand promenade, or garden mall. We have added dotted lines to the Plan, to show the triangular relationship between these three structures. The base of the triangle shows the line of the mall, with the Capitol Building at the intersection of the base with the hypotenuse. The White House is at the other end of the hypotenuse, at the apex of the triangle. The Plan departed from a straightforward grid by having strong diagonal boulevards criss-crossing the city and meeting at open squares or circles. Roads were to be wide and the scale of the whole was to be monumental, with large public spaces punctuated by suitable structures to anchor vistas (courtesy of Library of Congress, Department of Maps)

was constructed slightly away from its planned location at the junction of axis lines from the White House and the Capitol because of the unsuitability of the soil for such a large structure at the intended site. Some 555 feet (169 metres) high, the monument was of masonry construction, with a sharp-pointed aluminium cap, but had a separate, internal wrought-iron structure which served as the lift shaft and support for the staircase; it was the tallest iron structure of its time (Condit, 1982, pp.90–92). The Mall, instead of being developed as a grand avenue, was given rustic treatment as a romantic park. A more drastic change from the L'Enfant Plan was the building of a railway station on the Mall, with railway tracks running across the park at street-level. It was not until the early stages of the City Beautiful movement, at the turn of the twentieth century, that efforts were made to realize L'Enfant's Plan more fully and create a truly monumental city. To some extent, the realization of the new plan of 1902 was possible only because of earlier engineering projects undertaken for the purposes of sanitary reform (Peterson, 1979, p.96): a fetid creek was incorporated into the drainage system by means of a culvert along the Mall's northern boundary, and an area of malarial swamp-land was reclaimed along the edge of the Potomac, where it was proposed to terminate another vista with a new monument. Although in other areas of the city there were overflowing slums (Hall, 1988, p.178), this high-profile and elegant series of public improvements in the centre of the federal capital was an important boost for the City Beautiful movement: it incorporated landscape-architecture and municipal improvement in civic design, and, at the same time, it established an important role for architects in both civic design and city planning (Wilson, 1989, p.69).

Chicago – the Burnham Plan

Daniel Burnham was a noted Chicago commercial architect who worked on the 1902 Plan for Washington DC, and who subsequently worked on City Beautiful plans for many cities in the United States and abroad; it was for his home town of Chicago, though, that he produced in 1909 the planning masterwork that came to be known by his name: the Burnham Plan. Sponsored by the Commercial Club, a private group of Chicago businessmen, the Burnham Plan was a comprehensive scheme for the refurbishment of the city and, importantly, its surrounding region; it also included some plans for the future growth of the city. Its aim was unashamedly to promote Chicago – to make of it an attractive and therefore prosperous city, whose new wealth should eventually benefit all its citizens. The Plan aimed to restore harmony to the ethnic chaos that was Chicago by 'cutting new thoroughfares, removing slums, and extending parks' (Hall, 1988, p.179), and to tackle Chicago's notorious congestion problem, which threatened to strangle business (see Chapter 3). Not only was transport movement in the city centre considered but also the restructuring of the rail system into and around the city, and the improvement of port and transhipment facilities. Parks and open spaces within the city, and in the region more widely, were also important features. In addition, a uniform building style along grand Haussmannite boulevards was encouraged and there was an emphasis on providing centres of intellectual and civic life (Wrigley, 1983, pp.62–3). In terms of planning for the future, though, the Plan was surprisingly backward-looking – both architecturally and in its use of land. Other than the central business district of The Loop, no areas were specifically designated for commerce or manufacturing. Nor was systematic attention paid to the need for housing for workers, to the provision of schools or to sanitation requirements (Scott, 1969, pp.107–9; Sutcliffe, 1981, pp.106–10).

The publication of the sumptuously illustrated (see Figure 3.16) Burnham Plan was a great public relations success. Practical or not, it won a great following in the city. The Commercial Club continued to sponsor it into the implementation phase with the establishment of the Chicago Plan Commission. Under its chairman, businessman Charles Wacker, and its executive director, Walter Moody, who was a professional organizer and manager, promotion was as much an activity of the Commission as gradual implementation funded by the floating of bonds. Moody saw promotion as a 'scientifically professional activity' equal in importance to the architectural and engineering aspects of the Plan, and therefore a fundamental aspect of professional city-planning (Schlereth, 1983, p.78). He targeted specially tailored publicity at each of the important power groups of the city, using it to counter their criticisms of the Plan.

One of Moody's masterstrokes was the publication in 1911 of what he called *Wacker's Manual of the Plan of Chicago*, to identify the Plan strongly with the leader of the implementation commission.[3] Extract 6.2, at the end of this chapter, is taken from the 1913 edition. The *Manual* was aimed in particular at schoolchildren in the eighth grade (about 14 years old); at the time, this was the highest year of schooling that the majority would reach and a stage when children were thought to be most receptive. It was expected, of course, that their parents would also read the *Manual*; indeed, children were encouraged to share it with them. Eventually, some 50,000 copies were printed, and it was used in the school system for more than a decade. Moody believed that schools were where Chicago's future citizens were formed, and they had a duty to educate their charges in what enlightened citizenship meant; in the city around them students could readily observe what they were studying in the classroom. In turn, as a reformer schooled in the Progressive Era, Moody believed that building a better city environment – creating not just the City Beautiful, but also the City Practical – would build better citizens. The ultimate implementation of the Burnham Plan depended on widespread voter approval of each of its separate proposals; with the *Manual*, Moody planned for the long haul. His task was to get public approval for implementation by the city of an essentially private plan. Aspects of the implementation of the Burnham Plan relating to transport technologies and new buildings are mentioned in Chapters 3 and 4 (see Figures 3.18, 3.19, 3.20, 4.27, 4.29 and 4.31).

New York City

The Plan of 1811

The first important city plan to be drawn up after L'Enfant's Plan for Washington DC was the New York Plan of 1811. Established by law to forestall problems with property owners, the plan was utilitarian, aiming to facilitate New York City's rapid growth as a commercial rather than a political centre; the grid imposed on the city made possible the rapid sale of plots and thus aided land development. The creation of a less-congested city for public-health reasons was also cited as a reason for drawing up the Plan (Spann, 1988, p.18). The Plan laid out, well in advance of settlement, a dense, rectilinear street system for almost the whole of Manhattan (see Figure 6.7 overleaf). The standard plot size, with a frontage of 25 feet (7.6 metres) and a depth of 100 feet (30 metres), would affect the nature of the buildings erected on it for many years to come. Later critics argued that the plots were too small for houses of the wealthy, but too large to provide inexpensive housing for the poor; their 'unsatisfactory' size led to multiple dwellings being built on a single plot, with consequent cramped housing and poor conditions (see Chapters 4 and 7).

---

[3]   This discussion of Moody's promotional activities draws heavily on Schlereth (1983).

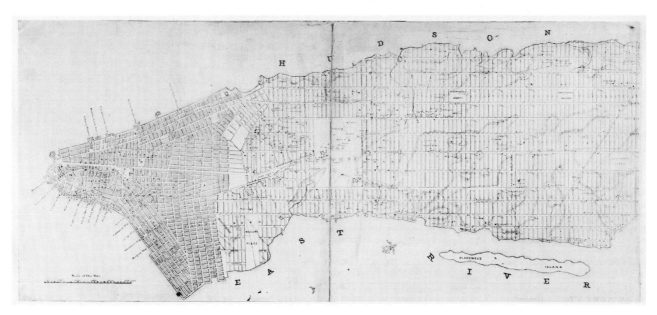

Figure 6.7 The 1811 Plan for
New York. This section of the
map shows the southern half of
Manhattan Island; the old,
unplanned city can be seen at
the tip of the island (courtesy of
Map Division, The New York
Public Library. Astor, Lenox
and Tilden Foundations)

Figure 6.8 Mid-Manhattan, 1864. This shows how some of
the original irregular property lines were superseded by the
grid. It also shows how the 1811 Plan was modified by the
creation of Central Park in the 1850s (Spann, 1988, p.32; map
courtesy of Map Division, The New York Public Library. Astor,
Lenox and Tilden Foundations)

The few reserved open spaces in the 1811 Plan were soon replaced by standard plots. However, settlement progressed so rapidly that pressure mounted for the establishment of public parks. In 1853, the state legislature authorized the city to acquire the necessary land in the centre of the island, amounting to 770 acres (312 hectares), for the construction of a park (see Figure 6.8). A competition held to design the new park was won in 1857 by Calvert Vaux and Frederick Law Olmsted, Sr; Olmsted had been much influenced by his youthful travels in England and would become famous nationally as a landscape-architect and an early exponent of city-planning. Central Park is still seen as a great achievement of civic design. Planned as the lungs of the city and an enhancement for neighbouring property values, the romantic, naturalistic, flowing design, using rather than overriding the island's topography, was an antidote to the rigidities of the grid. In Olmsted's phrase it was intended to provide 'conditions remedial of urban conditions' (quoted in Spann, 1988, p.30). Olmsted later promoted the idea of city parks plus parkways (attractively landscaped, unobtrusive roads such as those built through Central Park), linking planned garden suburbs into comprehensive, regional schemes of city design (White, 1988, p.97). The first of the US landscape-architects, Olmsted was an early representative of the wing of what would become formal 'planning' that saw planning as 'the potential instrument of a social-democratic urban regeneration'; this 'regenerationist' or 'communitarian' approach, in which schemes for wholesale change were proposed, can be contrasted with the 'meliorist' approach of many later planners, who worked incrementally with what existed (Simpson, 1985, pp.122, 129).

The zoning ordinance of 1916

Fired by the regeneration of Washington DC at the turn of the century, New York politicians moved to create a plan for the beautification of their city, newly expanded in 1898. Though stillborn at the time, a 1907 Plan expanded Olmsted's ideas and stimulated Charles Dyer Norton to establish a 'Committee on the City Plan' in 1914 (Johnson, 1988, p.170[4]). Norton was an insurance executive, newly arrived from Chicago where he had been involved in the Commercial Club's planning activities. One result of his committee's deliberations was the New York City zoning ordinance of 1916. Some of the earliest legislation affecting land-use dates from 1913, when the states of Minnesota, Wisconsin and New York permitted the designation of dedicated residential districts. New York City's is generally recognized as the first comprehensive zoning law in the USA. Chapter 4 described its regulations for the height of skyscrapers – long a concern of US cities, and particularly pressing in Manhattan where the blocking of light and air to the city's streets and neighbouring buildings by skyscrapers had become an area of conflict (Wilson, 1983, pp.90–94). The 1916 ordinance was also important because – for the first time in the USA – it tackled systematically the issue of land-use, defining separate districts for commercial, residential and unrestricted use (an 'unrestricted' district was one in which factories were permitted).

New York's zoning regulations were devised to solve a particular set of problems at a particular time, in a manner that would be legally robust and enforceable against challenges from property owners. Their weakness was that they were not devised in the context of a comprehensive plan for future developments, or even of population calculations. For example, the provisions of the zoning ordinance would have permitted the accommodation of the entire population of the United States in 1900 within New York City. Furthermore, they were made less effective because, despite

4   An abridged version of Johnson's essay appears in Roberts (1999), the Reader associated with this volume.

the delimitation of separate districts for distinct land-uses, mixed use continued to be permitted in all but the residential zones, thereby perpetuating many of the old problems. Such a misfit between zoning and planning would be characteristic of many cities' efforts in the inter-war years (Jackson, 1984, p.338; Scott, 1969, pp.153–63). Zoning proved to be a practical and attractive tool and New York's regulations stimulated measures elsewhere. In 1924, Herbert Hoover, secretary of commerce, who had trained as a mining engineer, published national guidelines on zoning in the form of a proposed standard act (a legislative framework that can be adapted to specific local circumstances); by this date, more than 218 municipalities had adopted zoning ordinances (Scott, 1969, p.194).

The Regional Plan of 1929

Norton, however, felt that the zoning ordinance was an inadequate outcome of the work of the committee; he therefore formed a private pressure group to promote the formulation of a plan for the whole region of New York. The idea was taken up by the philanthropic Russell Sage Foundation in 1921, and a team was put together to prepare a Regional Plan. The team was headed by Thomas Adams, a Scotsman who had worked in the British Garden City movement and become an adviser on urban planning to the Canadian government in 1914. The Plan, based on an extensive survey of regional social and economic trends and physical conditions, was published in 1929 when a Regional Plan Association (RPA) was incorporated to foster it. The Regional Plan allowed for even greater population growth than eventually occurred, and, like Chicago's Burnham Plan, it tackled the transport system; its aim was to knit together the whole vast region into an efficient unity, by constructing an effective rail and highway system and an integrated rail and harbour system. The Plan favoured a balanced, 'recentralized' city (that is, the aim was to reverse the tide of decentralization), opposing suburban sprawl, while at the same time fostering a radial highway network. In this way, though it did not anticipate the scale of motorization that occurred, the Plan made some attempt to take into account the implications of the motor car. The RPA, though influential, was a private pressure group which had to rely on the separate decisions of many different responsible authorities for the implementation of its plans. For example, Chapter 3 shows how Robert Moses, as commissioner for parks, was able to influence the infrastructural development of the city and, reciprocally, how he made use of the favourable climate created by the Plan.

It had weaknesses but, unlike the Burnham Plan, the New York Regional Plan linked the social, the technical and the economic:

> No longer could engineering or architectural works be promoted solely on narrow grounds of individual project efficiency. Relationships between public investments in infrastructure and the distribution of jobs and people would have to be taken into account. The plan and the regional survey persuasively demonstrated the need to consider complex connections among basic regional systems.
>
> (Johnson, 1988, p.192)

Planning in the suburbs – Radburn

The Regional Plan was not without its critics; foremost among them was Lewis Mumford, a journalist, publicist and founder in 1923 of the Regional Planning Association of America (RPAA), a loose coalition of like-minded architects (Clarence Stein and Henry Wright), engineers, economists (such as Stuart Chase) and sociologists (such as Catherine Bauer). In contrast to the meliorist approach of the New York Regional Plan (Johnson, 1988), the RPAA

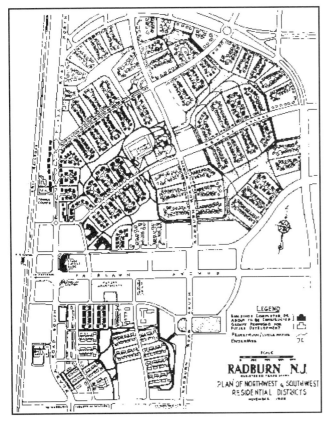

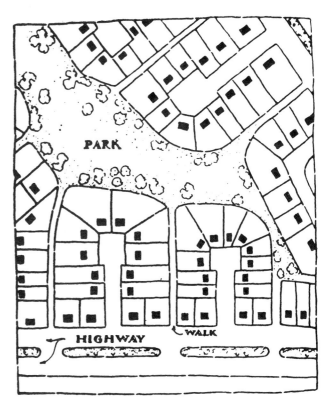

Figure 6.9 The Radburn Plan, 1928. The final plan for Radburn reveals the hierarchical circulation system (courtesy of The Radburn Archives)

Figure 6.10 The Radburn Plan, detail, as reprinted in the 1939 *FHA Manual* (reproduced from Birch, 1983, p.139, plate 7.7; courtesy of The Radburn Archives)

drew on Olmsted's communitarian principles (Simpson, 1985). Seeing town and country as interdependent, Mumford, who favoured the Garden City ideal of smaller-scale, decentralized development, vehemently objected to the 'motorization' of the city and its 'recentralization'. He believed these ideas to be destructive of his ideal – the organic, human-scale city network within the context of a humane regionalism – and felt that a regional plan should have regard to the requirements of the region as a whole, rather than start from what the metropolis required of the region (Wilson, 1983, pp.111–13). Adams's meliorist response was that he was interested in the art of the possible rather than attempting to implement an ideal (Johnson, 1988, pp.178–86).

Even as the New York Regional Plan was being prepared, the RPAA was fostering, with Mumford's involvement, a Garden City experiment at Radburn, near Paterson (NJ), just across the George Washington Bridge from Upper Manhattan (see Figure 6.9). The group consulted a wide range of people, including social scientists, as it formulated the plan for Radburn. The plan was therefore not primarily an architectural product, but 'reflected a multidisciplinary synthesis of the most current data and expert advice' (Birch, 1983, p.126). Radburn was designed as a town for the motor age, with the strict separation of vehicular traffic and pedestrians (see Figure 6.10). It consisted of a series of neighbourhood-like culs-de-sac,

> grouped in a super-block around a central park. The traffic highways border the super-block. The houses face the front yards and parks rather than the streets. The cul-de-sac roadways are service drives and give access to the rear of houses. Traffic passes by rather than among houses.

(Birch, 1983, p.139)

Furthermore, it was to be controlled by a community organization, the Radburn Association, which would 'administer the public lands, enforce restrictions, and supply supplementary municipal services such as recreation and day care activities'; these services and activities, which were to take place in purpose-built community centres, were intended to form so-called 'neighbourhood units', providing focal points for local identity (Birch, 1983, p.128). The timing was bad, as the Great Depression intervened and only about one-tenth of the scheme was built. However, as planning became a national issue under the New Deal (Franklin D. Roosevelt's uncle, Frederic Delano, who had been involved with both the Burnham Plan and the New York Regional Plan, was brought to Washington to take charge of federal initiatives), the Radburn model proved influential – including on the planners working for the Tennessee Valley Authority, the great New Deal energy, drainage and settlement project. One of them commented:

> Radburn stands out singly not because it is the biggest or most beautiful of cities but because it is the first tangible product of a new urban science … that seeks to make the places of man's habitation and industry fit the health requirements of his daily life … Radburn is not a theory, it is a demonstration.
>
> (Tracy Augur, quoted in Birch, 1983, p.129)

Aspects of the model were taken up by the New Deal's 'Greenbelt Towns' experiments under its Resettlement Administration, as well as in a later generation of private new-town developments in America.[5] Ironically, in the post-war period, the super-block neighbourhood concept of Radburn was applied to public housing, but as high-rise tower blocks in areas of slum clearance in a different legislative, economic and political context.

6.5 Summary

It has been possible to give only a few examples of the way in which technology became involved in the governance of cities during the nineteenth and early twentieth centuries. It was both regulated by and a tool of the new service-oriented city governments. In some areas, through regulation for common benefit, city governments promoted technological innovation. In others, technologies became part of the mechanism of government. And in one case, the practices and expertise developed in the course of the emergence of the technological infrastructure stimulated the establishment of a whole new context for the subsequent development of technologies in the city: planning.

Extracts

6.1 Freitag, J.K. (1912) *Fire Prevention and Fire Protection as Applied to Building Construction: a handbook of theory and practice*, London, Chapman and Hall, pp.11–23

Comparative losses in United States and Europe

… First, a comparison may be made between the conservation exhibited by European countries and the reckless waste permitted in the United States. From special reports of United States Consuls in Europe, it has been shown by the committee on statistics of the National Board of Fire Underwriters that the average per capita loss in six European countries for a period of five years was $0.33 …

---

5   And also in Britain, in Milton Keynes, the new town where The Open University is based (Clapson, 1999).

The total per capita fire loss in the United States for the five years ending with 1907 was $3.02, or nearly ten times as much as the European average quoted above.

Or, again, comparing cities with cities, the result in thirty foreign cities gave an average per capita loss of $0.61 as compared with $3.10 in the five years' average of 252 cities in the United States ...

Had the United States a per capita loss of $0.33 as given above for European countries, instead of an actual per capita loss of $2.51 for the year 1907 (based on a population of 85,532,761), then the total fire loss in the United States in that year would have amounted to only $28,623,290, or a saving, in fire waste alone, of $186,461,419 ...

Frequency of fires in United States and Europe compared

In such large cities in the United States as New York, Boston, Philadelphia, etc., the annual number of fires has been steadily increasing year by year, but in far greater proportion than the growth of population ... The frequency of fires has also increased of late years in such foreign cities as London, Berlin, and Paris, probably due to the increasing complexities of modern living; but whereas the total American fire losses have increased out of all proportion to city growth or expansion, fire losses in Continental cities have not materially increased. Thus the average fire loss in Boston is now about $2,000,000 while in an average European city of equal population the fire loss will seldom be found to range over $150,000, and this in spite of the usually marked superiority of our fire-fighting facilities.

The committee on statistics of the National Board of Fire Underwriters found that the number of fires per 1000 of population averaged 4.05 in cities of the United States, compared with 0.86 for similar cities in Europe.

Extent of fires in United States and Europe compared

In the investigations carried out by the Technologic Branch of the United States Geological Survey, it was found that a prominent cause of the tremendous fire waste in the United States was due to fires extending beyond the limits of the buildings in which they started. 'Exact figures as to the losses due to exposure were not obtainable, but the most conservative estimate indicates that at least 27 per cent of the losses resulted from fires extending beyond the building of origin.' [*] On the other hand ... it will be found that in such cities as Havre, Rouen, Milan, Rome, Brussels, Antwerp, and Leeds, Sheffield, and Bristol in England, every fire in the year 1890 was confined to the building in which it originated; while in Dresden, Florence, Vienna, and other cities, every fire was confined to the *floor* on which it originated ...

It must also be borne in mind that many of these results are obtained in spite of what Americans would consider the most inadequate fire-fighting facilities. Thus in Rome, where, in 1890, 328 fires were practically all confined to the *room* of origin, the extinguishment of fires was thus described by Consul-General Bourn:

> Buckets and fire extinguishers are chiefly used for extinguishing fires. If these are not sufficient, small hose, perhaps $1\frac{1}{4}$ inches in diameter, are brought into service. But the force of the water in many parts of the city is not great, although the supply is very abundant. If the hydrant pressure is not sufficient, small, portable fire engines are used, and in cases of great emergency there is one steamer, but, as it is so seldom required, no proper arrangements exist for bringing it into service. The last time the steamer was called out it was over two hours before it was ready to throw water on the fire ...

Causes of foregoing differences

The striking contrasts between the losses, frequency, and extent of fires in the United States as compared with European countries as given above, are due to four principal causes.

First: Differences in the view-point and in the civic responsibilities of the individual in the United States and in Europe, and the consequent laws or regulations which govern the individual.

[*] Herbert M. Wilson, *Transactions Am. Soc. C. E.*, vol.LXV, p.277.

Second: Differences in general character of buildings outside of congested areas.

Third: Differences in thoroughness of construction and maintenance.

Fourth: Differences in regulations and their enforcement regarding especially hazardous materials and conditions.

A candid enquiry into the first-mentioned differences, *viz.*, the individual view-point and responsibility, will disclose the fact that our national fire losses are principally caused by the moral attitude of the individual toward the phenomenon of fire waste.

European cities long ago learned the lesson that safety to the individual means safety to the whole community, and *vice versa*.

> They have learned that fire waste emanates in larger part from either criminal indifference or criminal intent, and that to this extent it is preventable through laws which go directly to the root of the evil by holding the individual citizen to a rigid accountability for every act of omission or commission which tends to increase the danger. In all parts of Europe where the Code Napoleon prevails, the law of Voisinage holds the landlord responsible for his negligence to all concerned, tenants or neighbors, and if fire originates from carelessness of tenant, he is held responsible to all concerned, landlord or neighbors. This law places the responsibility where it belongs and works automatically in making everyone interested in having his premises as safe as they can be made by human foresight. This is not only strictly logical, but in harmony with the attitude of every civilized government in dealing with the spread of contagion. †

It is just some such civic responsibility which is needed, and needed very soon, in all American cities and towns. Responsibility of the individual to the community, which will cause the individual to contribute to the public safety in matters of building construction by erecting structures which will not prove a menace to his neighbors; and responsibility of the community to the individual, in that those investors who improve their land by the erection of more costly and permanent structures shall not be allowed to suffer constant hazard through irresponsible neighbors who have no thought or care of their civic duties.

It is now a trite saying that 'fireproof buildings must stand in fireproof cities', but this statement contains the whole truth of the matter of fire protection. If American cities are not to suffer such conflagrations as have occurred at Chicago, Boston, Paterson, Baltimore, and San Francisco, besides many other lesser ones; if the realization of this tremendous financial drain is once grasped in an effort to lessen it; if it be admitted that isolated buildings surrounded by severe risks cannot withstand conflagration conditions, then the achievement of fire-resisting cities (or at least the congested areas therein) must be made possible by uniform fire-resisting construction throughout.

In the United States we are so prone to consider the rights of the individual that we are apt to overlook the rights of the aggregation of individuals. It is not denied that municipal building regulations adopted by any American city, requiring uniform fire-resistive building construction after any fixed date, would give rise to seeming injustices and hardships; but if laws requiring the remodeling of present risks were also rigidly enforced, in addition to laws covering the erection of new buildings, the hardships would soon be equalized, and benefit accrue to the community in the way of reduced fire losses, reduced insurance premiums, reduced expenses for maintaining fire-fighting equipments, and added security to life and property interests.

The second great cause of our excessive fire loss is to be found in the materials of construction employed in localities outside of the congested areas of large cities. Nearly all of our large American cities now have fire limits defining the congested areas within which frame buildings may not be erected, but, save in years when conflagration sweeps over some city, it is found that such congested areas do not contribute the greater proportion of the fire loss. Thus in the year 1907, when the actual fire losses to buildings and their contents in the United States amounted to about $215,084,709, the loss in brick, concrete, or other slow-burning construction totaled only $68,425,267, while double that amount, or $146,695,442 was on losses in frame buildings. ‡ In that year the total urban and rural losses were practically the same, but while the loss on *contents* was naturally greater in the urban property, still the loss on *buildings* was greater in the rural districts …

---

† Mr A. F. Dean, *National Fire Protection Association Quarterly*.

‡ The number of fires in brick, iron, and stone buildings was 36,140, while the number of fires in frame buildings was 129,117.

Throughout nearly all European countries, save in Norway and Sweden where wooden construction is prevalent, the erection of frame buildings is prohibited in all municipalities, and few are erected in rural districts. It is seldom that any considerable number of frame buildings are to be found, while a whole community of inflammable structures (as is common enough with us) is almost unknown …

Of course the conditions noted above are primarily due to the relatively high cost of lumber in European countries; while in the United States lumber has been available, cheap, and most readily adaptable to building uses.

Regarding the third cause of our fire waste, *viz.*, lack of thoroughness of construction and maintenance of fire protection appliances, it cannot be denied that, while our buildings are generally higher and larger than in European countries, yet they are more carelessly constructed and less efficiently inspected by proper authorities … while the maintenance and inspection of our fire protection auxiliary appliances is generally very perfunctory.

The fourth cause under discussion, namely the differences in regulations and their rigid enforcement regarding special hazards, may be admirably illustrated by comparing our recklessness and carelessness regarding such matters as lighting, heating, the care and storage of paints, highly inflammable liquids, and explosives, with the conditions obtaining, for instance, in Berlin, Germany, as given by the Consul-General to that city:

> Another important factor in the case is the strict and carefully enforced regulations concerning the storage, handling, and transportation of highly inflammable substances and explosives. The scrutiny of the building police extends to every detail of apparatus for heating and illumination. The wires of electric lighting plants must be inclosed, wherever they may be located inside a building, in non-combustible sheaths or tubing, with every practicable provision against breakage or short circuits. The construction and setting of stoves, the thickness of walls and floor foundations in proximity to stoves, furnaces, and fireplaces of all kinds, the construction of flues, ash bins, and chimneys, are all carefully regulated and subject to periodical inspection by the police. Gas stoves must be supplied with gas through fixed iron pipes; rubber tubing may not be used for that purpose. If any flexible tube is used it must be sheathed with asbestos. Finally, every chimney, whether in use or not, provided it is connected with an inhabited building, must be periodically cleaned by a member of the authorized force of chimney sweeps. The net result of the whole enforced system of construction, maintenance, and constant inspection is the practical immunity of Berlin from serious conflagrations and the important economies thereby secured in losses by fire and expenses of insurance …

Waste of structural materials

'It is evident that something must be done to stop the unnecessary waste of structural materials. Certain of these materials, such as wood and iron, are not inexhaustible by any means, but are even approaching exhaustion. In order to obtain the best use from these materials in the future, they must be used with a less lavish hand. Waste means increased cost in the very near future.

'… [A] study of the causes of waste of structural materials is evidently of prime necessity. The first source of such waste has been shown to be fires. A second source, and one closely related to fire losses, is that due to waste of iron and steel placed underground in city water mains or in pumping plants, on account of fire and conflagration protection.

'… It is believed that through dissemination of information as to the local availability of cement-making materials, of gravel and sand suitable for concrete construction, of clay suitable for brick- and tile-making, and through tests and investigations which will show the most appropriate method of mixing and proportioning these materials, and of designing them with the minimum amount of each material which may suffice its purpose, the cost of constructions will be reduced and the use of such materials be encouraged.

'Within the past few years marvelous strides have been made in the substitution of iron and steel for wood, due to the investigations of engineers, physicists, and chemists into the properties of these materials and the great amount of attention given to their fabrication by manufacturers and architects. More recently the engineering and technical professions have advanced to a great extent the uses of cement in concrete manufactures, but in a vastly greater period practically nothing has been done toward ascertaining the physical and chemical properties and the better modes of manufacture and use of the products of clay and stone. With these objects in view, the Government, as the largest consumer of such materials, is undertaking such tests and investigations as may develop the most suitable of these less perishable building materials for each particular use and locality. These tests have in view the establishing of the physical properties of these materials, the suggestion of improved methods of manufacture with a view to economy, improved methods of mining and marketing in order to improve the quality, reducing the quantity and cost, and extending the life of such materials. The investigations include the assembling of information relative to the most fire-resisting and fireproof forms of construction, the former for the prevention of conflagrations due to secondary or exposure fires and the latter for the prevention of the destruction of the buildings in which the fires originate.' §

6.2 Moody, W.D. (1913) *Wacker's Manual of the Plan of Chicago: municipal economy*, Chicago, Chicago Plan Commission, pp.95–112

The plan of Chicago: its purpose and meaning

The Plan of Chicago, as it has been worked out, is a plan to direct the future growth of Chicago in a systematic and orderly way. Its purpose is to make Chicago a real, centralized city instead of a group of overcrowded, overgrown villages. It means, when it is carried out, that Chicago will hold her position among the great cities of the world, that Chicago is to be given opportunities for indefinite growth in wealth and commerce, and that Chicago is to become the most convenient, healthful and attractive city on earth. History shows that this work will give to us, the owners and builders of Chicago, world-wide fame that will be everlasting …

Above everything else, the Plan of Chicago is concerned with our vital problems of congestion, traffic and public health. The plan will do away with congestion in the city and its streets, and so promote the health and happiness of all. It will make traffic easy and convenient, and so make it easier and cheaper to carry on business, thus increasing the wealth of the city and its people faster than will be possible otherwise. The plan will give Chicago more and larger parks and playgrounds, and better and wider streets, and thus make the whole people more healthy and better able to carry on the work of commerce and civilization in our great city.

All over the world today cities are growing as they never did before. Steam and electric transportation have made it easy to transport food for multitudes. Modern manufacturing methods draw large numbers of men together in cities to cheaply produce clothing, machinery and the varied supplies men need in their daily lives throughout the world. No country in the world, however, has given rise so rapidly to large cities as the United States, where it was shown by the census of 1910 that forty out of every one hundred people now reside in cities, and, of these, twelve reside in the three cities of New York, Philadelphia and Chicago.

… [M]en of science, devoting their lives to a study of the effect of city life upon humanity, declare to us that the physical condition of people in the cities, as compared with the people of the open country, is deteriorating. City life, they say, saps the energy of men, and makes them less efficient in the work of life. The remedy for this, they tell us, lies in providing increased means of open-air recreation, better sanitation in city houses, and more light and air in city streets. The Plan of Chicago provides for complying with this imperative demand. To preserve ourselves and our city by meeting this call for better health conditions is an aim of the Plan of Chicago …

In drawing the Plan of Chicago, the architects constantly kept in mind the needs of the future city in the three great elements of congestion, traffic and public health. They took the city as it has grown up and applied to it the needs of the future in transportation, in recreation and in hygiene.

Because we are a commercial people, and live in a great commercial city, first thought was given to transportation. The architect's first care, therefore was to create a proper system of handling the business of Chicago in its streets, and upon its street railways, its

---

§ Herbert M. Wilson, *Trans. Am. Soc. C. E.*, vol.LXV, p.287.

steam railroads and its water courses. The greatest part of the plan, then, refers to improving the existing streets, to cutting new ones where necessary, to arranging the city's railway and water terminals most effectively, and to the quick and cheap handling of all the business of Chicago.

This plan of transportation completed, the architects set about a plan of making Chicago more attractive, of providing parks for the people in the places where they should be provided, of giving the people recreation grounds, both within the city and in the outer district nearby, of improving and beautifying the lake front of the city, and so arranging all things that the future people of Chicago may be strong and healthy, and so ambitious to extend the fame and the commerce of their city.

Finally, in their planning, the architects recognized the need of giving the people of Chicago a way to express in solid form their progressive spirit. The people of Chicago have always been proud of their city, of its importance and its power. The architects strove, therefore to provide a means whereby the civic pride and glory of Chicago could be shown to the world in imposing buildings of architectural grandeur. Thus they provided a civic center upon a vast scale, to be improved with towering buildings serving as the seat of city government, uniting and giving life to the whole plan of the metropolis, and standing as a notice to the world of the tremendous might and power of a city loved and revered by its millions of devoted and patriotic citizens …

Solving Chicago's transportation problems

It has been seen how modern cities are governed in their growth and development by their facilities for transportation, and how Chicago has come to be a big city only because of its fine location in regard to the carrying to and fro of merchandise. We must recognize, then, that Chicago has become great largely by virtue of the railroads, and that upon the railroads it is dependent for its future growth and prosperity. Chicago is now the greatest railroad center in the world. Railway lines extend from the city in every direction. The problem, then, is to make these roads more effective in commerce, to bring them all together as one great machine in the service of the city.

Under modern conditions … the city which has the best and cheapest railroad service has an advantage in trade over every other city … Men operating great railroads are agreed that their greatest problem is to provide quick and cheap means of handling traffic in the great cities …

The great difficulty in moving freight in Chicago today arises from the lack of organization for handling the merchandise coming into the city over the various railroads, but intended for use in the country, or in other cities. Every month thousands of cars of goods are hauled into central Chicago over the various roads, switched to other railroads and drawn out of Chicago again unopened. Other thousands of cars are shipped into Chicago, unloaded in the center of the city and their contents carted through the streets to warehouses, only to be again carted away, loaded into cars, and shipped out of Chicago by wholesale merchants a few days or weeks later.

Under the Plan of Chicago, as drawn by the architects, all this wasteful effort and crowding within the center of the city will be ended. This is to be done by establishing upon the broad prairies southwest of Chicago a great freight and warehousing center. After this center is established all merchandise not intended for constructing buildings, for food or to be sold and used by the people of Chicago will be kept out of the city itself. The railroads will quickly and cheaply interchange traffic at the outside center. In the warehouses there all merchandise to be sold at wholesale and shipped to other towns and cities will be stored until time for shipment arrives, when it will be reloaded and started on its way without delay.

The advantages of this great common unloading and reloading station, where much of the work now done by manual labor downtown can be done by machinery, and where crowding of industries will not interfere with the business for which the center is planned, are apparent. All interests will benefit; the railways by increasing the effective use of their equipment by rapid loading and unloading, the merchants by avoiding the expense and delay of carting and handling products at the city's center, and by being enabled to more quickly and effectively serve their customers outside the city; and the city at large by relief from the crowding in the streets due to the teaming [sic] congestion, by a saving upon its pavements, and by a cessation of the noise and smoke nuisance growing out of the removal of so many freight engines to the new freight handling locality.

The freight traffic of Chicago has been carefully studied by experts, and it has been found that ninety-five per cent is handled by the railroads, and five per cent is done by water. For the care and development of water transportation the Plan of Chicago provides for the building of two great systems of docks, one at the mouth of Chicago river, the other at the mouth of Calumet river, adjoining South Chicago. The Chicago river docks are planned to care for vessels bearing cargoes of package freight, such as furniture, sugar and manufactured products. Those at South Chicago would be more extensive, and would care for the vessels carrying bulk cargoes, such as coal, grain and ore.

Connecting the two harbors and the outer freight center, and running through the center of the city that it may serve the merchants and manufacturers, there is provided in the plan an underground freightway to be operated by electricity. This railway would bind together and make complete all the parts of the great machine of traffic intended to make limitless the possibility of business expansion for Chicago. If it be made an underground way, it may be connected with the present tunnel system serving the central part of Chicago to carry out a complete system of underground distribution.

In drawing the plan of Chicago as affecting transportation, the designers did not neglect the detail of improving passenger service in and out of Chicago. Good order among the passenger stations was considered a necessity, and a pleasing arrangement of the stations was decided upon ... Thus the railway stations would be grouped in a semicircle about the heart of the city. An elevated, surface or subway loop railway could be built to connect the stations, and give easy transportation from any one of them to the others.

One of the great results of carrying out this plan of arranging the passenger stations will be to extend the present crowded business center of Chicago to as far south as Twelfth Street, and as far west as the river. The need of this extension is already great and is growing more acute every day. The unpleasant and distressing conditions of crowding already suffered in the loop district of Chicago can be relieved in no other way; and in no other way but undertaking a work for economy in trade and transportation can Chicago fulfill her destiny as mistress of the commerce of half a continent. To gain this relief, and to provide a certain way by which the railways which have made Chicago great may give the people of Chicago the greatest possible amount of service in future, is a powerful reason urging our immediate adoption of the Plan of Chicago.

Perfecting our street system

... One of the first needs of the future city is a perfect street system. There must be enough streets to easily accommodate the traffic of the millions who are to live in the city. They must be wide enough to insure comfort in traversing them, and they must run in the right directions to enable the people to go from place to place quickly. We must realize that lifetimes are made up of minutes, and that to save minutes means to lengthen life. Thus we can justify the spending of millions of dollars today if it means saving time for millions of people in years and centuries to come ...

The architects, in their Plan of Chicago, have prepared for great changes in the street plans of the city. They have provided for wider streets throughout the city, for widened and improved boulevards, and they have laid out, as absolutely necessary to a properly arranged and permanent city, a large number of new streets and ways, in the creation of which it will be necessary to destroy or remove hundreds of buildings in the crowded parts of the present city.

Circuits – An idea of the plan is to establish several circuits of existing thoroughfares and to improve them so traffic can move freely and directly about the city's center.

Quadrangle – The first constructive work of the Chicago Plan Commission – the foundation for all that is to follow – is to carry out the circuit idea by completing the great quadrangle formed by Twelfth street on the South, Halsted street on the West,

Figure 6.11 Proposed Twelfth street improvement at its intersections with Michigan avenue and Ashland avenue. [This is a typical illustration from the Burnham Plan showing block-like buildings, wide through-streets punctuated by a diagonal boulevard and provision for the railways] (copyrighted by the Commercial Club; reproduced courtesy of Chicago Historical Society ICHi-29514)

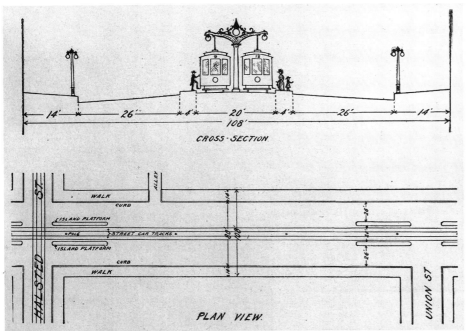

Figure 6.12 Plan of the new Twelfth street. Top diagram shows sidewalks 14 feet and 26-foot roadway on either side, with a 20-foot strip in the center for double street car line and bracket trolley poles. Plan below shows 'islands' on both sides of the car tracks at the intersection of each street for safety in entering and leaving street cars. Dimension of island 4 by 60 feet. Car tracks at street level, with free access for traffic to either side (prepared for the Chicago Plan Commission; reproduced courtesy of Chicago Historical Society ICHi-29513)

Chicago avenue on the North and Michigan avenue on the East. These four streets are destined to bear the heaviest traffic of any streets in Chicago ...

Twelfth street, the first section of the quadrangle, is being developed under the Chicago Plan. It is to be widened and arranged to bear easily a heavier traffic than that which now makes it a badly congested street. The widening of Twelfth street is the initial step in the constructive work of developing the plan as a whole and bears a relationship to the whole scheme of street construction and street widening.

The necessity for the improvement of that street lies in the fact that it is the only through thoroughfare ... connecting the west side with the downtown district. The actual heart of the city's population today is a little north of the corner of Twelfth and Halsted streets. Traffic and the city's growth are gradually moving in a southwesterly direction. Adequate provision must be made for a suitable outlet from that district to the present business center of the city ...

It is not intended to boulevard the street but to make it a clean, wide, business thoroughfare with a double, rapid-transit surface street car line down the center, and on it might be established stations of all the great railroads entering the city from the east, south and southwest. It is hoped that the railroads may be induced to locate terminals south of Twelfth street between State street and the river …

The improvement might properly be designated as both a 'local improvement' and a 'general benefit.' The Chicago Plan Commission has made a strong recommendation for a large 'general benefit' in order that a large percentage of the cost of the improvement shall be borne by the whole city, in which case the matter of a bond issue to defray the city's part of the cost will have to be referred to the people in a referendum.

Public sentiment generally and the united support of the press is back of this movement. Every citizen of Chicago should aid with his influence and vote at the proper time in the realization of this improvement, thus insuring the success of the first practical step in carrying out the Plan.

Halsted street, a section of the quadrangle, it is predicted, will, in time to come, carry an enormous traffic. It is so situated that its usefulness, already great, may be very much increased. It is selected as, next to Michigan avenue, the most important north and south traffic thoroughfare. Under the Chicago Plan the street would be widened, paved properly and developed as one of the great central business streets of the future city.

Chicago avenue, a section of the quadrangle, already one hundred feet wide, will serve for a long time the traffic it will be made to carry. Crowding of vehicles is not so great upon the north side of the city and is not increasing so fast as in other sections. It will connect with the proposed Michigan boulevard extension at Pine street, completing the first circuit of improvement in our streets.

Next to the quadrangle, by far the most important in the plans for streets are those relating to the time and distance saving diagonal thoroughfares which Chicago needs so badly. The city is fortunate in having, as a foundation for this system of diagonal streets, a large number of such thoroughfares … These thoroughfares, for the most part, are the routes followed for hundreds of years by the Indians, whose wide trails were developed first into country roads leading to the settlement at Chicago, and gradually became city streets as Chicago extended its limits with its growth.

The aim of all the present diagonal streets is to bring all traffic to the center of the city. The effect of this, in the present city, is to produce congestion and crowding that is fast growing unbearable. It is apparent, then, that the city's great need now is for diagonal streets to give more direct routes throughout the city, and so stop the crowding of traffic into the city's business center.

The architects found, in studying the street system of Chicago, that the greatest need is for diagonal streets to connect the widely extended west side of the city with the north and south sides. Ways must be opened, it is seen, by which the people of the various parts of Chicago may go quickly and conveniently to other parts. Ways must be created by which the people of the great west side may go directly to the lake front parks on either the north or south sides, and thus have greater freedom in recreation.

The Plan of Chicago, as completed, provides a complete system of diagonal streets which, if they were in existence today, would be used by hundreds of thousands of people with a saving of time and effort which cannot even be estimated. Millions of people will use these streets in the future. Their creation will remove every limitation now existing to prevent the city's growth in population.

Two of the great diagonal streets the architects have proposed will, upon the adoption of the Plan, be cut through the central part of the city … The function of these thoroughfares will be to give traffic which now crowds into the business center of the city a direct route between the north and south central districts and the central west side territory. The second street described will provide, also, a direct route connecting the northwest and southeast districts of Chicago …

Besides cutting the new diagonals and widening the principal thoroughfares within the city, there will be constructed thoroughfares along both sides of Chicago river and its branches. This work, in all probability, will be the finishing labors of the city in its street transformation plans. The water fronts of the great European cities are thus improved and beautified. Broad ways, according to the Plan of Chicago, will surely line both banks of the river branches … This street construction is to be on a plan so laid as not to interfere with the use of the river in commerce and trade, the driveways being elevated and running above the roofs of low warehouses and wharves lining the edges of the stream.

The city's streets would be linked together and unified by the wide semi-circular boulevard drive described in the next chapter as more properly a part of the vast park system by which the city is, according to the Plan of Chicago, to become the most attractive and healthful great city the world has ever known.

A system of outer roadways and highways encircling the city to connect the various parts of Chicago with each other, with the center of the city and with the outlying sections, is considered a great need. With the exception of five per cent, a perfect system of outer highways – called 'turnpikes' in the old days – now exists. Partly disconnected roads form ninety-five per cent of the proposed system today.

References

BIRCH, E.L. (1983) 'Radburn and the American planning movement: the persistence of an idea' in D.A. Krueckeberg (ed.) *Introduction to Planning History in the United States*, New Brunswick, NJ, Center for Urban Policy Research, Rutgers University, pp.122–51.

BOUMAN, M.J. (1987) 'Luxury and control: the urbanity of street lighting in nineteenth-century cities', *Journal of Urban History*, vol.14, pp.7–37.

CLAPSON, M. (1999) 'Technology, social change and the planning of a post-industrial city: a case-study of Milton Keynes' in D.C. Goodman and C.W. Chant (eds) *European Cities and Technology: industrial to post-industrial city*, London, Routledge, in association with The Open University, pp.279–300.

CONDIT, C.W. (1982, 2nd edn) *American Building Materials and Techniques from the First Colonial Settlements to the Present*, Chicago, University of Chicago Press.

FREITAG, J.K. (1912) *Fire Prevention and Fire Protection as Applied to Building Construction: a handbook of theory and practice*, London, Chapman and Hall.

GREENBERG, A.S. (1998) *Cause for Alarm: the volunteer fire department in the nineteenth-century city*, Princeton, NJ, Princeton University Press.

HALL, P. (1988) *Cities of Tomorrow: an intellectual history of urban planning and design in the twentieth century*, Oxford, Blackwell Publishers.

HAMMACK, D.C. (1988) 'Comprehensive planning before the Comprehensive Plan: a new look at the nineteenth-century American city' in D. Schaffer (ed.) *Two Centuries of American Planning*, London, Mansell, pp.139–65.

HOLLERAN, M. (1996) 'Boston's "Sacred Sky Line": from prohibiting to sculpting skyscrapers', *Journal of Urban History*, vol.22, 552–85.

JACKSON, K. (1984) 'The capital of capitalism: the New York Metropolitan Region, 1890–1940' in A. Sutcliffe (ed.) *Metropolis 1890–1940*, London, Mansell, pp.319–53.

JOHNSON, D.A. (1988) 'Regional planning for the great American metropolis: New York between the world wars' in D. Schaffer (ed.) *Two Centuries of American Planning*, London, Mansell, pp.167–96.

KRUECKEBERG, D.A. (1983) 'Introduction' in D.A. Krueckeberg (ed.) *The American Planner: biographies and recollections*, New York, Methuen, pp.1–34.

MONKKONEN, E.H. (1988) *America Becomes Urban: the development of US cities and towns 1780–1980*, Berkeley, Cal., University of California Press.

MOODY, W.D. (1913) *Wacker's Manual of the Plan of Chicago: municipal economy*, Chicago, Chicago Plan Commission.

NYE, D.E. (1992) *Electrifying America: social meanings of a new technology, 1880-1940*, Cambridge, Mass., MIT Press.

PETERSON, J.A. (1979) 'The impact of sanitary reform upon American urban planning, 1840–1890', *Journal of Social History*, vol.13, pp.83–103.

PLATT, H.L. (1991) *The Electric City: energy and the growth of the Chicago area, 1880–1930*, Chicago, University of Chicago Press.

REPS, J.W. (1965) *The Making of Urban America: a history of city planning in the United States*, Princeton, NJ, Princeton University Press.

ROBERTS, G.K. (ed.) (1999) *The American Cities and Technology Reader: wilderness to wired city*, London, Routledge, in association with The Open University.

ROSEN, C.M. (1986) *The Limits of Power: great fires and the process of city growth in America*, Cambridge, Cambridge University Press.

SCHLERETH, T.J. (1983) ' Burnham's *Plan* and Moody's *Manual*: city planning as progressive reform' in D.A. Krueckeberg (ed.) *The American Planner: biographies and recollections*, New York, Methuen, pp.75–99.

SCHULTZ, S.K. (1989) *Constructing Urban Culture: American cities and city planning, 1800–1920*, Philadelphia, Pa., Temple University Press.

SCHULTZ, S.K. and McSHANE, C. (1978) 'To engineer the metropolis: sewers, sanitation, and city planning in late-nineteenth-century America', *Journal of American History*, vol.65, pp.389–411.

SCOTT, M. (1969) *American City Planning since 1890: a history commemorating the fiftieth anniversary of the American Institute of Planners*, Berkeley, Cal., University of California Press.

SHOWALTER, W.J. (1923) 'The automobile industry', *National Geographic Magazine*, vol.44, pp.337–414.

SIMPSON, M. (1985) *Thomas Adams and the Modern Planning Movement: Britain, Canada and the United States, 1900–1940*, London, Mansell.

SPANN, E.K. (1988) 'The greatest grid: the New York Plan of 1811' in D. Schaffer (ed.) *Two Centuries of American Planning*, London, Mansell, pp.11–39.

SUTCLIFFE, A. (1981) *Towards the Planned city: Germany, Britain, the United States and France, 1780–1914*, Oxford, Basil Blackwell.

TARR, J.A. (1989) 'City at the point' in S.P. Hays (ed.) *Essays on the Social History of Pittsburgh*, Pittsburgh, Pa., University of Pittsburgh Press, pp.213–63.

TARR, J.A. (forthcoming) 'Transforming an energy system: the evolution of the manufactured gas industry and the transition to natural gas in the United States, 1807–1954' in O. Coutard (ed.) *The Governance of Large Technical Systems*, New York, Routledge.

WERMIEL, S. (1996) 'Nothing succeeds like failure: the development of the fireproof building in the United States, 1790–1911', Ph.D. dissertation, Massachusetts Institute of Technology.

WHITE, D.F. (1988) 'Frederick Law Olmsted, the pacemaker' in D. Schaffer (ed.) *Two Centuries of American Planning*, London, Mansell, pp.87–112.

WILSON, W.H. (1983) 'Moles and skylarks' in D.A. Krueckeberg (ed.) *Introduction to Planning History in the United States*, New Brunswick, NJ, Center for Urban Policy Research, Rutgers University, pp.88–121.

WILSON, W.H. (1989) *The City Beautiful Movement*, Baltimore, The Johns Hopkins University Press.

WRIGLEY, R.L., Jr (1983) 'The Plan of Chicago' in D.A. Krueckeberg (ed.) *Introduction to Planning History in the United States*, New Brunswick, NJ, Center for Urban Policy Research, Rutgers University, pp.58–72.

Chapter 7: THE 'CAR CRISIS' IN THE LATE TWENTIETH-CENTURY CITY

by Philip Steadman

7.1 Introduction

In 1903 Dr H. Nelson Jackson and his chauffeur Sewall K. Crocker made the first ever coast-to-coast crossing of the United States by car. They drove a brand new Winton automobile, and travelled from San Francisco to New York in sixty-three days (see Figure 7.1). Today one could drive it in four or five. Or one could hop on a plane, and be there in the day.

The first telephone line between the Atlantic and the Pacific coasts was built in 1915. A radio-telephone link between the USA and Europe was inaugurated in 1928; and by the 1930s the oil magnate J. Paul Getty could run his California business from the comfort of his European hotel suite. But the first trans-Atlantic cable, laid in 1956, had only the capacity for some ninety simultaneous conversations between the two continents, and it could take a day or more to get a connection (Cairncross, 1997). Today a subscriber can call up any of more than 700 million telephones world-wide, and be connected in most cases almost instantly. The cost of a long-distance call within the USA is not greatly different today from that of a local call in many other countries, and it is expected that these very low prices will soon apply across the globe.

In the twentieth century, the world shrank. And two key inventions in particular helped it to shrink – the petrol engine and the telephone. Many commentators expect that the most recent explosion of innovation in telecommunications – satellite links, mobile telephones, computer networks,

Figure 7.1 Dr H. Nelson Jackson, heir to the Payne's Celery Tonic fortune, and his chauffeur Sewall K. Crocker at a stop on their journey from San Francisco to New York in 1903, the first coast-to-coast crossing of the continent by car (reproduced from Flink, 1988, p.94; photograph: courtesy of the National Museum of American History, Smithsonian Institution, Washington DC)

the Internet, digital television, and the 'convergence' of these various devices and systems into all kinds of hybrids – will serve only to accelerate the shrinking process. Geographers and economists are even speaking of the imminent 'death of distance' (Cairncross, 1997; Couclelis, 1996).

In this and the following chapter we will look at some of the ways in which these two clusters of technologies – road transport and telecommunications – affected the form of cities in the twentieth century, and could mould them in the future. It might be thought that both had a predominantly dispersing and decentralizing effect – allowing suburbs to spread, firms to split their operations between many sites, and cities to become ever larger, until the boundary between town and country became blurred into continuous 'subtopia' or 'urban sprawl'. The truth seems to be more complicated, as we shall see.

Car travel has had hugely liberating effects and has brought to many people opportunities for employment, a wider choice of shops, and many kinds of entertainment and leisure pursuit previously only within reach of a privileged few. But it has not been an unmixed blessing, as we all know. The physicist Freeman Dyson says he was greatly fascinated as a boy by E. Nesbit's story *The Magic City*, a place where if you wish for anything, you can have it (Dyson, 1979). There is a special rule, however, applying to people who wish for a piece of machinery: they are obliged to keep it and go on using it for the rest of their lives. Dyson sees this as a larger allegory of technological 'progress': 'Nikolaus Otto plays for a few years with a toy gasoline engine and – bingo! – we all find ourselves driving cars.' We all wished for them, and now we cannot wish them away again.

With the rapid growth in car-ownership and road travel in many parts of the world during the last half century have come the risks of injury and death from accidents, continuing congestion, noise, air pollution and – most recently recognized of all these penalties – the threat of global warming. Some writers have argued that new computer and communications technology can play a part in solving these problems – by helping to manage and control road traffic, or by making it unnecessary to travel in the first place. So the two groups of technologies might have not reinforcing but complementary and counteracting effects. We will examine the merits of these arguments.

The place where the world shrank first was the USA. This is the country where the telephone and the Internet began; where automobiles were first mass-manufactured and where their use spread fastest; where commercial air travel was pioneered. So we will concentrate on the American story. These technologies and their impact have since spread to many parts of the world, of course, and cities around the globe are now connected one to another by ever-increasing flows of people, goods and information. So we will also turn briefly from time to time to survey the international scene.

7.2   Changes in the density of cities in the twentieth century

Some major European cities were able to grow in the late nineteenth century by extending into the countryside along tramways and suburban railway lines. Earlier chapters in this volume have described similar processes at work in US cities – trends greatly accelerated by the growth in car-ownership from the 1920s. Typically, as population migrated from the crowded city centre to the more spacious suburbs, so residential densities (numbers of people per square mile) dropped in the centre and rose on the periphery. In the 1950s the economist Colin Clark made one of the first comparative statistical analyses of urban population densities, an exercise which, before computers, required long and laborious measurements (Clark, 1951).

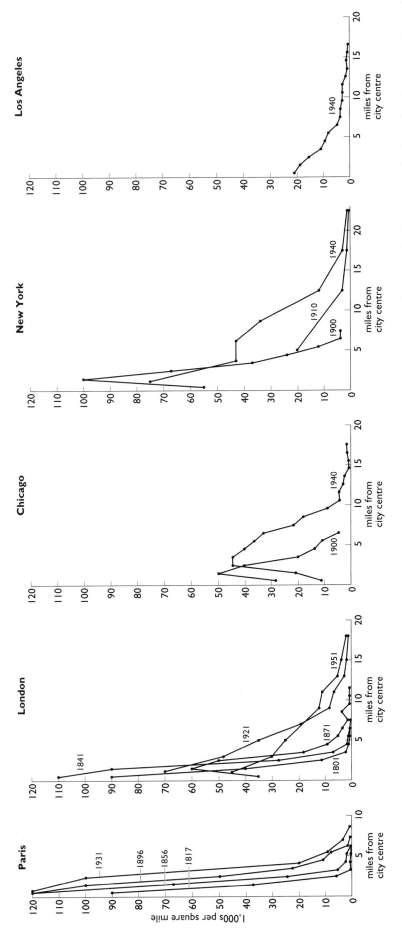

Figure 7.2 Population densities in five cities from the early nineteeth to the mid-twentieth century. The horizontal scale shows the distance in miles from the city centre; the total diameter of the city is therefore obtained by doubling this distance. The vertical scale shows population density in thousands per square mile. (In Clark's originals the latter are plotted on a logarithmic scale) (adapted from Clark, 1951, pp.492–3; by permission of Blackwell Publishers)

Clark counted the total numbers of people living in a series of concentric rings around the centre, in different cities world-wide, and for various dates from the early nineteenth century to the 1950s. His results for Paris, London, Chicago, New York and Los Angeles are illustrated in Figure 7.2. Take the example of London: in 1801 the urban area was just six or seven miles in diameter, and the density at the centre was about 90,000 per square mile. By 1871 the residential density at the centre had dropped to well under half this figure, as the centre emptied out and London had spread across more than twenty miles. By the 1950s the capital had a diameter of more than thirty miles. As Clark says, it is possible to make two generalizations about his statistics. First:

> In every large city, excluding the central business zone, which has few resident inhabitants, we have districts of dense population in the interior, with density falling off progressively as we proceed to the outer suburbs.

And second:

> In most (but not all) cities, as time goes on, density tends to fall in the most populous inner suburbs, and to rise in the outer suburbs, and the whole city tends to 'spread itself out'.
>
> (Clark, 1951, p.490)

Clark takes something of a technological determinist position, in interpreting these results, as we might expect, in terms of the costs and speeds of different means of urban transport. (For him, transport is the 'maker and breaker of cities', as he puts it in the title of another of his very influential papers: Clark, 1957.) If transport costs are high or mechanized transport is simply unavailable, then there is overcrowding at the centre. In Paris or London in the early 1800s, most people travelled by foot, and this effectively limited the radius of the city to the three or four miles that could be covered in an hour. There were some horse-drawn bus services, but they were not extensively used. The effects are visible in Clark's statistics for both Paris and London: densities *rose* in both cities in the first half of the nineteenth century, as more and more people packed into their restricted land areas.

There is a dearth of reliable statistics on urban densities before the nineteenth century and the institution of regular censuses. But for what they are worth, such records as do exist seem to suggest that densities in London and Paris have never been much greater than 150,000 per square mile (about 60,000 per square kilometre) in the entire history of those cities. The highest figures reached in the nineteenth century were in the Lower East Side of Manhattan around 1900, where the density rose – just in this small area of New York – to some 350,000 people per square mile. In Manhattan as a whole, around 40,000 tenements housed more than one and a half million people. Congestion was worse even than in central Bombay. Such densities have only since been exceeded in certain areas of Hong Kong in the 1980s where the housing is in skyscraper blocks of apartments (Hall, 1996).

It is not easy to appreciate quite what these extraordinary numbers meant for the quality of everyday life. A typical figure for the average density of a US city today might be 5,000 people per square mile, or less. Lower Manhattan in 1900 was therefore some sixty to seventy times denser than this. How were people packed in so tightly?

Except for the streets, no open land was left unused for building. Ramshackle multi-storey tenements were built as tall as they could be without going to the expense of elevators. The buildings were also extremely deep, and were pushed as close to their neighbours as the constraints of access allowed. Often several families shared one apartment, and many people slept in each room. Running water and shared water-closets were only provided outside the blocks (see Figure 7.3). In some of the most wretched lodging houses, the homeless paid a few cents to sleep in dormitories on suspended strips of canvas. 'All-night restaurants' offered refuges where the completely destitute slept, seated in chairs.

Figure 7.3 Rear alley of a block of tenements in Elizabeth Street, Manhattan, 1903, in a part of the city where population densities rose to some of the highest levels ever recorded anywhere in the world. Running water was available only from hydrants in the street. The roof on the right covers a row of forty-six communal water-closets (photograph: the Jacob A. Riis Collection, Museum of the City of New York)

Figure 7.4 (overleaf) illustrates a notorious type of New York apartment block known from its plan shape as the 'dumb-bell'. The dumb-bell was by no means the worst of its kind. Indeed it represents a relatively late stage in the evolution of the tenement type, when philanthropic and public-health concerns had already brought about a few reforms: it was the winner of a competition for 'improved' designs. The dumb-bell had narrow slots down either side to provide the only daylight and 'fresh' air to ten of the fourteen rooms on each floor. These shafts quickly became dumps for rubbish. One can hardly imagine what it must have been like to try to sleep in these airless rooms in the sweaty heat of a New York summer.

Figure 7.4 'Dumb-bell' tenements in New York. *Left* slot at the side of the block, the only source of daylight. *Above* block of dumb-bell tenements (reproduced from de Forest and Veiller, 1903, pp.10, 14 (opposite); photograph: courtesy of The Bodleian Library, University of Oxford, 247554.d.2)

The need to be within walking distance of the workplace had severe consequences for the form of working-class housing in the great cities of the USA and Europe. I have described the situation in Manhattan in some detail to emphasize this point. New York was the most extreme case, but the 'back-to-back' houses of Britain's northern industrial cities were hardly much better; nor were the *Mietskasernen* ('rent barracks') of Berlin.[1]

We know some of the attractions that pulled people out to the suburbs once the tram or the car gave them the chance: the possibility of houses instead of apartments, with gardens and greenery, clean air, peace and quiet, personal privacy. These statistics on central-area densities at the turn of the century serve to remind us of the powerful forces that were pushing people from behind – the terrible overcrowding, the insanitary and disease-ridden conditions, the crime, the noise. It was not the unhappy poor who escaped first, but the upper- and middle-income groups who could afford larger houses and the cost of commuting journeys. But as incomes rose in the USA through the century, the flow to the suburbs became a flood, especially in the years following the Second World War. Although the central-area slums were demolished, the poor remained behind. But even they were rehoused in private or public developments at significantly lower densities than before. So the process of 'spreading out' that Clark describes was further accelerated.

To go back to transport economics: only when the average speed of urban transport rose, and the costs fell, could the areas of cities begin to be enlarged. In the second half of the nineteenth century, the growth of London, New York, Chicago (and other cities) was made possible first by horse-drawn buses and tramcars, then by steam or electric commuter railways – on the surface in Chicago, on both surface and underground in New York and London. Notice from Figure 7.2 (p.203) how the middle of London in 1871 had a *lower* density than the ring 1–2 miles from the centre. A similar pattern is to be seen in the 1900 statistics for New York, and in both sets of statistics for Chicago. This is because office buildings and shops on high-value land had begun to force residents out of the very centres.

---

[1] See Chapter 6 of Goodman and Chant (1999), a companion volume in this series.

Clark thought that the comparatively low density at the centre of Los Angeles in 1940, and its very flat 'density gradient', reflected its origins as a city that grew up in the automobile age, and had always relied on the automobile. We have seen this view challenged in Chapter 3, which showed the early dependence of Los Angeles on public transport. Nevertheless the flattening out of the density gradients for London and all three US cities in the 1940s and 1950s, and their spread across a diameter of thirty or forty miles, must surely signal the growing use of the car.

It would be good to have equivalent figures to Clark's for dates since 1950. Clark introduced his 1951 paper by remarking that 'this branch of geography does not yet seem to have received adequate quantitative study'. Unfortunately it does not appear to have received much since. One purely theoretical difficulty is that, as cities have sprawled, it has become ever more difficult to define their precise geometrical centres; indeed they have in many cases developed multiple sub-centres, as neighbouring towns and villages have been overtaken and absorbed. So the simple, circular, single-centred model of urban form no longer fits.

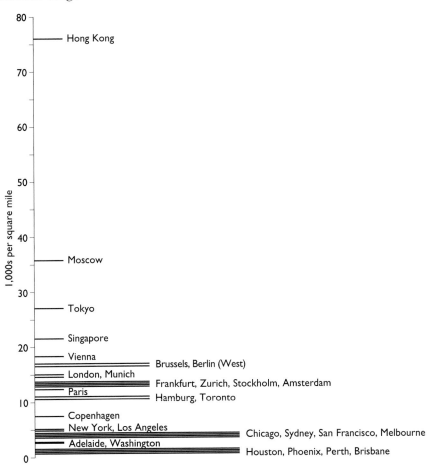

Figure 7.5 Average population densities in twenty-nine cities world-wide (adapted from Newman and Kenworthy, 1989a)

However, a study was made in the 1980s of the *overall* population densities of major cities across the world (Newman and Kenworthy, 1989a). Here the higher densities of 'central' areas are averaged out with the lower densities of the suburbs, but comparisons with Clark's results are still illuminating. They confirm what has been mentioned already: that many larger US and Australian cities now have average densities under 5,000 people per square mile (see Figure 7.5). These are cities where surrounding land has been cheap and

Figure 7.6 Typical detached housing in the USA; photograph: © Aerofilms Limited.

plentiful, and whose transport systems are dominated by the private car. The housing here would typically take the form of detached single-storey dwellings on large plots, with garages and gardens (see Figure 7.6).

We can conceive, then, of a spectrum of urban densities, from 1900 Manhattan at one end, to late twentieth-century Houston or Phoenix at the other. As the century progressed, so US housing tended to move along this spectrum, towards ever greater space-standards per person, both inside and around the dwelling.

With the exception of Copenhagen, the population densities of the European cities in Figure 7.5 lie between 10,000 and 20,000 people per square mile, reflecting historical legacies of higher-density housing and more effective public-transport services. Of the cities shown, the highest densities are in: Tokyo, where houses are small and tightly packed; Moscow, where car-ownership in the 1980s was still very low and a majority of the population lived in apartments; and Hong Kong and Singapore, both of which are held in by their small land areas and island sites.

7.3   Growth in car use in the USA and Britain

We saw in Chapter 3 that the USA came to lead the world in ownership and use of motor vehicles. By 1925 there was one car for every four Americans. (The entire population could, in theory, have taken to the road at once!) The countries closest behind the USA – Canada, New Zealand and Australia – were also places with plenty of land and dispersed, but relatively prosperous, farming populations. Europe still trailed far behind, with typical rates of ownership in the 1920s of around one car for every twenty or thirty people. There was a world-wide *fall* in ownership after 1929 as a consequence of the Great Depression. Even so, automobile registrations dropped in the USA only by 10 per cent between 1929 and 1932. People held on to their automobile while they sold everything else, since mobility held out some residual hope of getting work.

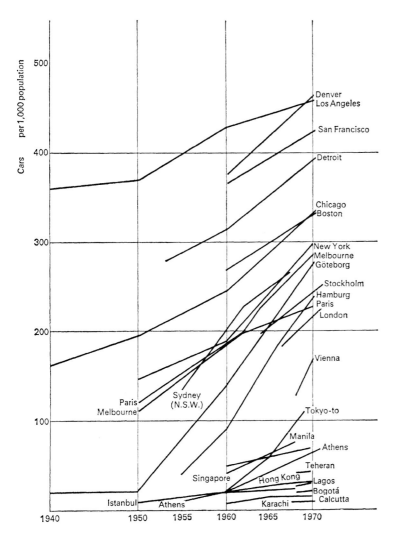

Figure 7.7 Rates of car-ownership in twenty-six major world cities, 1940–70 (reproduced from Thomson, 1977, p.80)

Figure 7.7 shows the growth in rates of car-ownership after the Second World War in a selection of twenty-six major cities around the world. (The sample is not exactly the same as in Figure 7.5, but there is extensive overlap.) For most of the period 1940–70, Los Angeles, which by the mid-1920s led US cities in car ownership, retained its premier position – being just overtaken by Denver in 1970. Overall, the cities with the highest levels of car-ownership are those of the American South-West, followed by the great Eastern cities. Levels in European cities are typically half of those in Los Angeles or Denver. The relationship of these statistics to the density levels in Figure 7.5 is very clear. They mirror each other: the higher the car-ownership, the lower (in general) the density. Note the position of Tokyo in Figure 7.7, where car-ownership remained low until the mid-1960s. Only since then, with the growing strength of the Japanese economy and its dependence on car manufacture, has the picture changed. This helps to explain Tokyo's high density today, and the great difficulties it has had in adapting rapidly to the car.

Increasingly in the second half of the twentieth century, the car came, in the USA and Europe, to displace other modes of land transport. Rail-passenger traffic reached a peak in the USA in the 1920s and has been in decline ever since, except for a brief revival during the Second World War. Commercial air travel was only just beginning before the war. Figure 7.8 (overleaf) shows changes in means of passenger transport in the whole country between 1960 and 1990.

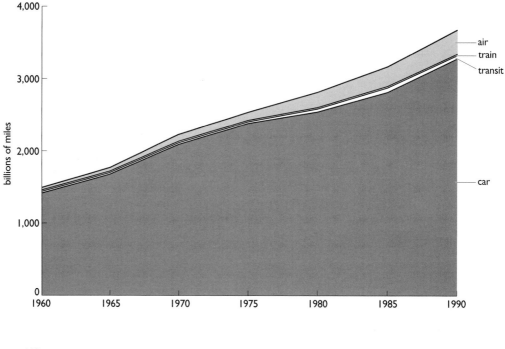

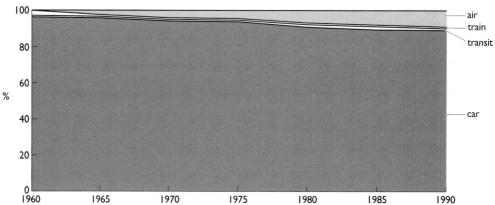

Figure 7.8 Total travel in the USA in passenger miles by mode, 1960–90. ('Passenger miles' are obtained by multiplying the mileage travelled by a vehicle, by the number of passengers it carries; for example, a car with three occupants travelling ten miles counts as thirty 'passenger miles'.) 'Car' here includes motorcycles, vans and small trucks; 'transit' is mostly bus and light rail; 'train' is intercity/Amtrak. The two graphs show (top) actual distances in billions of miles and (bottom) the same figures as percentages (plotted from data in United States Department of Transportation, 1999)

(Rural and inter-city traffic is included, as well as trips within cities.) Even in 1960 more than 95 per cent of all mileage travelled in the USA was by car. Since that date the percentage taken by train has continued to shrink, with a minor resurgence in the 1980s. There was also some modest growth in the use of urban public transport – bus and light rail – in the 1980s. But these effects are almost too small to see in the graphs. Meanwhile the long-distance journeys previously made in the USA by rail have been made increasingly by plane.

Over the thirty-year period shown in Figure 7.8, the total distance travelled by car in the USA more than doubled. There was a slight faltering in the rate of growth of total mileage in the mid-1970s, which can be attributed to the Arab–Israeli War, the emergence of the Organization of Petroleum Exporting Countries (OPEC), and the resulting sharp rise in the international price of petrol. But throughout the period, the trend has always been relentlessly upwards.

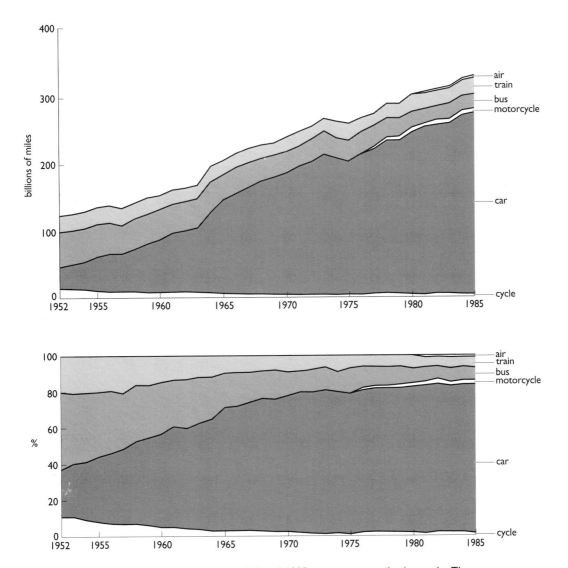

Figure 7.9 Total travel in Britain between 1952 and 1985, in passenger miles by mode. The two graphs show (top) actual distances in billions of miles and (bottom) the same figures as percentages (plotted from data in *Transport Statistics: Great Britain*, London, Her Majesty's Stationery Office)

Europe has followed the US pattern of growth in car use, but with a delay of perhaps thirty years. Figure 7.9 shows changes in passenger transport in Britain over a broadly similar period and in the same format as the US statistics of Figure 7.8. In the 1950s public transport still dominated in Britain. It was only around 1960 that the total distance travelled by car came to account for more than half of all mileage. Train travel declined somewhat during the 1960s following the 'Beeching cuts' and the closure of many rural lines. What remains are the intercity journeys and commuting by rail. The proportion of total mileage travelled by car in Britain has levelled off, since around 1975, at about 80 per cent – somewhat lower than the figure for the USA of 95 per cent. This difference can probably be explained by the fact that the densities of cities are typically higher in Europe than in the USA, as we have seen, and the population is not nearly so dispersed.

The most striking feature of Figure 7.9 is the extraordinary growth in car mileage in Britain – a more than fivefold increase over thirty years. No wonder that town centres are clogged, and that the road-building programme has failed to keep pace.

One caveat should be issued about the statistics in Figures 7.8 and 7.9: they ignore completely one means of transport which is, still, immensely important in cities – human legs. (The statistics for the USA also omit bicycling.) This is because they are collected by government departments whose main interests are in traffic engineering. Precisely because pedestrian movements are less often the subject of surveys, it is more difficult to know their numbers. But, for example, Thomson (1977) estimated that around half of all trips in large cities were then made on foot or by bicycle. (Since they are almost all short, they represent on the other hand only a small fraction of total 'passenger miles'.)

7.4 Transport technology and urban form

In 'Transport: maker and breaker of cities', Clark (1957) argued that successive transport technologies have allowed cities to grow, and have made possible distinctively different patterns of land use. But in turn the form of the city created by one means of transport has often made it difficult to introduce its successor. That is to say, land uses and transport technologies have got out of step. The basic patterns of roads and buildings are changed more slowly than are the mechanical characteristics and performance of vehicles. Thus, for example, many European cities have found it difficult to adapt to the car, and have pedestrianized their historic centres, so returning these small areas in effect to their pre-industrial mode of transport. The higher densities of European cities compared with those in the USA and Australia (Figure 7.5, p.207) provide some crude evidence of these historical constraints.

Influenced by Clark's analysis, the planning theorist Peter Hall claimed that cities throughout the world had experienced 'four successive crises of transport technology and urban form, of which the fourth is still in progress' (1994, p.s79). These issues have been described in a previous volume in this series (Goodman and Chant, 1999) and in previous chapters in this volume. The first crisis was the constraint on growth, and the extreme urban densities, resulting from the absence of mechanical transport. This was overcome with horse trams and steam railways, and resulted in the spread, from 1870, of 'streetcar suburbs' (Warner, 1978). Hall refers to the US and European cities that developed in this form as *early-public-transport cities*. He distinguishes them from cities that reached populations of more than a million by the 1890s, and relied on electric trams and underground or surface electric railways to help solve the problems of carrying huge flows of commuters in and out of their centres. London and New York were the first examples, by the 1920s and 1930s, of this type of *late-public-transport city*. Berlin and Paris followed.

Hall's third 'crisis' was experienced first by Los Angeles, where the unprecedented urban form created by the electric railway network paved the way – almost literally – for the complete takeover of the city's transport system by the car. (Many of today's freeways – urban motorways – follow the old railway lines.) Los Angeles, which grew rapidly only from the turn of the twentieth century, has always lacked the kind of strong, dense central area that characterized the old 'walking cities' of Europe and the US East Coast. The original town of Los Angeles – now the 'downtown' of the metropolis – remained, certainly, the principal focus of employment and commercial activity until the late 1920s. But from its very beginnings the emerging supercity consisted of a wide scattering – an archipelago – of small separate towns, which were linked by electric railways. For a time during the 1910s and 1920s, the land in the 'holes' of this network remained undeveloped, and housing was concentrated within walking distance of the railway lines. But the intervening areas rapidly became filled with low-density, single-family houses, once rising car-ownership made them accessible.

The ever-growing numbers of Angeleno automobile-drivers then began to converge, from the early 1920s, on workplaces and shops downtown, and created insoluble problems of congestion (Bottles, 1987). The situation was made worse by the fact that electric trains and automobiles shared the same road space, and were not separated on different tracks or different levels. Department stores, speciality shops and manufacturing began to decentralize and move to the suburbs, where they had room to expand and to provide parking, and were more easily accessible by car. Meanwhile, as a corollary, the electric trains were losing patronage. They were slowed by the downtown congestion, and found it increasingly difficult to serve the very dispersed patterns of residential and then commercial and industrial development.

The city responded, first with a programme of improvement of the existing roads and then, in the years immediately before the Second World War, with an ambitious plan for a complete new network of high-speed motorways, with restricted access and separated from the remainder of the street system. A start was made on the freeways in the 1940s, but their real growth came in the 1950s and 1960s. Los Angeles became the prototype, then, of what Hall terms the *auto-oriented city*. Other cities in the American West – including Dallas, Houston and Salt Lake City – have followed LA's lead.

The fate of the electric railways in Los Angeles foreshadowed the difficulties experienced by many urban public-transport systems, once in competition with the car. Bus services and commuter rail services are most effective where they can pick up riders along relatively densely-populated corridors and deposit them in places where many workplaces or shops are concentrated. They have greater problems in serving lower-density suburbs, where the origins and destinations of people's trips are widely scattered. A positive feedback effect is created, whereby the use of cars allows lower densities of development; public-transport services become ever less competitive, with fewer routes, fewer vehicles and higher fares; more passengers shift to cars; use of public transport declines further; and so on. At the same time lower densities encourage people to use their cars for journeys – to local shops, taking children to school – that would previously have been made by bicycle or on foot.

There is a further problem for public transport. Its main role has traditionally been to carry commuters in the morning and evening rush hours. This means that transit authorities must provide a vehicle fleet sufficient to carry passengers without overcrowding at the peaks – a fleet that then sits idle for the remainder of the day. Some off-peak demand is created by shoppers, students, retired people and others without cars. But as car-ownership has grown, so this demand too has been increasingly met by private transport, thus further exacerbating the imbalance between peak and off-peak use of bus and rail.

In a comprehensive review, *Great Cities and their Traffic*, Michael Thomson (1977) classified the land-use patterns and transport networks of a number of generic types of late twentieth-century city. These are schematic diagrams and fail to do justice, of course, to the geographical and historical complexities of real individual cases. They can serve a useful purpose nevertheless for making broad comparisons. Thomson's archetype of 'full motorization' is shown in Figure 7.10 (overleaf: top). Complete dependence on the car for passenger transport is possible, he says, without great technical difficulties in cities of up to 200,000 or 300,000 people – by upgrading radial roads, building inner ring roads, and providing central car-parking. Oldenburg (Germany) and Norwich (England) are examples of historic European towns that have adapted in this way, and Milton Keynes an example of a new town designed in effect for full motorization.

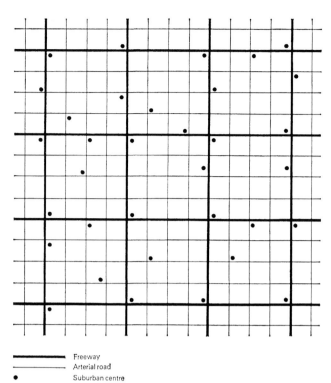

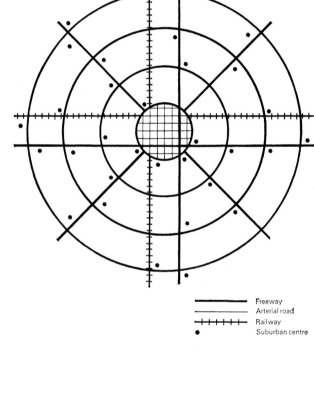

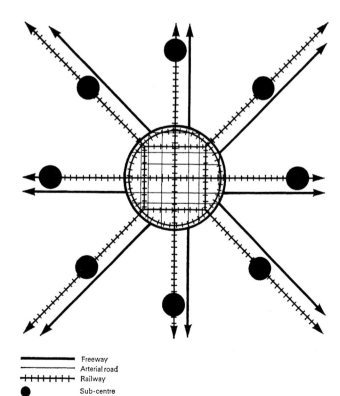

Figure 7.10 Archetypal forms of late twentieth-century cities and their transport networks, according to Thomson (1977). *Left* 'full-motorization' archetype. *Below* 'weak-centred' archetype. *Bottom* 'strong-centred' archetype. Note that 'freeways' are urban motorways without crossings or traffic-lights, while 'arterial roads' are high-capacity streets or boulevards, usually with several lanes in each direction (reproduced from Thomson, 1977, pp.100, 131, 163)

But in larger cities of this full-motorization type, employment, shopping and entertainments must be dispersed, and served by a network of motorways, together with a secondary system of high-capacity arterial roads (as shown in Figure 7.10, top). As Thomson points out, although Los Angeles is famous for its freeway network, by the 1960s Chicago had more miles of freeway in a smaller land area, and New York had three times the length of the LA network in an area only 40 per cent greater:

> Los Angeles's claim to be the automobile city rests more on her secondary roads than her freeways. The traffic capacity of her tight grid of six- or eight-lane arterial highways is possibly unrivalled.
>
> (Thomson, 1977, p.104)

Thomson's 'strong-centred' archetypal form (Figure 7.10, bottom) relates to those cities that had large, well-established centres before 1900, such as Paris, New York, Tokyo and (to a lesser extent) London. They have retained high concentrations of employment and services in these centres: typically they have more than one million downtown office workers, concentrated in ten square miles (twenty-five square kilometres) or less. For this reason a strongly radial transport network is still required. It would be quite impossible for anything but a small fraction of the central-city workers or shoppers to gain access by car. Far too many lanes of motorway would be needed, and even if this were feasible, there would be no room for sufficient exit ramps or parking. The commuter flows can only be accommodated by rail – up to 40,000 people per hour on one track – where the total space occupied by a single traveller is tiny compared with that needed for a car. If such a city is to retain and strengthen the dominant position of its centre, it can probably not allow an outer orbital such as London's M25, which draws development towards the edges – although it can allow an orbital motorway close-in, to serve the centre, such as Paris's *Boulevard Périphérique*. If sub-centres are to be allowed in the strong-centre city, they should ideally lie on the radial routes, as Thomson's diagram indicates (Figure 7.10, bottom).

Between the poles of full motorization and strong centres, lie those cities that Thomson characterizes as 'weak-centred'. Figure 7.10 (middle) illustrates his schematic diagram. A weak-centred city typically has between 250,000 and 500,000 downtown office workers, and brings them in by a mixture of car, bus and light rail, with perhaps some radial commuter train services. Thomson gives several examples – Melbourne, Copenhagen, San Francisco, Chicago, Boston – but there would be many other regional centres in the USA, and minor European capital cities, that would fit this description. The weak-centred city, according to Thomson, is not a stable form, and has a historical tendency to move, as it grows, either towards the strong-centred type or towards full motorization.

In US cities whose history goes back before this century, their centres – whether 'strong' or 'weak' – continue to be significantly served by public transport. However, their outer suburbs have increasingly become as dependent on automobiles as Houston or Los Angeles. Robert Cervero (1985, 1989) distinguished three 'waves' in the process of US suburbanization. First was the movement of the upper- and middle-income population to largely residential suburbs in the first half of the century, which I have already described.

Second was the migration of shops and industry to the urban periphery in the decades after the Second World War, in order to be nearer the suburbanites who made up their clientele and workforce. In this way the suburbs were transformed from dormitories to places with most of the facilities of the traditional city – albeit much more dispersed. With cheaper land, and catering to a car-borne population, out-of-town shopping centres and commercial 'strip' developments were typically built no higher than a single storey, and surrounded by acres of parking. Airports, again sited on the urban periphery because of their demand for land and their noise nuisance, attracted warehousing and hotels.

Cervero's 'third wave' has been the arrival in the suburbs, from the 1980s, of large numbers of high-tech industries and office workers, especially in the financial, insurance and property sectors. This third wave has arrived with particular rapidity: in 1980, 57 per cent of all office space in the USA was in city or town centres and 43 per cent in the suburbs. Just six years later the position was reversed, with 60 per cent in suburbs, 40 per cent in the city centres (Cervero, 1989). Some of the new office space has been in high-rise buildings, sprouting improbably along what only twenty or thirty years previously were leafy backroads (Figure 7.11). Other developments have been built in low-rise, landscaped 'campus-style' settings (Figure 7.12). Cervero refers to such buildings as 'horizontal skyscrapers' (perhaps 'groundscrapers' would be better).

Figure 7.11 An 'edge city': Tysons Corner, Virginia. *Top* at about the time of the Second World War (photograph: Fairfax County Library, Virginia Room Archive). *Bottom* Tysons Corner in 1988 (photograph: Craig Herndon copyright © 1988 *The Washington Post*)

These new suburban commercial and office developments began to coalesce into major sub-centres that could almost qualify as cities in their own right – except that they had no mayors or formal civic identities, and in some cases even lacked names. Joel Garreau (1991) called them 'edge cities'. His rough-and-ready definition of an edge city required that it have five million square feet or more of lettable office space, 600,000 square feet or more of lettable retail space, more jobs than bedrooms, and had been nothing like a 'city' thirty years ago. On these criteria he identified around 200 existing or rapidly emerging edge cities in the USA in 1988. Some examples are provided by the area around Route 128 and the Massachusetts Turnpike near Boston, the Schaumberg area west of Chicago's O'Hare Airport, and Irvine in Orange County south of Los Angeles. Many edge cities are larger than the downtowns

Figure 7.12 'Horizontal skyscraper' (in the middle distance) providing offices for over 5,000 workers, surrounded by low-density suburban housing, in San Ramon, California (reproduced from Cervero, 1989, p.35 by permission of Professor R. Cervero)

of existing major cities: for example, the oxymoronically named Perimeter Center outside Atlanta is bigger than downtown Atlanta.

Suburban commercial centres and edge cities have brought about new patterns of automobile travel, first seen in Los Angeles, where cross-town shopping and leisure trips, and 'cross-commuting' between sub-centres, have come increasingly to replace the radial in-and-out flows typical of the old 'strong centres'. This is Hall's ongoing fourth 'crisis': road congestion has spread from downtown to the suburbs. Motorized and weak-centred cities such as Houston, San Francisco and Denver came, in the 1990s, to experience solid traffic moving at 10–12 miles per hour on orbital and cross-town routes. Greatly increased volumes of traffic have been shifted on to what were previously minor rural or residential roads. The fact that edge cities are generally being built at very low densities obliges their inhabitants to travel exclusively by car. Although all of the elements of the traditional city are present, they tend to be zoned into large single-use shopping centres, 'office parks' and housing estates, rather than being mixed at the fine scale typical of the high-density 'strong centre'. This means that even a trip from the office at lunch-time to go shopping or eat out means getting into the car.

7.5 Growth in world mobility

It might be thought that the enormous growth in car use across the world in the last half-century has resulted in people spending more of their waking hours travelling, but curiously this is not the case. Traffic planners and sociologists make 'time budget' studies in which they ask people to fill in diaries recording the amounts of time they spend on different activities including travel. As one would expect, there are large differences between individuals, with travelling sales representatives and long-distance lorry drivers

Figure 7.13 Time spent travelling by people in a number of continents and sizes of settlement, as shown by studies made between the mid-sixties and early nineties. The vertical scale shows hours per day spent in travel as an average per person. The horizontal scale shows income (US dollars). The 'travel-time budget' is highly stable and does not vary with income (adapted from Schafer and Victor, 1997a, Figure 2; by permission of Massachusetts Institute of Technology)

at one extreme, and housebound people at the other. But when averages are taken across whole populations – and provided walking and cycling are included along with motorized travel – then it turns out that the 'travel-time budget' is remarkably stable, working out at just over an hour per day (Marchetti, 1994; Zahavi, 1981). This figure has stayed much the same since such studies were first made in the 1960s. It does not vary significantly with income: it applies to African village-dwellers and US city-dwellers, to Europe, Latin America and the Far East. Figure 7.13 shows results from a selection of international studies, in which hours spent travelling per day have been plotted against average incomes. In nearly all societies, whatever the average income, people travelled for between one and one and a half hours per day.

What happens, as incomes rise, is that people spend roughly the same amount of time travelling, but they go *faster* and *further*. Instead of walking, they buy bicycles; they then give up the bicycles and go by bus; they buy cars and abandon the bus. During the twentieth century the same modes of transport themselves got faster as mechanical improvements were made to vehicles, and road surfaces and rail track were upgraded. The post-war automobile on the USA's Interstate Highways was capable of much higher average speeds than the Model T. The European and Japanese high-speed trains run at more than twice the speed of the fastest conventional expresses.

The US transport analysts Andreas Schafer and David Victor brought together historical statistics on travel patterns for eleven regions around the world, to create an extraordinarily illuminating picture of the growth of 'global mobility' from 1960 to 1990 (Schafer and Victor, 1997a, 1997b). Figure 7.14 shows their statistics for total distance travelled per person per year by all motorized modes – car, bus, train and aircraft. They call this quantity 'traffic volume'. The graph of Figure 7.14 plots traffic volumes in different groups of countries against average income per person, for the thirty years in question. It is clear that in all parts of the world, traffic volume has increased in proportion to income. Thus in 1960 the average North American earned just under $10,000 and travelled 7,500 miles (12,000 kilometres). By 1990 both income and total distance had roughly doubled.

The relationship between traffic volume and income is more variable in developing countries. For instance, in China average income tripled between 1960 and 1990, but the total distance travelled went up (it seems) tenfold. This however, according to Schafer and Victor, is partly a consequence of the way in which the statistics are collected. In these countries the poorer people, who in the past only walked or cycled, are increasingly travelling instead by bus or

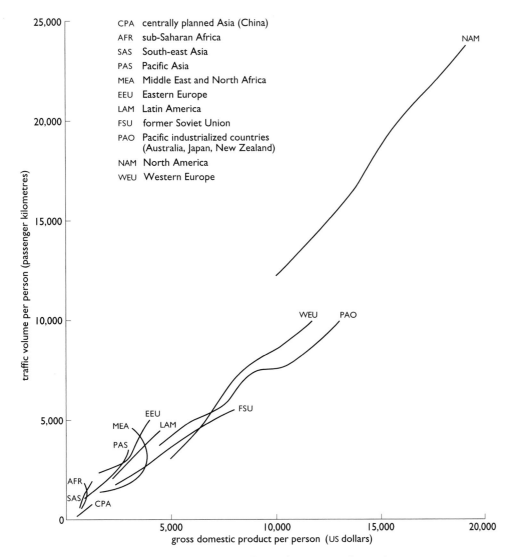

Figure 7.14 Total distances travelled per year by car, bus, train and aircraft, in various regions of the world (vertical scale), compared with average income in US dollars (horizontal scale), 1960–90. There is a close relationship: as incomes rose, so distances travelled increased in proportion (adapted from Schafer and Victor, 1997a; by permission of Massachusetts Institute of Technology)

train. Travel surveys are notoriously bad, as already mentioned, at recording distances travelled on foot and by bike. For this reason these non-motorized modes are omitted from Schafer and Victor's figures. Hence the apparent large increases in distances travelled.

These international findings confirm what transport economists, notably Zahavi (1981), had previously argued: that people tend to spend a roughly predictable fraction of their incomes on travel. In developing countries, where most travel is by public transport, by bicycle or on foot, this proportion is typically 3–5 per cent. In the richer, industrialized countries the figure rises, with growing use of cars, to stabilize somewhere at 10–15 per cent of income, at levels of car-ownership around one for every five people. Nearly all member countries in the Organization for Economic Co-operation and Development (OECD) have now made this 'automobile transition'.

We therefore find a global pattern of behaviour, in which people continue to devote a roughly constant amount of time to travelling; but as their incomes rise they use the extra money, and the same amount of time, to travel at ever

Figure 7.15 How total motorized travel has been divided between different modes of transport, as average incomes have changed in selected regions of the world, 1960–90. For each region, total distances travelled have been divided between four modes – cars, buses, conventional trains, and a 'high speed' category that combines high-speed rail with air travel. These shares are expressed as percentages, on the vertical scales (adapted from Schafer and Victor, 1997a, Figures 9, 14 and 15; by permission of Massachusetts Institute of Technology)

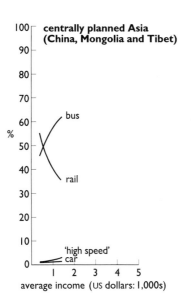

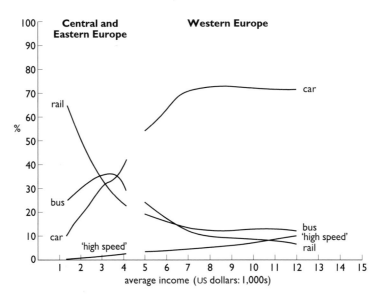

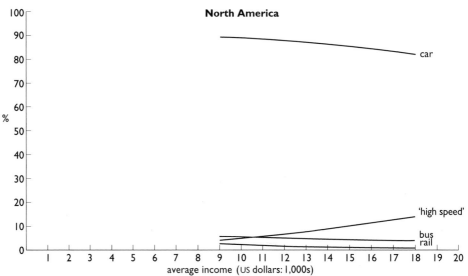

higher speeds. When, as in Figure 7.15, Schafer and Victor's data are separated by different regions and into different transport modes, they reveal what appears to be a common process of change. Of the four cases illustrated, North America with the highest average income is the furthest along the path; Western Europe follows; then come Central and Eastern Europe; while China, Mongolia and Tibet are at the earliest stages in the process. In effect the four regions are at different stages along broadly similar historical paths (which is not to say that there is any historical inevitability requiring that future patterns of mobility in China become like those in the USA).

The modes of transport in Figure 7.15 are divided into four categories – cars, buses, conventional low-speed railways (shown just as 'rail' in the figure), and a 'high speed' category that combines air travel with high-speed rail travel. In percentage terms, bus and rail have been in decline in recent decades in the USA, as we know, while 'high-speed' travel – in the USA exclusively by air – has started to take a share of traffic volume from cars. In Western Europe the same transition is occurring from cars to 'high-speed', but in this case to high-speed trains as well as to planes. 'Low-speed' rail travel continues to take a large but fast-decreasing share in Central and Eastern Europe; bus travel has peaked and is going into decline; and car use is growing very rapidly. In China, Mongolia and Tibet the total volume of traffic is more or less equally divided between rail (declining) and bus (still growing), while car use has barely started to take off. In Schafer and Victor's description:

> At low incomes (below $5,000 per capita), motorized travel is dominated by buses and low-speed trains that, on average, move station-to-station at approximately 20 to 30 kilometers per hour. As income rises, slower public transport modes are replaced by automobiles, which typically operate door-to-door at 30 to 55 kph and offer greater flexibility. (These average speeds, which vary by region, are lower than the posted speed limits because of congestion and other inefficiencies.) The share of traffic volume supplied by automobiles peaks at approximately $10,000 per capita. At higher incomes, aircraft and high-speed trains supplant slower modes. At present, aircraft supply 96 percent of all high-speed transport, flying airport-to-airport at about 600 kph.
>
> (Schafer and Victor, 1997b, pp.37–8)

Although there are certain similarities in these processes of change that have gone on in countries around the globe, there are nevertheless some marked regional differences. Schafer and Victor attribute the most significant of these to differences in geography and the typical densities of cities. In the USA, as we saw in Figure 7.8 (p.210), car travel grew to the point, in the 1960s, where it accounted for 95 per cent of all mileage by motorized modes. This was because the country had plentiful supplies of land for cities to grow and roads to grow with them. The percentage of all travel represented by cars has never risen to such a level in Western Europe, where population densities are higher and cities more constricted. Here the car's share of 'traffic volume' has peaked at about 70 per cent and 'is poised to decline'. Schafer and Victor predict that in the richer Asian countries, notably Japan, the peak in the car's share of all travel will be even lower, at around 55 per cent, and public transport will continue to provide the remainder. This is because of typical urban densities as much as three times higher than in Western Europe (see Figure 7.5, p.207).

Schafer and Victor found that the total volume of passenger traffic in the world was nearly five times higher in 1990 than in 1960, and was divided in the following proportions: cars 53 per cent, buses 29 per cent, 'low-speed' railways 9 per cent, and 'high-speed' 9 per cent. On the assumption of constant travel-time budgets, and using projections of income by region, they go on to make some alarming predictions of future growth in traffic volume up to the year 2050. But this is a history book, not an exercise in future-gazing, so at this point we must part company with them.

7.6   *Motor vehicles and photochemical smog*

All this car travel has had its costs, and not just in money. During the 1940s a brownish-yellow haze started to appear in the blue skies over Los Angeles, especially on still and sunny days in summer. The problem became serious towards the end of the decade. The authorities put the blame on oil-burning factories, and shut down many of these during a crisis in 1947. Urban air pollution was already well known to Londoners as 'smog', a term coined by the *Daily Graphic* at the turn of the century to mean a combination of smoke and fog. London's smogs were caused mainly by coal smoke from open fires in houses. They too became acute in the late 1940s and early 1950s, culminating in the winter of 1952 when 4,000 were killed.

Meanwhile, back in Los Angeles at the California Institute of Technology, a team led by Dr Aarlie Hagen-Smit had discovered that industry was not the only culprit, and that exhaust fumes from vehicles were playing an increasing role. The particular geographical and meteorological conditions of Southern California, and the high levels of Angeleno road traffic, created a specially unfortunate combination. The sunshine and high temperatures that had attracted so many to California in the first place, were themselves part of the cause of the air pollution.

The action of sunlight on nitrogen oxides and unburnt hydrocarbons (from vehicles and industry) produces ozone, mixed with a cocktail of other gases and dust particles. This is 'photochemical smog' (Elsom, 1996). The pollutants can irritate the eyes and throat, cause headaches and respiratory problems, and – as has since been discovered – exacerbate asthma (although perhaps not cause it). Ozone is harmful not just to humans and animals, but can stunt the growth of plants, and may be one cause of pollution damage to forests.

Under certain conditions, especially in river valleys or plains (as in Los Angeles), the pollution can become trapped beneath a 'temperature inversion'. If the sky is clear at night, the slopes of the hills radiate heat and become cooler, chilling the air around them. This cold air then slips down the slopes and lies in the valley, producing a layering of cold air below and warmer air above. The cold air is denser, the warm air less dense. This means that the smog created at ground level cannot rise upwards and become dispersed, as it otherwise would be.

Figure 7.16   Smog in Manhattan, August 1970 (photograph: UPI/Popperfoto)

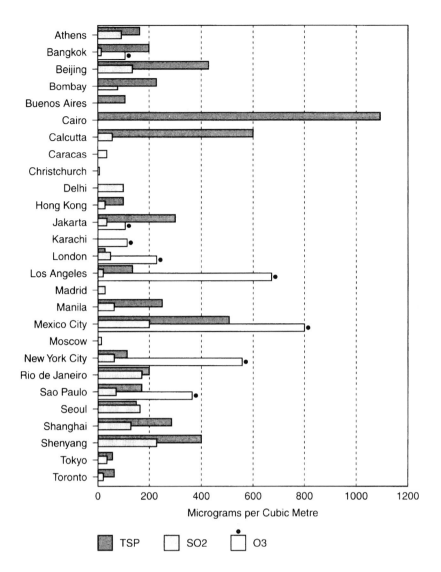

Figure 7.17 Urban air pollution across the world. The bars show levels of total suspended particulate matter (TSP), sulphur dioxide (SO$_2$) and ozone (O$_3$) in twenty-seven cities; data on ozone were available only for eight cities (reproduced from Elsom, 1996, p.8, based on United States Environmental Protection Agency, 1994, p.136; by permission of Earthscan Publications Ltd)

The city of Athens has had recurrent smog problems for this reason, as has Mexico City. (Oil burned for domestic heating has added to the problems caused by cars in both these cities.) In Athens in the 1980s an estimated six to ten people – mainly the very old and very young – were dying each day in the summer from the effects of the *nefos* or smog. In Mexico City the authorities responded in 1990 by requiring that only cars with odd-numbered licence plates be used on one day, and only those with even-numbered plates the next. During a specially serious crisis in 1991, all schools and factories in the city were closed (McShane, 1997).

Since the 1960s smog has become endemic in many cities in both developed and developing countries. In the 1990s it was estimated that air quality was bad enough to affect the health of about one half of the world's urban population (Elsom, 1996). Figure 7.17 shows levels of three air pollutants in the late 1980s in a selection of cities world-wide. The high ozone levels in Los Angeles, Mexico City, New York and São Paulo stand out. Many Asian cities are especially polluted.

At one time it was thought that London might remain immune from photochemical smog. Ironically, the Clean Air Act of 1956 got rid of the 'pea-souper' fogs by banning the domestic burning of coal, so making the capital much more sunny and preparing London to join the other victims of car-based air pollution. Meanwhile it has emerged that other emissions from vehicles can also be injurious to health, in particular fine particles from diesel exhausts, which have been implicated in lung disease and cancer.

Car manufacturers in the USA responded only slowly to the emissions problem, as to the issue of vehicle safety, until forced by government (Flink, 1988). California was the first state to introduce legislation to control vehicle emissions, requiring modifications to crankcases in all new cars sold from 1963, and controls on exhausts from 1966. The Motor Vehicle Air Pollution and Control Act of 1965 resulted in national standards comparable to those set in California. By the 1970s most US car manufacturers were fitting catalytic converters for controlling exhaust emissions. This is a technology adapted to large conventional petrol engines; smaller, imported cars could meet US standards in other ways. Cleaner petrol, more efficient engines and better fuel performance can also contribute to cutting emissions. But in many countries the growth in car use has wiped out much of the gain from these various 'technological fixes'.

In California today the controls on all sources of air pollutants, not just motor vehicles, are among the most stringent in the world. As a result the emissions *per person* are low compared with those in other parts of the developed world. But even so the *total* quantity of hydrocarbons put into the atmosphere per day in the Los Angeles air basin was about the same, in the late 1980s, as it had been in 1940 (see Figure 7.18). The controls have only just kept pace with the city's continuing growth in population and vehicle numbers.

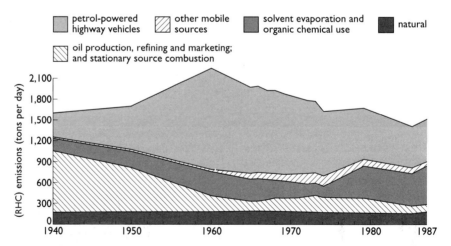

Figure 7.18 Quantity of reactive hydrocarbons (RHC) put into the atmosphere (tons per day) in the Los Angeles air basin, 1940–87. The effects of controls on the petroleum industry from 1940, and on motor vehicles from 1960, are clearly visible. But total emissions have remained at much the same level over the half century (reproduced by courtesy of Glen Cass, California Institute of Technology)

Several car firms were working in the 1990s on experimental or prototype production cars with very low or even zero emissions (Zetsche, 1995). These included lightweight cars powered by petrol engines of extremely high efficiency, and cars powered by fuels other than petrol or diesel. Both methanol and natural gas are cleaner than petrol. Cleanest of all would be hydrogen, whose only combustion product is water. Hydrogen could be burned directly in a combustion engine, or converted to electricity in a fuel cell. The obstacles seem to be not so much in the mechanical design of hydrogen-powered vehicles, as in the costs of making and distributing the gas.

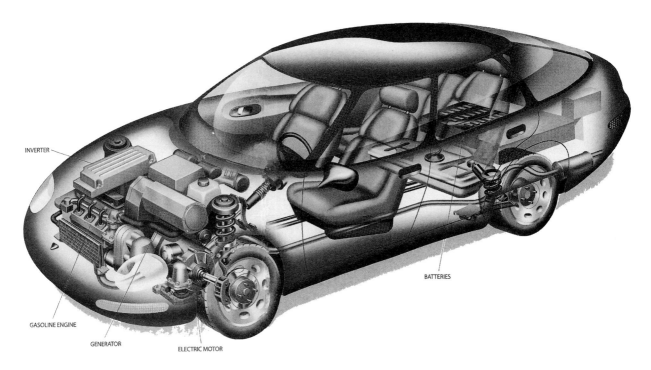

INVERTER

BATTERIES

GASOLINE ENGINE

GENERATOR

ELECTRIC MOTOR

Figure 7.19 Hybrid electric car, with combined petrol engine, electric motor and batteries (reproduced from Wouk, 1997, pp.44–5; *Scientific American*, October 1997, copyright © George Retseck)

Electric vehicles have existed since the late nineteenth century; indeed in the 1880s and 1890s it remained unclear, as between electric-, petrol- and steam-powered cars, which would turn out to be the dominant technology. The weakness of electric cars has always been the inadequacy of their batteries, which limit their power and range. From an air-quality point of view they are nevertheless attractive, since the vehicles themselves are totally non-polluting. (They may cause pollution indirectly, depending on how the electricity to charge the batteries is generated, and how the batteries are finally disposed of.)

To get the best of both worlds, the electric motor and the internal combustion engine can be combined in the 'hybrid electric vehicle' (HEV) (Figure 7.19), a technology that has excited much interest since the 1970s (Wouk, 1997), although the first patent was issued as long ago as 1905. One form of propulsion or the other is used, depending on road conditions. The combustion engine can serve part of the time to recharge the electrical batteries. Volkswagen, Mitsubishi and Toyota were all developing HEVs in the early 1990s, and some HEV buses were on the road.

7.7 *The contribution of transport fuels to global warming*

The greater part of the petrol and diesel fuel burned in motor vehicles ends up as carbon dioxide gas (CO_2). This is not a direct hazard to health, but is the main contributor to a global phenomenon – man-made change to the earth's climate – that may prove to be more damaging than all the other dangers of road traffic put together. Over the last two centuries the concentration of carbon dioxide in the atmosphere has risen, from about 280 parts per million in 1800 to 350 parts per million at the time of writing. Half of the rise has occurred since 1960. It is now the consensus among climate scientists that this increase (together with the release of other 'greenhouse gases' such as methane) has resulted in a small rise in the average temperature of the earth. The greenhouse gases serve to trap some of the long-wavelength radiation that would otherwise escape from the atmosphere, with the result that temperatures are raised near the earth's surface. In the words of the international scientific body charged with studying global warming, the Intergovernmental Panel on Climate Change (IPCC),

> there has been a real but irregular increase in global mean surface temperatures of 0.3 to 0.6°C over the past 100 years, a marked but irregular recession of the majority of mountain glaciers and the margin of the Greenland ice sheet, and a rise in the average sea level of between 1.0 and 2.0 mm per year.
> (Houghton *et al.*, 1990, p.365)

The rise in sea levels is not primarily due to the ice melting, but to higher temperatures causing the sea to expand.

How this process might develop in the future is the subject of vigorous debate. But studies by the IPCC using computer models indicated that if levels of carbon dioxide in the atmosphere were to double, as they might do on present trends by 2050, then an average warming of between 1.5 and 4.5°C could result. By the standards of daily fluctuations in temperature these might seem small changes. But, as is well known, the consequences are predicted to be catastrophic: many thousands of square miles of low-lying coastal land – where large parts of the world's population are concentrated – could be permanently flooded. Effects on the weather, causing droughts in some places and increased rain and storms in others, could have severe impacts in turn on agriculture. Already the increased strength of the El Niño weather phenomenon, responsible for storms in South America and drought in the Asian Pacific, is being blamed on global warming.

The prospect has been taken very seriously by many governments, although oil-producing nations and some industrial interests in the USA have argued against taking early precautionary measures that would damage trade. At the Climate Change Convention in Rio de Janeiro in 1992, the industrialized nations agreed to stabilize carbon dioxide emissions at their 1990 levels by the year 2000. However, at the follow-up meeting in Kyoto in 1997, deadlines for action were pushed back, and only some weak compromises agreed. There is moral and political pressure on the nations that were first to industrialize – Western Europe and the USA in particular – to make larger cuts than other countries, since they have released more of the carbon dioxide in the past. In the 1990s carbon dioxide emissions from North America were, for example, eight times greater per capita than those from China.

Road traffic is by no means the sole cause of global warming. Between 1850 and 1950 most of the additional carbon dioxide came from coal burned in industry and houses, and used in power stations to generate electricity. Since then, however, the world's energy supplies have become dominated by natural gas and oil – and much of that oil is burned in road vehicles. Figure 7.20 shows changes in fuel-use between 1960 and 1992 in the UK (Department of Trade and Industry, 1993). In 1960 some 17 per cent of all energy was used in transport. Petroleum made up one-quarter of all fuels used. Thirty years later transport was accounting for nearly one-third of all energy use in the UK, and petrol made up 42 per cent of all fuels. These figures for petrol use are dominated by road vehicles, which account for about 80 per cent of total energy use in the transport sector (Hughes, 1991). Air travel represents a significant and growing fraction of the remainder.

Figure 7.21 indicates that there are large differences in the efficiencies with which different kinds of vehicle use energy. It compares typical vehicles in the UK, with their fuel consumption expressed in equivalent units. (The 'fuel' in the case of walking and cycling is food.) Two values are given for each type of vehicle: 'typical loading' is the value for a typical number of people per vehicle; 'maximum loading' is the value for the vehicle fully loaded. It is assumed, for example, that the typical number of people per car is just under two.

Two messages stand out very clearly from the statistics in Figure 7.21. The first is that there is large scope for improving fuel efficiencies by filling vehicles more

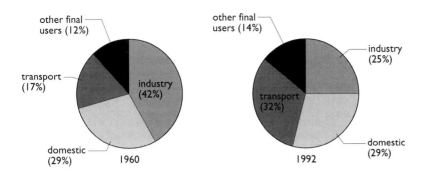

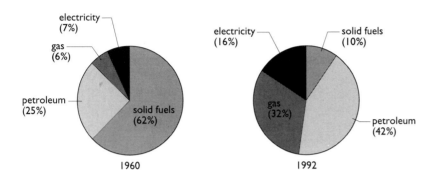

Figure 7.20 Changes in delivered energy (top) and in fuels used (bottom); UK, 1960–92. ('Delivered energy' excludes losses in distribution and all uses of fuels in the energy industries themselves) (reproduced from Alexander, 1996, Figure 1.7)

nearly to capacity. The second is that private cars and aeroplanes are much less efficient per passenger kilometre than are trains or buses. Perhaps the most remarkable statistic is that a large petrol-powered car, with a driver and no passengers, can be less fuel-efficient in these terms than a Boeing 737 jet aircraft. We can understand therefore why governments in industrialized countries are concerned about the trends in travel modes which we have already looked at – away from trains and buses towards cars and aeroplanes. Not only are total distances travelled increasing, but these are being covered in types of vehicle that are less fuel-efficient.

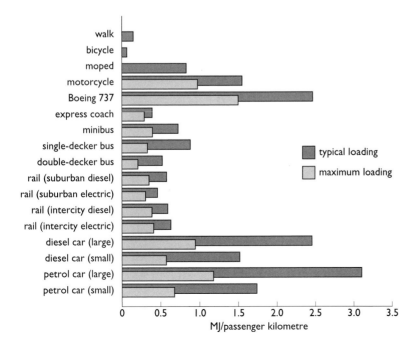

Figure 7.21 Energy efficiency of different transport modes in the UK. The bars show energy used (in megajoules) per passenger kilometre. Thus the longer the bar, the less efficient the mode. Two figures are given in each case – one for a typical number of passengers per vehicle, the other for the vehicle fully loaded (reproduced from Blunden and Reddish, 1991, p.103)

By no means all motorized traffic is within cities, of course. But in most industrialized countries a majority of the population now lives in urban areas, while there is rapid urbanization going on in the developing world. So if global warming is to be mitigated, then we need to change the ways in which houses, offices, shops and factories are heated, lit and air-conditioned, since buildings typically account for around one half of all fossil fuel used in developed countries. But car travel and public transport must change also, either through developments in the technology of vehicles themselves, or through reductions in traffic volumes brought about by the re-planning of cities and their infrastructure.

The various design changes that would cut local air pollution from cars, discussed in the last section – improved engines, alternative fuels, electric propulsion – could also play their part in reducing the emissions of carbon dioxide that contribute to global warming. Meanwhile, so long as vehicles are powered by fossil fuels, there are other measures that can cut emissions – ensuring that cars and buses are more fully loaded, shifting trips from cars to public transport or better still to cycling or walking, and (perhaps most difficult of all) reducing the total distances that people travel in motor vehicles. All these measures would help at the same time to reduce traffic congestion.

7.8 *Getting people out of their cars*

The Californian authorities have used financial incentives to encourage car pooling, and have forbidden the use of some freeway lanes to cars with only a single occupant. This lead has been followed elsewhere, as for example in Washington DC in the 1990s where one freeway was restricted in peak hours to so-called 'high occupancy vehicles'.

Persuading people back on to the buses may be more tricky, given the relentless trend in the twentieth century towards the car, which we have been

Figure 7.22 A 'superbus' in Leeds, UK, running on a guided busway (photograph: Marcus Enoch)

following in this volume. Many cities and bus companies have nevertheless been making vigorous efforts. They have tried to make buses more comfortable, faster and more responsive to demand.

One way to increase speeds and so compete on journey times with the car is to provide separate lanes with priority for buses. If these are converted from part of the existing highway, they can have the compounding effect of taking capacity away from car lanes and slowing the cars. To prevent bus lanes being abused, and to avoid hold-ups at junctions, the lanes can be completely segregated, as in the New Towns of Runcorn in the UK and Evry in France. The city of Curitiba in Brazil has a much-admired and highly effective bus system running on special lanes down the central reservations of dual carriageways. To increase speeds yet further, some cities have introduced a technology that is somewhere between a bus and a train or tram – the 'guided busway'. The bus runs on ordinary rubber tyres between raised rails, within which it is guided by smaller horizontal wheels pressing on the rails (see Figure 7.22). The O-Bahn system in Adelaide uses this principle, and there are also guided busways in Essen in Germany. The advantage is that the vehicles can run on the busways at high speeds, like a light railway, along heavily used corridors; but can then leave the tracks and run on ordinary roads to pick up riders in the lower-density suburbs.

Some bus companies and city authorities have tried to tackle the most outstanding disadvantage of bus travel – the time spent waiting at the stop. There seem to be two ways to improve this situation: either to make services so frequent that the waits are always short, as in many Dutch cities; or to let the passengers summon the bus. One difficulty with a very dispersed pattern of demand for travel in low-density cities is that there are not the numbers of passengers along many routes to justify the traditional multi-seater bus. In these circumstances there might be some merit in a road-based public-transport system that combines the characteristics of buses and of taxis, perhaps based on small vans or the types of limousine that take passengers to US airports. Information technology such as radio links and satellite positioning systems could allow a central office to know the positions of all these vehicles, and direct their movements minute by minute depending on demand. Passengers might use the same technology to summon the vehicles.

There is something paradoxical, as Hall remarks, in the way in which so many of the 'innovations' in transport now being put into practice

> do not use notably new technologies … The odd fact is that we have acquired virtually no new urban transport technology for a century. Electric trains, subways, light rail, even the private car were all around by 1890. It is surely about time for a change.
>
> (Hall, 1994, p.s88)

Hall's own proposal for a '21st-century metropolitan transport system' is one based on small vans, powered by electricity or 'hybrid electric' propulsion, running on special guideways along high-volume routes, and then spreading out along existing streets to serve local areas. On the guideways the vehicles might be physically linked like carriages in trains, or coupled electronically to travel separately but with very short gaps between them, in order to maximize the rate of flow.

Ironically, there are even some historical precedents for Hall's vision, at least so far as the 'communal taxi' element is concerned. Between 1914 and 1916 there was a short-lived craze in several US cities for 'jitneys' – private cars, usually Model Ts, operated as freelance taxis offering shared rides at very low fares – until public transport operators and licensed cab-drivers managed to get them banned (McShane, 1994). Some jitneys carried regular commuting passengers, like van pools today. Modern-day counterparts of the jitney continue to flourish in many cities around the world, as for example the 'jeepneys' of Manila and the 'collectivos' of Caracas.

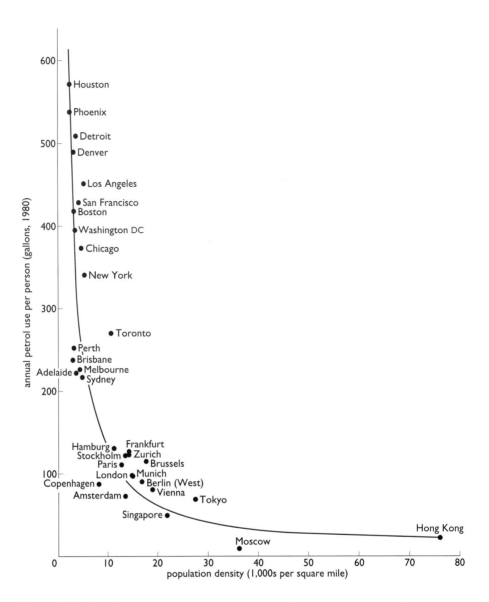

Figure 7.23 Petrol use, in gallons per person per year, in thirty-two cities world-wide, compared with their population densities in people per square mile (reproduced from Newman and Kenworthy, 1989b; by permission of The American Planning Association)

As a final way of cutting car travel and moving passengers back to public transport, bicycles or their own feet, some planners have argued for measures that would try to turn round the trend in the densities of cities that we looked at earlier in this chapter. They point to some striking findings by Newman and Kenworthy (1989b), whose statistics on urban density were presented in Figure 7.5. These authors also collected data on the use of petrol per person in the same cities. The relationship between the two sets of figures is illustrated in Figure 7.23. The message is clear: the lower the average urban density, the greater the petrol consumption. This is no doubt attributable to longer car trips, and fewer trips by public transport, by bicycle or on foot. The moral appears to be that if we could somehow make cities dense again, through land-use planning, we might cut fuel-use in transport. (This could not be done quickly, since we are talking about undoing the work of the twentieth century.)

It should be said that Newman and Kenworthy's conclusions – although not their data – have been challenged, notably by Gordon and Richardson (1989) who argued that market forces would act more effectively to conserve petrol,

Figure 7.24 The centre of Kentlands, a planned community in southern Maryland, laid out according to the tenets of the 'new urbanism' (reproduced from Cervero, 1995, p.93; photograph: copyright © Alex S. Maclean)

and that in any case such Canute-like policies could not possibly stop the tide of suburbanization. They pointed out that the cities with the lowest densities and highest fuel-use were those with the highest average incomes and the highest levels of car-ownership. (This is undeniably true: the question is, what chain of causes might link these phenomena?) The controversy did not deter the Commission of the European Communities (1990) from arguing in their *Green Paper on the Urban Environment* for planning strategies to contain Europe's cities and make them more compact; nor did it deter the British government from following suit in planning and policy guidance to local authorities (Department of the Environment, 1994).

So far the closest that town planning has come to realizing this 'compact city' ideal is the 'new urbanism' of the US architects Andres Duany, Peter Calthorpe and followers, exemplified in some communities in Florida, Maryland and California (see Calthorpe, 1993). Figure 7.24 shows the centre of Kentlands near Gaithersburg, Maryland. These developments are characterized by somewhat higher-density housing than is usual in the USA. There are paths for pedestrians, local shops and public-transport stops near by. Offices and light industry are mixed with housing, not zoned separately, with the intention of encouraging shorter journeys to work. Getting people out of their cars is not the only aim. These 'neotraditional' schemes seek to recover some of the social ambience, civic sobriety and visual charm of the small towns of the nineteenth century, for which Americans feel so much nostalgia. It remains to be seen whether they can actually start to turn back the clock of twentieth-century urban change.

References

ALEXANDER, G. (1996) 'Overview' in G.A. Boyle (ed.) *Renewable Energy: power for a sustainable future*, Oxford, Oxford University Press, in association with The Open University, pp.1–12.

BLUNDEN, J. and REDDISH, A. (eds) (1991) *Energy, Resources and Environment*, London, Hodder and Stoughton.

BOTTLES, S. (1987) *Los Angeles and the Automobile*, Berkeley, University of California Press.

CAIRNCROSS, F. (1997) *The Death of Distance: how the communications revolution will change our lives*, London, Orion.

CALTHORPE, P. (1993) *The Next American Metropolis: ecology, community, and the American dream*, Princeton, Princeton University Press.

CERVERO, R. (1985) *Suburban Gridlock*, New Brunswick, Center for Urban Policy Studies, Rutgers University.

CERVERO, R. (1989) *America's Suburban Centers: the land use–transportation link*, Boston, Unwin Hyman.

CERVERO, R. (1995) 'Why go anywhere?', *Scientific American*, vol.273, no.3, pp.92–3.

CLARK, C. (1951) 'Urban population densities', *Journal of the Royal Statistical Society*, series A, vol.CXIV, pp.490–96.

CLARK, C. (1957) 'Transport: maker and breaker of cities', *Town Planning Review*, vol.28, pp.237–50.

COMMISSION OF THE EUROPEAN COMMUNITIES (1990) *Green Paper on the Urban Environment*, Brussels, CEC (EUR 12902).

COUCLELIS, H. (1996) 'The death of distance', editorial, *Environment and Planning B: Planning and Design*, vol.23, pp.387–9.

DE FOREST, R.W. and VEILLER, L. (eds) (1903) *The Tenement House Problem*, New York, Macmillan.

DEPARTMENT OF THE ENVIRONMENT (1994) *Transport Planning Policy Guidance Note 13 (PPG 13)*, London, Her Majesty's Stationery Office.

DEPARTMENT OF TRADE AND INDUSTRY (1993) *Digest of United Kingdom Energy Statistics 1992*, London, Her Majesty's Stationery Office.

DYSON, F. (1979) *Disturbing the Universe*, New York, Harper and Row.

ELSOM, D. (1996) *Smog Alert*, London, Earthscan/Kogan Page.

FLINK, J.J. (1988) *The Automobile Age*, Cambridge, MIT Press.

GARREAU, J. (1991) *Edge City: life on the new frontier*, New York, Doubleday.

GOODMAN, D. and CHANT, C. (eds) (1999) *European Cities and Technology: industrial to post-industrial city*, London, Routledge, in association with The Open University.

GORDON, P. and RICHARDSON, H.W. (1989) 'Gasoline consumption and cities – a reply', *Journal of the American Planning Association*, vol.55, pp.342–6.

HALL, P. (1994) 'Squaring the circle: can we resolve the Clarkian paradox?', *Environment and Planning B: Planning and Design*, vol.21, pp.s79–s94.

HALL, P. (1996, rev. edn) *Cities of Tomorrow*, Oxford, Blackwell.

HOUGHTON, J.T., JENKINS, J.G. and EPHRAUMS, J.J. (eds) (1990) *Climate Change: the IPCC scientific assessment*, Cambridge, Cambridge University Press.

HUGHES, P. (1991) 'The role of passenger transport in CO_2 reduction strategies', *Energy Policy*, vol.19, no.2, pp.149–60.

McSHANE C. (1994) *Down the Asphalt Path: the automobile and the American city*, New York, Columbia University Press.

McSHANE C. (1997) *The Automobile: a chronology of its antecedents, development, and impact*, London and Chicago, Fitzroy Dearborn.

MARCHETTI, C. (1994) 'Anthropological invariants in travel behaviour', *Technological Forecasting and Social Change*, vol.47, pp.75–88.

NEWMAN, P.W.G. and KENWORTHY, J.R. (1989a) *Cities and Automobile Dependence: a sourcebook*, Brookfield, Gower.

NEWMAN, P.W.G. and KENWORTHY, J.R. (1989b) 'Gasoline consumption and cities: a comparison of US cities with a global survey', *Journal of the American Planning Association*, vol.55, pp.24–37.

SCHAFER, A. and VICTOR, D.G. (1997a) *The Future Mobility of the World Population*, Cambridge, Center for Technology, Policy and Industrial Development, MIT (Discussion Paper 97–6–4).

SCHAFER, A. and VICTOR, D.G. (1997b) 'The past and future of global mobility', *Scientific American*, vol.227, no.4, pp.36–9.

THOMSON, J.M. (1977) *Great Cities and their Traffic*, London, Gollancz.

UNITED STATES DEPARTMENT OF TRANSPORTATION (1999) 'National transportation statistics' <http://www.bts.gov/ntda/nts> (February 1999).

UNITED STATES ENVIRONMENTAL PROTECTION AGENCY (1994) *National Air Quality and Emissions Trends Report 1993*, Research Triangle Park, North Carolina, USEPA (EPA454/R94026).

WARNER, S.B. (1978) *Streetcar Suburbs: the process of growth in Boston, 1870–1900*, Cambridge, Harvard University Press.

WOUK, V. (1997) 'Hybrid electric vehicles', *Scientific American*, vol.277, no.4, pp.44–8.

ZAHAVI, Y. (1981) *The UMOT–Urban Interactions*, Washington DC, US Department of Transportation (DOT-RSPA-DPB 10/7).

ZETSCHE, D. (1995) 'The automobile: clean and customized', *Scientific American*, vol.273, no.3, pp.76–80.

Chapter 8:
TELECOMMUNICATIONS AND CITIES SINCE 1840

by Philip Steadman

8.1 Introduction

In a rueful passage in *Civilization and its Discontents* (first published in 1930), Sigmund Freud lamented the way in which mechanical transport in the twentieth century allowed families to drift apart, something which for him was only partly mitigated by the possibility of keeping in touch electronically:

> Is there, then, no positive gain in pleasure, no unequivocal increase in my feeling of happiness, if I can, as often as I please, hear the voice of a child of mine who is living hundreds of miles away…? [But] if there had been no railway to conquer distances, my child never would have left his native town and I should need no telephone to hear his voice …
>
> (quoted in Fischer, 1992, p.15)

In this chapter we look at the interactions between transport and telecommunications technologies, and the effects the latter have had on the form and functioning of cities. With a few notable exceptions, urban historians have not given these subjects the attention they surely deserve. As Graham and Marvin (1996) remarked in a theoretical review of telecommunications and the city, there has been excessive emphasis on transport technologies in urban studies, to the detriment of any equivalent analysis of telecommunications. The fact that telephone and computer networks are less bulky and less visually obtrusive than transport infrastructure might possibly have something to do with this relative neglect.

Among planning theorists who provide exceptions to this rule, one leading figure is Melvin Webber, who in the 1960s began to comment on the emergence of new kinds of settlement patterns in which physical proximity of activities was less important than in cities throughout history. Webber (1963) coined the phrase 'community without propinquity' to encapsulate the idea that people might maintain networks of social or business contacts, via the car and the telephone, which were spread much more widely than ever before. The result would be the emergence of communities of interest, or patterns of association, not defined by spatial boundaries: he called these communities 'non-place urban realms'. Such changes could be observed in progress in the south and west of the United States. Webber's thinking was influential on the planning of Milton Keynes, the British new town of the 1960s, with its low density, its network of buried cabling and its 'neutral' rectangular grid of widely spaced roads.[1] Webber, though, did not imagine that trends in transport and communication technologies would universally act as agents of urban dispersal: 'They could push urban spatial structure towards greater concentricity, toward greater dispersion, or, what I believe to be most likely, toward a very heterogeneous pattern' (Webber, 1963, p.44).

---

[1] Milton Keynes is discussed in Goodman and Chant (1999), a companion volume in this series.

The coming together of telephone technology with personal computing, most visible in the spectacular growth of the Internet, has spawned a rich literature of futuristic speculation about the possible consequences for buildings and cities. Writers whose principal interests are in the technology itself have predicted that cities will be transformed, and might even dissolve completely. We can take just two examples. The computing specialist James Martin (1981) envisaged a future for cities where the need to travel would be much reduced. Most people would, he believed, work from home via computer terminals and videophones. He predicted that computers and television would provide interactive entertainment and films on demand at home, and create the channels for distance education on the model of the UK's Open University. Medical examination would be carried out remotely, and patients monitored and even treated at a distance using video and robotic instruments.

In his *City of Bits*, the architect William Mitchell (1995, p.47) argues that, 'Increasingly, telecommunications systems replace circulation systems, and the solvent of digital information decomposes traditional building types'. He predicts that, if documents of all kinds are available for consultation and printing at home via the Internet, the traditional library will become redundant. There will be no need to travel to the museum or the art gallery, when the world's collections can be viewed in comfort from home. Convicted criminals can be tagged and supervised electronically, so prisons will disappear. When consumer goods can be inspected on screen, ordered and paid for in electronic money, then the only remaining physical embodiment of shops will be in their warehousing and their delivery vehicles. Neither Martin nor Mitchell go so far as to envisage the disappearance of cities altogether; but others have imagined a completely dispersed rural society of 'electronic cottages' which would be a Utopia, a 'no-place' in the literal sense.

If such a vision were ever to be realized, or even if trends were to go in this direction, then many aspects of the 'car crisis' described in the previous chapter – road congestion, air pollution, the death toll from accidents – would simply fade away.

Other writers, treating these issues from economic and social science perspectives, have argued by contrast that telecommunications, far from replacing the need for physical movement of goods and people, might rather result in an *increase* in travel. Thus Netzer (1977), for example, suggests – although with little detailed evidence – that this may have been true historically of the telephone:

> there is a texture and subtlety in three-dimensional face-to-face communications that cannot be reproduced in any other way, so much so that past advances in telecommunications technology appear to have increased, not substituted for, some aspects of the demand for face-to-face communication.
>
> (Netzer, 1977)

Netzer's claim is not implausible. The telephone allowed people to conduct business over larger areas, and to maintain social contacts at a distance with friends and family (like Freud and his son). These people then *travelled further* to visit their far-flung acquaintance. The question of possible causality is complicated by the fact that use of the telephone spread simultaneously with use of the motor car (see Figure 8.1 overleaf), at a time when oil prices were also low. Should the real costs of transport fuels rise sharply in the future, and should the technical quality of telecommunications improve to the point of giving the participants a convincingly realistic illusion of being present together in the same 'virtual' space, then this relationship might change, and perhaps electronic communication really would come to replace much physical movement.

Another economist, Frances Cairncross (1997), has imagined that the effect of the 'information revolution' and the Internet might be not so much to reduce

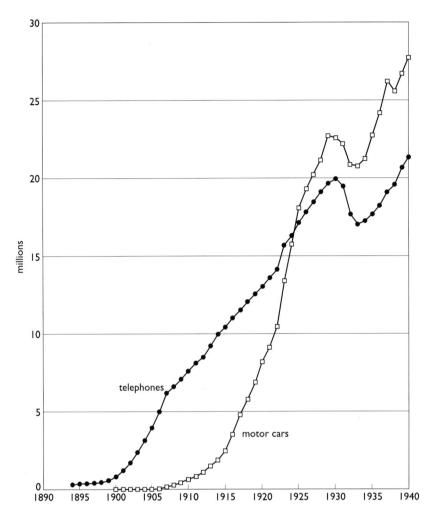

Figure 8.1 Growth in numbers of telephones and motor cars in the USA, 1894–1940; the dips from 1929 mark the effects of the Depression (data from the United States Bureau of the Census, 1975; graph reproduced from Fischer, 1992, p.44, figure 2; © 1992 The Regents of the University of California)

travel as to generate a *different kind* of travel. There might come to be fewer marked differences between homes and workplaces, and between work and recreation. Cities could become predominantly centres of entertainment and culture, and offices might change into something more like social centres. So trips would be made for new purposes and between new kinds of activity.

All this, however, is speculation, and we are concerned here with urban history. The Internet is young, and only its technological origins have so far attracted the attentions of historians. As Cairncross says, the consequences for the city may indeed turn out to be as far-reaching as those of the motor car. But in the late 1990s, computer communications had reached the same stage as the car in 1910: at that point it was a machine more or less fully developed in its engineering design, but with all its secondary effects on employment, urban form and the countryside yet to come.

In this chapter, then, I will be raising two key questions that I cannot hope to answer fully: I will simply marshal evidence on either side. The first question is: do electronic telecommunications serve in general to help activities become more separated and dispersed, so leading to lower densities and urban sprawl? The second is: do electronic telecommunications cause people to travel less? I will try to illuminate both issues by looking not forward, but back at those technologies whose relationship to cities has at least been studied a little by historians: the telephone and its predecessor, the electric telegraph.

8.2 'What hath God wrought?' The electric telegraph

There were telecommunications networks before the invention of the electric telegraph. The word 'telegraph' was first used to describe a system consisting of a line of hill-top towers, in sight of one another, with flag-like wooden arms with which messages could be sent in semaphore (see Figure 8.2). The French government began setting up such a visual telegraph for military communications in the late eighteenth century. By the 1840s Paris was linked to twenty-eight other cities (Attali and Stourdze, 1977). A message took between ten and twenty minutes to get from Paris to Toulon on the Mediterranean coast. A similar telegraph built in Massachusetts in 1800 by Jonathan Grout brought news of incoming vessels and cargoes, from Martha's Vineyard to shipowners and merchants in Boston – an early example of the kind of commercial application to which its electric successor was put (Garratt, 1958; Wilson, 1954). The place-name 'Telegraph Hill' is now the only relic of this vanished technology.

The first proposal for an electric telegraph was made as early as 1753 in an anonymous article in the *Scots Magazine*. In the succeeding one hundred years a number of European scientists experimented with a variety of designs (Garratt, 1958). Some of these employed static electricity, others depended on electrochemical or electromagnetic effects. The messages were conveyed either by rotating needles which pointed to letters on dials, or else in different forms of code. The first commercially successful electric telegraph was developed in England by W.F. Cooke and Charles Wheatstone, who formed a partnership and secured a patent in 1837. They went on to produce several electromagnetic instruments with needle-type displays (see Figure 8.3 overleaf). The advantages over the visual telegraph were speed, freedom from dependence on good light and good weather, and lower cost – since fewer operators were required.

Figure 8.2 The visual telegraph, used to send semaphore signals over long distances in the early nineteenth century (courtesy of Picture Collection, New York Public Library)

The earliest customers, in Britain as later in the USA, were the railway companies: they used the telegraph to communicate with signalmen and determine the positions of trains, so improving both timekeeping and safety. Cooke and Wheatstone gave a demonstration to the directors of the London to Birmingham Railway, on the line between Euston and Camden Town stations, a distance of some 5 miles (8 kilometres). Other companies were quick to show interest. By 1852, there were about 4,000 miles (6,400 kilometres) of telegraph lines installed in Britain. One sensational incident in 1845 brought the new technology to general public attention. A woman was murdered at Slough, and the suspect was seen boarding a train for Paddington. The railway company sent a full description of the man ahead by telegraph, and the police were waiting for him in London. He was arrested and later hanged.

In the USA, the electric telegraph was brought to the point of commercial application by Samuel Morse. He designed an instrument with which an operator could open and close a circuit by pressing and lifting a key, so sending a series of bursts of current down a wire to a receiving instrument. The message was transmitted in long and short pulses following the code to which Morse gave his name. With a single battery the range of the instrument was limited. But with the introduction of the relay, which retransmitted the signal, it became possible to send messages over long distances.

In late 1842, the US government awarded Morse a grant for a test line between Washington and Baltimore, a distance of 40 miles (65 kilometres).

Figure 8.3 Cooke and
Wheatstone's five-needle
telegraph, 1837; the needles
deflect in pairs and point to the
respective letters (courtesy of
Science Museum/Science and
Society Picture Library)

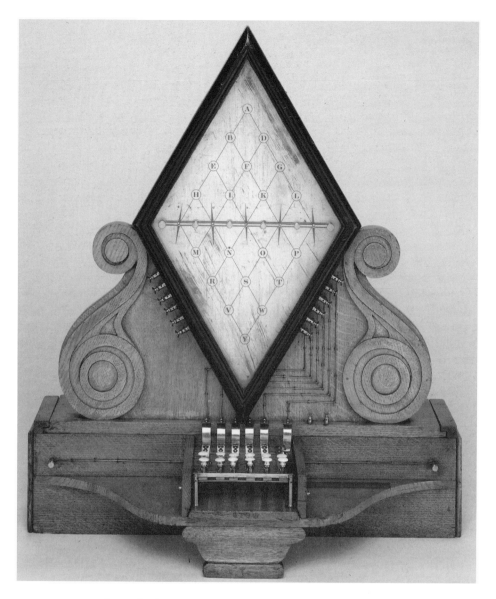

Morse was so pleased when he heard the news from the daughter of the
Commissioner of Patents, he told her she could send the very first message.
Her choice was a biblical phrase, taken from the Book of Numbers: 'What hath
God wrought?' Over the next eight years Morse's Magnetic Telegraph
Company installed more than 23,000 miles (37,000 kilometres) of wire
connecting 500 US towns and cities (Tarr *et al.*, 1987). In 1851 a cable was laid
linking England and France. The North American continent was crossed ten
years later. After several heroic attempts and dispiriting failures, the first fully
operational transatlantic cable was connected in 1866. *The Times* of London
published an article on 27 July that year, the eve of the cable's inauguration,
under the headline 'Shrinking world', a phrase that was to echo through much
of the later literature on telecommunications:

> So far as human foresight can judge, the Old and the New World will be in
> telegraphic communication before tomorrow night. The prospect opened to the
> world by this achievement is so marvellous that any attempt to describe it must
> give only a faint and feeble picture. The two most active and energetic nations of
> the world are placed in hourly communication. The Governments of England
> and the United States will be able to converse rapidly and freely … To the
> mercantile interests of both countries the gain must be immense … Indeed, now
> that the great enterprise is completed, there can be no doubt that in a few years
> the entire globe will be spanned by the telegraph wires, and the news of the
> planet will be given every morning in the London papers.

Figure 8.4 Dispatches from the Mexican War of 1848 were transmitted to newspaper offices by the 'lightning wire'; engraving, 1853, by Alfred Jones, after a painting by R.C. Woodville (Collection of the New-York Historical Society)

The Times was correct in its forecast. By the 1870s, the global network extended to India, Australia and South America.

Among the first US clients of the (inland) telegraph, after the railway companies, were the newspapers. Associated Press set up a telegraphic news service, and papers boasted about the speed with which dispatches from the Mexican War of 1848 reached them via the 'lightning wire' (see Figure 8.4). Reuter's followed, transmitting financial and economic news by telegraph between Berlin and Paris as early as 1850. Indeed, some of the most enthusiastic users of the new technology were brokers and speculators, for whom it provided instantaneous information on prices and markets. News services and railways operated *between* cities. It may well have been the use of the telegraph by the financial professions that began to affect cities themselves.

According to Joel Tarr *et al.* (1987), the telegraph played an important role in bringing the nation's security markets together in New York between 1850 and 1880. By 1860, Wall Street was connected by telegraph to all major cities across the USA. Within the cities, the information supplied by the telegraph was distributed by messengers (see Figure 8.5 overleaf). So, as Tarr *et al.* point out, communication *between* cities was at this stage much faster than communication *within* cities. Later, the invention of the 'stock ticker', by which commodity and share prices, transmitted by telegraph, were printed directly on to 'ticker tape', made it possible for private offices and hotels to be linked to the stock exchanges.

Figure 8.5 Messenger boys from the American District Telegraph Company, a part of Western Union, posed on the steps of the Sub-Treasury Building, Nassau Street, New York. The boys delivered telegrams from the telegraph office to their final destinations (reproduced from The Children's Aid Society, 1978, pp.20–21)

In this respect, then, it appears that the telegraph had the effect of *concentrating* financial activities into New York and other centres. In manufacturing, where the telegraph was also introduced early on, it seems more than likely that the opposite effect was encouraged – that is, a *dispersal* of activities – although the question has apparently been little studied. In the mid-nineteenth century, much manufacturing was in the centres of cities. Typically, a factory and its associated warehousing and offices were all combined on a single site. The telegraph allowed the office headquarters to be separated from the production areas. The former could stay in the city centre, close to markets, clients and sources of finance. The latter could be moved out to cheaper land, where factories could be built on a single storey and have room to expand. If there are doubts as to how much the telegraph permitted these developments, there are none about the effects of its successor, the telephone, as we shall see. By 1900, the modern 'central business district' was emerging, with the offices of manufacturing companies, financial institutions such as banks and insurance firms, and department stores and other specialized shops, all grouped together.

The electric telegraph was also used in cities to support the fire brigade and the police. William F. Channing, a Boston doctor who was interested in electricity, drew analogies between the telegraph and the human nervous system. Unlike the long-distance telegraph, which was essentially linear, Channing envisaged an urban telegraph service that would be something more like a web, and would 'cover the surface of the municipal body ... as thickly ... with telegraphic signalizing points as the surface of the human body is covered with nervous extremities or papillae' (Channing, 1855). In response to a newspaper article published by Channing in 1845, the mayor of Boston decided to set up a telegraphic fire-alarm system. Call-boxes were placed throughout the city, from which the alarm could be passed to a central station, which in turn alerted the fire-fighters. The idea was soon adopted in other large American cities, and by 1900 around 750 municipalities had telegraphic fire-alarms (as mentioned in Chapter 6).

It was in Boston, too, in the 1850s, that the telegraph was first adopted by a municipal police force. The Boston chief of police, inspired by the example of the telegraphic fire-alarm, set up a similar network to connect officers on the beat, local precincts and headquarters (Tarr *et al.*, 1987). There was an important difference, though. The user of the fire-alarm needed only to transmit what was in effect always the same message: 'There is a fire near the location of this box.' He or she did not need to be trained in Morse code. The policeman on the beat was not trained either, so he was unable to send or receive complex information via the telegraph. Instead, the police boxes were equipped with dials, with which officers could transmit just certain standardized signals: 'Send help', 'Send an ambulance', and so on. The police force had to wait for the invention of the telephone to give them a truly flexible communications system (see Figure 8.6 overleaf). For these reasons, the main use of the police telegraph was in co-ordinating forces to combat the urban riots which beset American cities in the mid-nineteenth century.

The same lack of flexibility hampered the use of the telegraph in many applications. It required specialized operators and was best suited to simple messages. The telephone, by contrast, since it effectively conveyed the human voice directly and needed no special skills to operate it, could be used by anyone to transmit the most nuanced and intimate of communications. More important still, it was *interactive*: it allowed conversation. The telegraph industry itself created some of the preconditions for the telephone's success: not just the great commercial companies, such as Western Union, which began in one technology and moved to the other, but the actual telegraph wires themselves, which were taken over wholesale for the telephone network.

Figure 8.6 Chicago police telephone and patrol system in the 1880s (reproduced from *Scientific American*, 23 April 1881, p.255)

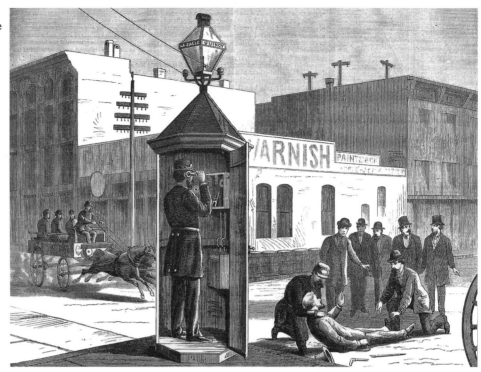

It should not be imagined that the telephone entirely and immediately displaced the telegraph. Domestic telegrams may have disappeared. But the two technologies continued in operation side by side throughout the twentieth century, especially for long-distance commercial communications. Although intercontinental telephony was introduced in the 1920s, it relied on short-wave radio, and its success was limited. It was only after the Second World War that the first oceanic telephone cables were laid. Even so, there was for example an explosive growth in high-speed dialled telegraphy from the 1950s, greatly stimulated by the introduction of the Telex machine, which produced printed records of every message and was used extensively by businesses and the news services (Cherry, 1978). Modern facsimile (fax) machines and electronic mail (e-mail), both of which transmit messages in digitally encoded form, can be seen as remote descendants of the original telegraphs of Cooke and Wheatstone, and of Morse.

8.3 'Watson, are you there?' Bell and the telephone

Alexander Graham Bell first showed his telephone in public at the Centennial Exhibition in Philadelphia on 25 June 1876 (Wilson, 1954). Bell read a paper describing a telephone conversation he had held over a long-distance telegraph wire between Boston and Portland (ME), about a hundred miles (160 kilometres) away. Conditions at the Centennial were not ideal for the live demonstration that followed, and to make himself heard Bell had to shout to his assistant through the instrument, 'Ahoy, ahoy, Watson, are you there?' (Wilson, 1954, p.284). The visiting public were not greatly impressed, but Bell's scientific audience saw the potential. The telegraph industry did not immediately appreciate the nature of the competition, however. Wanting to pay off his debts and get married, Bell offered his invention to the Western Union Company for 100,000 dollars. They refused it, so Bell kept his patent to himself. It turned out to be one of the most lucrative ever issued. (Later, Western Union realized their error and tried to commercialize a rival instrument designed along different principles by Thomas Edison, but failed and were forced to merge with Bell's backers.)

In its improved version of 1877, Bell's instrument consisted of a thin metal diaphragm set in front of one pole of an electromagnet. Speaking into the 'transmitter' caused this diaphragm to vibrate, so inducing a continuously varying current in the magnet (see Figure 8.7). This current was transmitted down the wire to the 'receiver' – identical in design to the transmitter – where it caused the receiving diaphragm to vibrate and emit sound again. Where the telegraph signal was *digital* (that is, the current was transmitted in separate pulses), the telephone signal was *analogue* (the current varied continuously with the sound waves that produced it).

Bell's subsequent demonstrations of the telephone emphasized its potential in providing entertainment: for example, it could allow listeners to hear operas and concerts from home. Indeed, the Electrophone Company, set up in 1894, achieved some modest success supplying such services (Briggs, 1977). From 1893, a 'telephone newspaper' operated in Budapest, transmitting poetry, lectures and music as well as news. These were *broadcast* services, from a single central transmitter to numerous receivers.

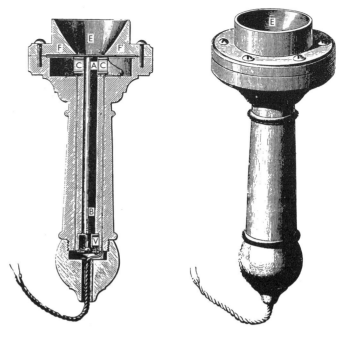

Figure 8.7 The commercial form of Bell's telephone, cut away to show the electromagnet, marked AB, and the vibrating diaphragm, marked FF′. The same instrument served as transmitter and receiver (reproduced from Wilson, 1954, p.281, where the picture is credited to Figuier, 1883–6)

Bell himself realized, though, that the real potential of the telephone was in one-to-one exchanges between individuals. (Notice how his first message at the Centennial was addressed to a specific person, Mr Watson, inviting him to *reply*.) He saw that its success as a universal means of distant communication was dependent on creating central switchboards, since it would be completely impractical for large numbers of instruments to be connected each one directly to every other. In a letter to his 'capitalists' in 1878, Bell made what one of the leading histories of the subject described as a 'sensationally good forecast' of the telephone's future:

> At the present time we have a perfect network of gas pipes and water pipes throughout our large cities. We have main pipes laid under the streets communicating by side pipes with the various dwellings, enabling the members to draw their supplies of gas and water from a common source.
>
> In a similar manner it is conceivable that cables of telephone wires could be laid under ground, or suspended overhead, communicating by branch wires with private dwellings, counting houses, shops, manufactories, etc., uniting them through the main cable with a central office where the wire could be connected as desired, establishing direct communication between any two places in the city. Such a plan as this, though impracticable at the present moment, will, I firmly believe, be the outcome of the introduction of the telephone to the public. Not only so, but I believe in the future wires will unite the head offices of telephone companies in different cities, and a man in one part of the country may communicate by word of mouth with another in a distant place.
>
> (quoted in de Sola Pool *et al.*, 1977, p.156)

Notwithstanding its lukewarm public reception at the Centennial, the telephone proved to be one of the most immediately popular and fastest-spreading technological devices in the history of innovation. The first public telephone exchanges were set up in New Haven (CT) in 1878 and in London in 1879. At first the networks were almost entirely within cities, as Bell envisaged. The majority of users were businesses. Among professional groups, physicians were quick to install telephones, as they were to buy the first motor cars. The wires were strung on poles and buildings. In many cases, as mentioned, former telegraph lines were taken over for telephone traffic. Soon the proliferating cat's cradles in cities became eyesores. The wires caused problems for the installation of overhead cables for trams in some cities; and, conversely, the passing trams could cause interference in telephone reception. A heavy fall of snow in Manhattan in 1888 brought down many overhead wires (see Figure 8.8), and it was decided to reroute them all below ground.

Figure 8.8 Overhead telephone wires in New Street, Manhattan, looking towards Wall Street, in the great blizzard of 11–12 March 1888. Many lines collapsed, and all cables were soon rerouted below ground to avoid a repetition of the chaos (photograph: courtesy of Brown Bros., Museum of the City of New York)

There were technical problems with the attenuation of signals over long distances which limited the early growth of intercity links. But in 1880 the Interstate Telephone Company built the first long-distance line from New York City to Boston, and others soon followed. The telephone, like the motor car, was especially welcomed in rural regions of the USA, where it promised to relieve some of the loneliness of farming life. This confounded the telephone companies, who anticipated only urban markets. Of all states, Iowa had one of the highest levels of telephone ownership in the early years of the new

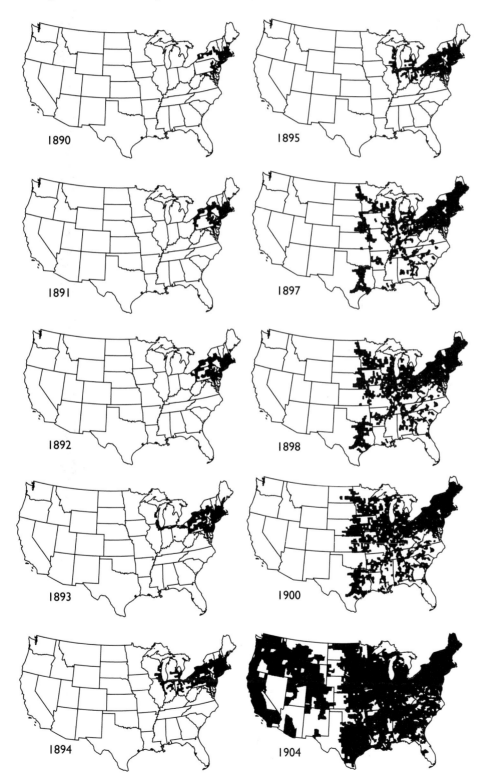

Figure 8.9 Growth of the American Telephone and Telegraph (AT&T) toll network between 1890 and 1904, by counties; a county is shaded if one or more towns were connected to the network in the year in question (reproduced from Abler, 1977, p.333; © The Massachusetts Institute of Technology)

century. In some places existing barbed-wire fences were used to carry signals across the prairies. By the 1920s, according to one writer, the telephone 'was a fundamental part of American farm life … the farmer's first aid' (quoted in Fischer, 1992). The instrument was too expensive for many private individuals to own, however: those who could not afford it used telephones at drug stores. Pharmacists had first installed telephones on their premises to speak to doctors, and then offered them for general use as a public service. By the turn of the century, the Bell system was operating public coin-in-the-slot telephones in most large cities (Fischer, 1992).

Some writers had predicted that increased use of the telephone would widen people's social horizons and encourage more long-distance contacts. The authors of *Recent Social Trends in the United States*, an influential report to President Hoover that was published in the 1930s, expressed some scepticism about this proposition, pointing out that 'Relatively, local contacts have increased more rapidly [via the telephone] than non-local contacts' (Willey and Rice, 1933, p.202). Modern communication, they said, may 'strengthen certain aspects of localism' (although perhaps this phenomenon was simply a reflection of existing community structures, reinforced by the difference at that time between the costs of local and long-distance calls, and the poor quality of long-distance reception).

Figure 8.9 shows the growth of the American Telephone and Telegraph (AT&T) network between 1890 and 1904. The first transcontinental link was made in 1915 (see Figure 8.10), and a complete national network established by 1920. The process had taken forty years.

Figure 8.10   The first coast-to-coast telephone call in the USA, January 1915 (photograph: courtesy of California Historical Society, North Blake Library, San Francisco FN-28803)

8.4 *The telephone and the office building*

The telephone came very rapidly, then, to connect cities and their inhabitants. What effects did it have on the actual *forms* of cities and buildings? The question of cause and effect is an especially problematic one here. Like some other urban technologies, the telephone has provided opportunities and freedoms of many different kinds. As Ithiel de Sola Pool has pointed out, these seem in many cases to have worked in opposite directions simultaneously:

> It saves physicians from making house calls, but physicians initially believed it increased them, for patients could summon the doctor to them rather than travel to him. The phone invades our privacy with its ring, but it protects our privacy by allowing us to transact affairs from the fastness of our homes. It allows dispersal of centers of authority, but it also allows tight continuous supervision of field offices from the center. It makes information available, but reduces or eliminates written records that document facts ... No matter what hypothesis one begins with, reverse tendencies also appear.
>
> (de Sola Pool, 1977, p.4)

We should therefore proceed with caution, and avoid simplistic diagnoses. So far as individual buildings are concerned, an argument has been made for the telephone being part of that cluster of innovations that made the high-rise office building possible in the first place. It was used at an early date in the construction of skyscrapers, to allow workers on the highest levels of the metal frame to keep in communication with their colleagues at ground level. It played a similar part from the turn of the century in those major construction projects and industrial enterprises that required the co-ordination of people and resources over large areas, for example in mining and the construction of the Panama Canal (Aronson, 1977).

Some commentators have suggested that the telephone made possible the very operation of large-scale office bureaucracies in multi-storey buildings. Offices had previously depended on armies of messenger boys. The passenger lifts of the new skyscrapers of New York and Chicago could, according to this argument, simply not have coped with movements on the scale required. Similar constraints applied to the big new high-rise hotels. Room-service telephones were first introduced around 1900. In 1904 the biggest concentration of telephones under one roof in the world was in the Waldorf-Astoria Hotel in New York (Aronson, 1977). John J. Carty, chief engineer of AT&T, was quoted in 1908 as saying:

> It may sound ridiculous to say that Bell and his successors were the fathers of modern commercial architecture – of the skyscraper. But wait a minute. Take the Singer Building, the Flatiron, the Broad Exchange, the Trinity, or any of the giant office buildings. How many messages do you suppose go in and out of those buildings every day? Suppose there was no telephone and every message had to be carried by a personal messenger. How much room do you think the necessary elevators would leave for offices? Such structures would be an economic impossibility.
>
> (quoted in de Sola Pool *et al.*, 1977, p.140)

The argument seems reasonable on the face of it. But it has been challenged by Claude Fischer (1992) in his book *America Calling*. 'This contention', Fischer says, 'lacks both evidence and plausibility' (p.27). As he points out, skyscrapers were being built in New York and Chicago in the 1870s, before telephone companies even existed. So it cannot be true that skyscrapers were *impossible* without telephone services. Perhaps we must conclude that the skyscraper created a pressing *demand* for the telephone, which meant that electric bells, speaking tubes and messenger boys were very rapidly superseded once it was introduced. Bell himself had foreseen this change.

Great Britain was slower than the USA to take up the telephone, in part because of the conservatism of public officials. In 1879 Sir William Preece, chief engineer to the British Post Office, gave evidence on the prospects for telephony to a House of Commons Committee:

> I fancy the descriptions we get of its use in America are a little exaggerated, though there are conditions in America which necessitate the use of such instruments more than here. Here we have a superabundance of messengers, errand boys and things [*sic*] of that kind ... The absence of servants has compelled Americans to adopt communication systems for domestic purposes. Few have worked at the telephone much more than I have. I have one in my office, but more for show. If I want to send a message – I use a sounder [a form of telegraph] or employ a boy to take it.
>
> (quoted in Dilts, 1941, p.11)

Such attitudes did not persist, however, once English businessmen and householders had experienced the technology for themselves. By 1890, there were 45,000 telephones in Britain; by 1920, nearly a million (Perry, 1977, p.91).

8.5 The telephone: an agent of urban dispersal or of urban concentration?

What of the effects of the telephone on the *location* of activities within and outside the city? We have already seen how the electric telegraph began to allow the office functions of industrial firms to separate from the manufacturing activities themselves, the former staying in the centre of the city, the latter moving to peripheral sites. As with the telegraph, the first clients for telephone services in the United States were businesspeople. Before telecommunications and motorized road transport, manufacturing firms and wholesalers had tended to group together within distinct zones in the city, even in single streets: the meat and grain markets, the jewellers' and goldsmiths' quarter, the garment district. Shippers and merchants were located next to docks or railway stations. Much business was done (then as now) over lunch or a drink in the restaurants and taverns. To be within walking distance was crucial to the exchange of commercial information, whether between competitors, collaborators, buyers and sellers, lenders and borrowers. Many writers on urban history have pointed to this kind of communication function as one important reason for the emergence of cities in the first place. Melvin Webber (1968, p.1097), for example, described the need for face-to-face contact as 'the glue that once held the spatial settlement together'. The new transport and telecommunications technologies of the twentieth century, he argued, have increasingly come to dissolve that glue.

Can the telegraph and telephone be said to have 'caused' the breakup and geographical dispersal of commerce and manufacturing? As with the question of whether the growth of suburbs was 'caused' by trams and trains, we have to be careful here. The telecommunications engineer J. Alan Moyer made a detailed study of the introduction of telephones into Boston (Moyer, 1977), which provides some concrete evidence. In 1850, Boston had been a densely built-up seaport where interdependent firms were clustered together, and employees lived close by, so that they could walk to work. The expansion of the city in the latter half of the nineteenth century was made possible by the new horse trams, electric streetcars and suburban railways (see Chapter 2). All this was before any telephone services existed. So clearly, Moyer concluded, the telephone was not crucial to the decentralization process at this stage.

On the other hand, the telephone does seem to have been instrumental in allowing retail-oriented businesses to flourish in Boston's new suburbs, where they could serve the more dispersed population, and still stay in touch with city-centre markets, wholesalers and services. As Moyer (1977, p.363) said, the telephone was used extensively by these businesses 'to transfer directions and orders, and to extend prevailing business practices'. Meanwhile, Moyer also diagnosed some effects of the growth in domestic telephones, in helping to encourage the move of residential development to the suburbs. This process accelerated with the connection of new telephone lines.

Notice, however, this *two-sided* effect of telecommunications, which we will meet again at the intercity scale. Small businesses, manufacturing, warehousing could all move to the suburbs or beyond and still remain in touch with the centre. But at the same time, this very change served to reinforce the dominant position of the city-centre headquarters. Here are the beginnings of a process of *dispersal* of those commercial and industrial functions for which space, cheap land, easy access to out-of-town transport routes or direct access to new suburban markets were important; this was balanced by the reverse process of *concentration* of those functions for which spatial proximity and face-to-face contact remained important, even at the cost of higher rents and traffic congestion. The two processes went on throughout the twentieth century. They were one cause of the characteristic transformation of great cities in the developed world in this period, from centres of manufacturing to centres for the service and 'information industries' – banking, insurance, accountancy, the law, publishing, advertising, broadcasting and entertainment.

8.6 *Growth in global telecommunications and information technology after the Second World War*

There was a huge growth in international telecommunications after the Second World War: growth in the volume of traffic, in the geographical spread of telecommunication links, and in the complexity of the supporting technology (Cherry, 1978). The post-war expansion of automatic telegraphy has already been mentioned. Transatlantic telephone cables were laid from 1956 onwards, and by the late 1970s the intercontinental network extended to South America, Africa, and across the Pacific to Australia and the Far East (see Figure 8.11 overleaf). This network allowed reliable high-quality international telephony for the first time. Businesses, governments and institutions were responsible for most of the early traffic. The new connections encouraged the spread of global companies with offices and factories in many countries. Private users were still relatively rare, and until the 1970s air mail – not the telephone – remained the primary medium of international personal communication.

The geographical pattern of these new telephone cables is perhaps significant for our general question as to whether telecommunications encourage or substitute for travel. Just as early telegraphs followed the railways, so this expanded telephone traffic followed the same routes as the heaviest air traffic, which itself grew rapidly during this period. As Colin Cherry (1978, p.90) said at the time: 'Much of today's [telecommunications *and* passenger] traffic, of various kinds, follows the general directions of traditional trade routes.' This certainly suggests that telecommunications and air travel were acting, by a kind of positive feedback effect, to promote each other's growth.

Earth-orbiting satellites were made possible by wartime advances in rocketry, and were put to peaceful use for civil communications from the 1960s. The science-fiction writer Arthur C. Clarke had been the first to imagine, in 1945, how an artificial satellite might be made to orbit the Earth at the same speed at which the planet rotates. It would thus be stationary relative to some fixed point on the Earth's surface ('geo-stationary') and could serve to relay radio and television signals. The International Telecommunications Satellite Consortium (Intelsat) was set up in 1964, and in the following year launched its first geo-stationary satellite, 'Early Bird', capable of carrying just 240 simultaneous telephone conversations and one television channel. Figure 8.11 shows the distribution of Intelsat satellites by 1977.

Since 1980 there have been two major transformations of the telephone network, both of them made possible by technological developments, both propelled commercially by the privatization and deregulation of an industry that had up to then been dominated by private or state-run monopolies. The first change was a huge increase in the capacity of the long-distance cable links, created by substituting optical fibres – bundles of thin strands of glass that use bursts of laser light to carry signals – for the traditional copper wires. The first transatlantic fibre-optic cable, laid in 1988, could carry 40,000 simultaneous conversations. Cables laid at the end of the twentieth century are each able to accommodate three million conversations. This staggering increase in the potential flow through the telecommunications 'pipes' had two consequences. By 1996 there was a glut on the long-distance routes, with two-thirds of capacity unused. And the real cost of carrying an international telephone call – as distinct from what companies actually charged their customers – was becoming so small as to be approaching zero. This is what telecommunications specialists mean when they speak of the impending 'death of distance': long-distance telephone calls that are almost free.

The second change was that telephone instruments escaped from the cable network and went on the move. Portable radio telephones were developed for military use in the Second World War. It was not until the necessary electronic components became small enough and cheap enough in the 1980s, however, that miniaturized mobile phones for business and private use became a

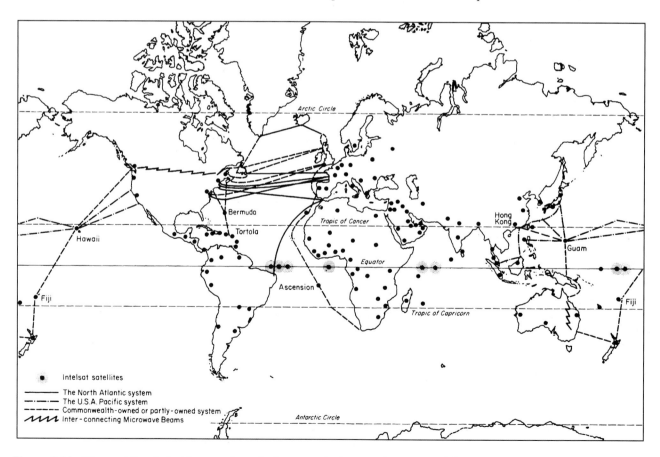

Figure 8.11   The world's principal intercontinental telecommunications trunk routes, and the approximate locations of Intelsat (International Telecommunications Satellite Consortium) stationary satellites and attendant ground stations, in 1977 (reproduced from Cherry, 1978, p.62, figure 3.1; © John Wiley and Sons Limited; reproduced with permission)

commercial proposition. At the time of writing, in the late 1990s, one new telephone subscriber in six in the world buys a mobile phone. Some commentators are again predicting significant spatial effects (for example Cairncross, 1996). The 'offices' of businesspeople can, for example, follow them wherever they go: back home, in the car, on the train, on the street. Mobile phone technology can have some specialized applications in regulating road traffic: for example, it can be used to direct cars on to less congested routes, to track stolen vehicles, or to charge drivers automatically and remotely for using streets or parking spaces.

Meanwhile – quite unnoticed by the great telephone corporations – developments in the rarefied worlds of advanced military and computing research were paving the way for yet another revolution in international telecommunications (Dyson, 1997; Hart *et al.*, 1992). In 1960, the RAND (Research and Development) Corporation commissioned a study of methods for protecting telephone and data links between US military bases and their command and control centres, in the event of nuclear war. The predicted cost of 'hardening' a system of fixed links and exchanges was astronomical. Paul Baran, an electrical engineer working for RAND, put forward an ingenious alternative. Instead of relying on a centralized – hence vulnerable – control and switching centre, Baran had the brilliant notion that *the messages themselves would know where they were going.*

A message would be broken into a series of small 'packets', each containing information about where it came from, where it was going to, and how it should be reunited with its fellow packets at the destination. When each packet reached a node in the network it would be kept moving along in roughly the right direction, according to what Baran called a 'hot potato' routing doctrine. A network of this kind would have many possible paths for any message, and would therefore be very robust. Should some links be knocked out, the packets would simply find alternative routes along other surviving links. Baran successfully demonstrated such a 'packet-switched data network' to RAND. It was never built for its intended military purpose. But packet-switching was to provide the key to the later success of the Internet. It created the basis for a telecommunications system without conventional exchanges, to which links could be added haphazardly at any point.

The Internet itself had its origins in the early 1960s in 'ARPANET', which was built to operate on Baran's principles, and linked up US military and university laboratories working on collaborative computing projects. The name derived from the Advanced Research Projects Agency of the US Department of Defense. ARPANET started with computers at four nodes. By 1983 it linked more than a hundred. It was used for transferring data, documents and software. The idea of sending electronic 'letters', or 'e-mail', occurred almost as an afterthought. Other specialist computer networks were joined up, and by the late 1980s ARPANET was buckling under the strain.

It was replaced by a new network funded by the National Science Foundation. A common set of 'protocols' was established to allow computers of many types to speak to each other. By 1992, this 'NSFNET' had linked together more than 5,000 smaller computer networks within separate institutions, most of them universities and government laboratories, and was beginning to extend overseas. The NSFNET 'backbone' of dedicated high-capacity cables (see Figure 8.12 overleaf) became interconnected in this way with existing telephone, fibre-optic and satellite systems, and the resulting 'network of networks' became known as the Internet. In the mid-1990s, the Internet was opened to commercial users, who had previously been excluded, and the National Science Foundation began to phase out the last government subsidies.

Up to this point, most traffic on the Internet was made up of separate documents consisting of numerical data or text. This changed with two programming innovations of the early 1990s (Anderson, 1996). Tim Berners-Lee at the CERN physics laboratory in Switzerland devised a way of connecting documents through something akin to electronic footnotes or cross-references. Marc Andreessen, an undergraduate at the University of Illinois, and others wrote a program called Mosaic, which made it possible to navigate along these so-called 'hypertext' links by simply pointing at and clicking on words or images on the computer screen using a 'mouse'. These developments made possible the World Wide Web, which has since spread over the Internet, and allows users to follow thematic links to 'surf' through what, by the late 1990s, were millions of pages of text, pictures, sound and animations. If the Internet is a giant library, as some have described it (though it is also a bazaar, and a huge poster site, and a craft fair, and many other things), then it is a library without librarians or a comprehensive catalogue, in which you must look in the books to find the locations of other books.

Statistics on the growth of the Internet are out of date by the time they are published. But between 1988 and the mid-1990s it doubled in size every year. By early 1997 there were around 57 million users worldwide, with another 13 million using just e-mail. Taken as a whole, the Internet must qualify as the biggest and most complex machine ever created.

The story of the Internet is an extraordinary one, not least because it was never really planned as such; it has evolved through a chapter of accidents, and its financing has seemed to defy the laws of economics. Anderson (1996) describes the recipe for its creation:

> Think up a universal way for networks to share data that will work with any kind of network, of any size, carrying any kind of data, on any sort of machine. Let anyone use it, for free, with no restrictions or limitations. Then just stand back.
> (Anderson, 1996)

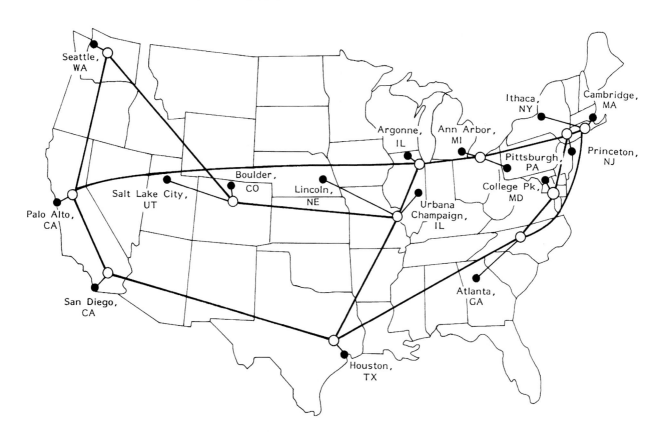

Figure 8.12 The NSFNET 'backbone' in 1992; most of the cities linked by the network are the sites of major universities and government laboratories (reproduced from Hart *et al.*, 1992, p.676, figure 5; with permission from Elsevier Science)

What is especially poignant is that while governments were announcing grandiose plans, and telephone and cable television companies were laying optical fibres, for a network of 'information superhighways' to be completed in the new millennium, a miscellaneous coterie of unworldly computer freaks beat them to it, under their very noses. Even the mighty Microsoft Corporation, itself a product of that same culture, was very late in appreciating the Internet's potential.

Part of the reason for its success and rate of growth is that the Internet has been in many ways parasitic on existing hardware, already in place for other purposes. Cheap microcomputers spread, from around 1980, first into businesses and then into homes; these computers were linked within firms and organizations by local networks, in order to share data and exchange e-mail; computers not already linked by dedicated cables were connected across existing telephone lines; and the international telephone network had enormous excess capacity because of the move to optical fibres.

8.7   *The death of distance prematurely announced?*

What, then, might be the effects of this new information and communications revolution on cities, and on how we travel in and between them? As Cairncross (1996) sagely warns: 'Predicting the future of a new technology is a mug's game … Guessing the social impact of a new technology is even more hazardous.' We can only cite conflicting opinions and point to trends.

The idea that telecommunications might allow office employees to work from home is as old as the telephone itself. H.G. Wells (1899) envisaged this use for the future videophone (his term was the 'kineto-tele-photograph') in his story *When the Sleeper Wakes*. A possible drawback is illustrated in Figure 8.13. Enthusiasm for 'telecommuting' grew in the 1970s, with the increased use of computers, and a renewed concern for saving fuel in transport prompted by the sharp rise in petrol prices that followed the Arab–Israeli war. (A 'commuter' was originally someone who commuted – that is, exchanged – daily public transport fares for a cheaper season ticket. A 'telecommuter' exchanges the costs of travel for the costs of telecommunication plus a saving in time.) The idea seemed to have many attractions, besides savings in fuel. It could make it easier for part-timers, disabled people and carers of small children to hold down jobs. From an employer's point of view, it shifted the costs of heating, lighting and accommodation to the telecommuter. On the other hand, it could make supervision more difficult. Telecommuters might spend some days at the office, some at home. An intermediate possibility was the creation of 'teleworking centres' in residential areas, where workers from different companies could share facilities and equipment.

Throughout the twentieth century, many people at both ends of the employment spectrum have operated from home with the help of the telephone. Professionals whose work involves processing information, such as writers, editors, financial advisers and computer programmers have traditionally worked in this way. There are also those who work at home on routine data-processing, directory enquiries, telephone sales and telephone services of still more

Figure 8.13 A cartoonist in 1929 envisages the future videophone and a possible drawback (from *Judge* magazine, reproduced in Durant, 1956, p.151)

disreputable kinds. The technology of videoconferencing, with full vision as well as sound, has allowed some business meetings that would previously have been face to face to be conducted at a distance. Many writers on the Internet are enthusiastic about its remote-working potential. Geographical addresses might become e-mail addresses in cyberspace. As Nicholas Negroponte (1995, p.165) asks rhetorically in his book *Being Digital*: 'If instead of going to work by driving my atoms into town, I log into my office and do my work electronically, exactly where is my workplace?'

One sector of industry that is entirely concerned with information, and that has become almost completely computerized, is banking and finance. These businesses have always been among the first to take up new telecommunications, as we have seen. Trading floors in France, Belgium, Spain and Canada have closed, to be replaced by screen-based trading. One of the world's biggest financial markets is now a 'virtual' one – NASDAQ (National Associated Automated Dealers Quotation System), which in the 1990s linked half a million dealers world-wide through telephone and fibre-optic lines (Warf, 1995). On the other hand, at the time of writing there are still financial markets where traders not only seem to find it necessary to be together in the same city, but even in the same room, shouting at each other. So face-to-face contact appears still to be crucial to activities that might seem to have the greatest potential for remote working.

Optimistic forecasts were made in the 1970s about the future numbers of telecommuters, but these have not been fulfilled. Patricia Mokhtarian (1997) has studied the subject for two decades and has become sceptical. She estimated that, in the late 1990s, only 16 per cent of the US workforce could *contemplate* telecommuting, and that on any one day probably around 2 per cent were actually doing so. One reason, she suggested, is that despite the stresses and strains, many office workers place high value on the social and professional interaction that goes on at the workplace. Mokhtarian's conclusion echoes those of other critics who have looked back at earlier telecommunications devices: 'Historically, transportation and communications have been complements to each other, both increasing concurrently, rather than substitutes for each other' (Mokhtarian, 1997, p.61).

People are shopping on the Internet in increasing numbers. According to the magazine *Business Week*, nearly one-third of all households connected to the Internet in the US used it for making purchases in the first half of 1998. Among the most popular items were computers and software, consumer electronics and music CDs. For those cultural conservatives who worry about computers displacing the printed word, it might be a matter of surprise, and some comfort, that Amazon, which sells exclusively over the Internet, is at the time of writing, the world's largest bookshop. Another surprise is the extent to which the Internet is used for selling *flowers* – until one reflects on the long-established success of the telephone-ordering service Interflora (see Figure 8.14). (Also, the Internet changes the balance of competition between major corporations and corner shops. The local florist can have a Web page which advertises gladioli globally.) Indeed, there are several areas in which the Internet is competing in retail and service markets with telephone and television: remote banking services, the sale of travel and theatre tickets, information services. In other areas it seems to be taking over business that would previously have been conducted through catalogues and mail order, a strong tradition in the United States since the nineteenth century.

Where Internet shopping may be slower to grow is in those sectors such as clothing, furnishings and food, where it is important for the shopper to see, touch and smell the real goods. In 1998, the British grocery chain Tesco offered a choice of 20,000 items over the Internet, for delivery in selected boroughs of London. But if remote shopping for food *does* turn out to be a success, it may

Figure 8.14 A Manhattan florist advertises on the World Wide Web, 1998 (by permission of Flower Fields, New York; ©1998 Webheads)

be more for standard goods than luxuries and specialities. Some earlier 'teleshopping' initiatives, such as the French Caditel service and the British Prestel system, have certainly had mixed fortunes. An ambitious system called Keyline, with 500,000 special terminals, was announced for launch in Britain in 1990 but never materialized. As with the social attractions of the office, one should not underestimate the extent to which going shopping has become a leisure activity, an excursion to be combined with eating out and entertainments, even for some a form of therapy. As Fischer (1992) pointed out, telephone shopping grew after the Second World War; but then 'so did in-person shopping, through the great expansion of suburban malls'.

While the Internet has been spreading in its anarchic way, the telephone and cable television companies have been busy installing local fibre-optic networks in cities, and corporations have been building private fibre-optic and satellite networks to link their offices and manufacturing facilities across the world. Because, as we have seen, the capacity of the optical-fibre bundles is so much greater than the old copper cables, they can be used to carry qualitatively different kinds of messages. They are 'broad-band' networks. Specifically, they can transmit video or film in digital form – something that is slow, if not impossible, over conventional telephone lines because of the sheer amounts of information involved by comparison with numerical data, text or sound. (It is because the Internet relies in many places on the existing telephone network that it has had difficulties, at least up until the late 1990s, in carrying moving images.)

These broad-band networks open up new possibilities for telecommunications: videoconferencing; collaborative working at a distance via video links, for instance by teams of designers; entertainment services where users can select films from libraries to view at home and have them delivered electronically ('video on demand'); distance education in which students can watch films, view live experiments and demonstrations 'in real time', or see, listen and talk to teachers at remote locations.

Authorities in a number of cities have seen investment in broad-band telecommunications as a key to future commercial competitiveness. Among the leaders has been Singapore, where the National Computer Board, set up in 1982, has overseen an ambitious plan to wire up the island city, its institutions, businesses and schools (Batty, 1990). A network to link and automate the import–export trading sector of the Singapore economy was launched in 1989, followed by other public-service networks to connect medical, education, traffic control and public transport services. Singapore is the world's largest port: the city's advanced Electronic Data Interchange network is used to control maritime shipping and the movement of cargoes (Warf, 1995). In Hong Kong an equivalent broad-band infrastructure has been built by the telephone and cable television companies, in a competitive free-market context which contrasts with the authoritarian central planning of Singapore. Smaller-scale experiments with broad-band networked services were going on around the world during the 1990s, especially in the USA, Scandinavia and Japan.

Messianic claims have been made for the liberating, transformatory prospects this new 'electronic frontier' is opening up. US Vice-President Al Gore, giving the first interactive computer news conference in 1994, spoke of the power of the new data highways to promote economic growth, solve social problems, foster democracy and 'link the peoples of the world' (quoted in Slouka, 1995, p.82). Heady visions indeed. Perhaps they should be tempered with some remarks made by Claude Fischer about similar predictions made for the telephone in its early days:

> Despite the romance of speaking over long distances, Americans did not forge new links with strange and faraway people. Despite experiments with novel applications of telephony, they did not attend concerts, get healed, hold town meetings, or change lifestyles via the telephone. As well as using it to make practical life easier, Americans – notably women – used the telephone to chat more often with neighbors, friends and relatives; to save a walk when a call might do; to stay in touch more easily with people who lived an inconvenient distance away. The telephone resulted in a reinforcement, a deepening, a widening, of existing lifestyles more than in any new departure.
>
> (Fischer, 1992, p.263)

So far from encouraging democracy and overcoming barriers between nations, there are indications indeed that the new broad-band networks are actually helping to sharpen existing geographical and economic divisions. We saw earlier how cities in the twentieth century changed from manufacturing centres to centres of service industry. It has been argued by some economic analysts that the new telecommunications technologies, especially private business networks, are acting to accelerate this process (see Moss, 1987; Warf, 1995). They are reinforcing the existing commercial dominance of a few 'world cities' – New York, London, Tokyo, Los Angeles, Hong Kong – in financial and information-intensive industries.

It is multinational companies in these fields that are investing most heavily in optical fibres and satellites. Some of these world cities have seen a notable boom in office-building during the 1980s and 1990s as a consequence – for example, Docklands in London. For decades Los Angeles lacked a single high-density office centre, until the city became an international financial centre doing business across the Pacific. Each of these world cities has, as Warf (1995, p.369) puts it, become 'more closely attuned to the rhythms of the international economy than the nation-state in which it is located'.

In the nineteenth century, the telegraph wires followed the railways. After the Second World War, the international telephone cables followed shipping routes and connected the same cities as the airlines. In the late twentieth century, the optical-fibre cables were being laid along existing railways and motorways, which provided convenient, established rights of way. Communications between existing cities were being further improved, while the rural would-be 'cybercottages' remained cut off. Many cities have built 'teleports' – office parks equipped with satellite stations and fibre-optic links – on the analogy of airports or seaports, but for information flows not cargoes. In the late 1980s there were more than fifty teleports in the world: the majority were in the United States, but the world's largest was in Tokyo.

Office work is increasingly being separated into two types of activity: decision-making, professional and creative activities; and routine data-processing and clerical activities – with the latter being pushed out of the big cities to premises in smaller towns or suburbs, just as manufacturing was separated from its administration and pushed out in the early twentieth century. As Moss (1987, p.534) describes the process:

> telecommunications technologies are leading to the centralization of business services in a small number of principal world cities, while simultaneously leading to the dispersion of routine information-based activities to the periphery of the metropolitan regions.

The 'front offices' in the city need face-to-face contact, and attract professional workers not just with highly paid and challenging jobs, but with the city's restaurants, clubs and theatres. The 'back offices' are moved out to cheaper sites and staffed with lower-paid workers. Typically, back offices are devoted to such tasks as data entry and office records; compilation of catalogues and directories; processing of payrolls, cheques and insurance claims; or processing subscriptions to magazines. Often they operate twenty-four hours a day, and exchange data continuously with headquarters over broad-band links.

This process has taken on an international dimension, as firms look for yet cheaper labour abroad. Several US insurance companies, for example, set up back offices around Shannon Airport in Ireland in the 1990s. The English-speaking parts of the Caribbean have attracted other US back offices, and the Dominican Republic has invested heavily in telecommunications infrastructure to encourage business in such areas as information services, publishing and the processing of credit-card transactions. The region around Bangalore in India has become a focus for software development because of the local supply of trained computer professionals who can work remotely for foreign firms – at lower rates of pay than Americans or Europeans – without the need to emigrate. Geography is still important in the global information economy. As Warf says:

> contrary to early, simplistic expectations that telecommunications would 'eliminate space', rendering geography meaningless through the effortless conquest of distance, such systems in fact produce new rounds of unevenness, forming new geographies that are imposed upon the relics of the past. Telecommunications simultaneously reflect and transform the topologies of capitalism, creating and rapidly recreating nested hierarchies of spaces technically articulated in the architecture of computer networks. Indeed, far from eliminating variations among places, such systems permit the exploitation of differences between areas with renewed ferocity … That the geography engendered by this process was unforeseen a decade ago hardly needs restating; that the future will hold an equally unexpected, even bizarre, set of outcomes is equally likely.
>
> (Warf, 1995, pp.375–6)

References

ABLER, R. (1977) 'The telephone and the evolution of the American metropolitan system' in I. de Sola Pool (ed.) *The Social Impact of the Telephone*, Cambridge, Mass., MIT Press, pp.318–41.

ANDERSON, C. (1996) 'The Internet' in The Economist (ed.) *Going Digital*, London, Profile Books, pp.95–128.

ARONSON, S.H. (1977) 'Bell's electrical toy: what's the use? The sociology of early telephone usage' in I. de Sola Pool (ed.) *The Social Impact of the Telephone*, Cambridge, Mass., MIT Press, pp.15–39.

ATTALI, J. and STOURDZE, Y. (1977) 'The birth of the telephone and economic crisis: the slow death of monologue in French society' in I. de Sola Pool (ed.) *The Social Impact of the Telephone*, Cambridge, Mass., MIT Press, pp.97–111.

BATTY, M. (1990) 'Intelligent cities: using information networks to gain competitive advantage', *Environment and Planning B: Planning and Design*, vol.7, pp.247–56.

BRIGGS, A. (1977) 'The pleasure telephone: a chapter in the prehistory of the media' in I. de Sola Pool (ed.) *The Social Impact of the Telephone*, Cambridge, Mass., MIT Press, pp.40–65.

CAIRNCROSS, F. (1996) 'Telecommunications' in The Economist (ed.) *Going Digital*, London, Profile Books, pp.15–57.

CAIRNCROSS, F. (1997) *The Death of Distance: how the communications revolution will change our lives*, London, Orion.

CHANNING, W.F. (1855) 'The American fire-alarm telegraph', *Ninth Annual Report of the Smithsonian Institution*, Washington, DC, Smithsonian Institution Press, pp.147–55.

CHERRY, C. (1978, rev. edn) *World Communication: threat or promise? A socio-technical approach*, Chichester, Wiley.

CHILDREN'S AID SOCIETY (1978) *New York Street Kids*, New York, Dover.

DE SOLA POOL, I. (ed.) (1977) *The Social Impact of the Telephone*, Cambridge, Mass., MIT Press.

DE SOLA POOL, I., DECKER, C., DIZARD, S., ISRAEL, K., RUBIN, P. and WEINSTEIN, B. (1977) 'Foresight and hindsight: the case of the telephone' in I. de Sola Pool (ed.) *The Social Impact of the Telephone*, Cambridge, Mass., MIT Press, pp.127–57.

DILTS, M.M. (1941) *The Telephone in a Changing World*, New York, Longmans Green.

DURANT, J. (1956) *Predictions*, New York, Barnes.

DYSON, G. (1997) *Darwin Among the Machines*, Harmondsworth, Allen Lane.

FIGUIER, L. (1883–6) *Les Nouvelles Conquêtes de la science*, Paris.

FISCHER, C.S. (1992) *America Calling: a social history of the telephone to 1940*, Berkeley, University of California Press.

GARRATT, G.R.M. (1958) 'Telegraphy' in C. Singer, E.J. Holmyard, A.R. Hall and T.I. Williams (eds) *A History of Technology*, Oxford, Clarendon Press, vol.IV, pp.644–62.

GOODMAN, D. and CHANT, C. (eds) (1999) *European Cities and Technology: industrial to post-industrial city*, London, Routledge, in association with The Open University.

GRAHAM, S. and MARVIN, S. (1996) *Telecommunications and the City: electronic spaces, urban places,* London, Routledge.

HART, J.A., REED, R.R. and BAR, F. (1992) 'The building of the Internet: implications for the future of broadband networks', *Telecommunications Policy*, November, pp.666–89.

MARTIN, J. (1981) *Telematic Society: a challenge for tomorrow*, Englewood Cliffs, NJ, Prentice-Hall (previously published as *The Wired Society*, 1978).

MITCHELL, W.J. (1995) *City of Bits: space, place and the infobahn*, Cambridge, Mass., MIT Press.

MOKHTARIAN, P.L. (1997) 'Now that travel can be virtual, will congestion virtually disappear?', *Scientific American*, October, vol.277, no.4, p.61.

MOSS, M.L. (1987) 'Telecommunications, world cities and urban policy', *Urban Studies*, vol.24, pp.534–46.

MOYER, J.A. (1977) 'Urban growth and the development of the telephone: some relationships at the turn of the century' in I. de Sola Pool (ed.) *The Social Impact of the Telephone*, Cambridge, Mass., MIT Press, pp.342–69.

NEGROPONTE, N. (1995) *Being Digital*, London, Hodder and Stoughton.

NETZER, D. (1977) 'The economic future of cities: winners and losers', *New York Affairs*, winter, vol.4, no.4, pp.81–93.

PERRY, C.R. (1977) 'The British experience 1876–1912: the impact of the telephone during the years of delay' in I. de Sola Pool (ed.) *The Social Impact of the Telephone*, Cambridge, Mass., MIT Press, pp.69–96.

SLOUKA, M. (1995) *War of the Worlds: cyberspace and the high-tech assault on reality*, London, Basic Books.

TARR, J.A. with FINHOLT, T. and GOODMAN, D. (1987) 'The city and the telegraph: urban telecommunications in the pre-telephone era', *Journal of Urban History*, vol.14, no.1, pp.38–80.

WARF, B. (1995) 'Telecommunications and the changing geographies of knowledge transmission in the late 20th century', *Urban Studies*, vol.32, no.2, pp.361–78.

WEBBER, M.M. (1963) 'Order in diversity: community without propinquity' in L. Wingo Jr (ed.) *Cities and Space: the future use of urban land*, Baltimore, Johns Hopkins Press, pp.23–54.

WEBBER, M.M. (1968) 'The post-city age', *Daedalus*, vol.97, no.4, pp.1091–1110.

WELLS, H.G. (1899) *When The Sleeper Wakes*, London, Nelson (revised 1910 as *The Sleeper Awakes*).

WILLEY, M.L. and RICE, S.A. (1933) 'The agencies of communication', *Recent Social Trends in the United States*, New York, McGraw-Hill, vol.1.

WILSON, M. (1954) *American Science and Invention: a pictorial history*, New York, Simon and Schuster.

Index

Page numbers in *italics* refer to figures

Acknowledgements

Grateful acknowledgement is made to the following for permission to reproduce material in this book:

pp.15, 17 Reps, J.W. (1965) *The Making of Urban America: a history of city planning in the United States;* copyright © 1965 by Princeton University Press, reprinted by permission of Princeton University Press.

pp.88–90 Excerpt from *Middletown in Transition* by Robert S. Lynd and Helen M. Lynd, copyright 1937 by Harcourt Brace and Company and renewed 1965 by Robert S. Lynd and Helen M. Lynd, reprinted by permission of the publisher.

pp.133–4 Reprinted with the permission of Scribner, a Division of Simon & Schuster, from *Skyscrapers and the Men who Build Them* by William A. Starrett. Copyright © 1928 by Charles Scribner's Sons.

Every effort has been made to trace all copyright owners, but if any has been inadvertently overlooked, the publishers will be pleased to make the necessary arrangements at the first opportunity.